BIODETERIORATION AND PRESERVATION IN ART, ARCHAEOLOGY AND ARCHITECTURE

AF776260

Dedicated to the memory of
Muriel Mitchell

BIODETERIORATION AND PRESERVATION IN ART, ARCHAEOLOGY AND ARCHITECTURE

Edited by Ralph Mitchell
and Jennifer Clifford

Archetype
Publications
www.archetype.co.uk

First published 2018 by Archetype Publications Ltd

Archetype Publications Ltd
c/o International Academic Projects
1 Birdcage Walk
London SW1H 9JJ
www.archetype.co.uk

Text © 2018 the authors and Archetype Publications
Images © 2018 the authors except where stated otherwise

ISBN: 978-1-909492-64-6

British Library Cataloguing in Publication Data
A catalogue record for this book is available from the British Library.

All rights reserved. No part of this publication may be reproduced, stored in a retrieval system, or transmitted, in any form or by any means, electronic, mechanical, photocopying, recording or otherwise, without the prior permission of the publishers.

The views and practices expressed by individual authors are not necessarily those of the editors or the publisher. The authors and publisher cannot take any responsibility for any harm or damage that may be caused by the use or misuse of any information contained herein.

Every effort has been made to trace copyright holders and to obtain their permission for the use of copyright material. The publisher would be pleased to rectify any omissions in future reprints.

Front cover illustrations
Fungal colonies on book covers and conidia and conidiophore of *Aspergillus halophilicus*.

Back cover illustrations
Left: Cellulose fibres that appear to be torn due to the enzymatic action of the fungus *Acrostalagmus* sp.
Right: Frescoes in the catacombs of Domitilla in Rome: detail from paintings in the Cubicolo of Apostoli piccoli (photo courtesy of the Pontificia Commissione di Archeologica Sacra, Rome, Italy).

Cover images courtesy of Dr Flavia Pinzari and Dr Clara Urzì.

Designed by Marcus Nichols at PDQ Digital Media Solutions Ltd.
Typeset by PDQ Digital Media Solutions Ltd, Bungay

CONTENTS

FOREWORD

The world's cultural heritage treasures are at increasing risk of deterioration from environmental threats such as climate change, air pollution and tourism. All forms of heritage materials are vulnerable, both external and internal, ranging from those found on archaeological sites to historic manuscripts. No country or institution is immune from these threats and if preventive action is not taken now, many of these materials may not survive into the next century.

Microorganisms, which grow both on the surface and interior of heritage materials, play a central role in these deterioration processes – metabolic products of the microbial processes are commonly responsible for the degradation. However, our knowledge of the microbial community's role in both the deterioration and preservation of a wide range of cultural heritage materials has improved significantly in recent years. New molecular methods have permitted early detection of biodeterioration and provided a deeper understanding of the mechanisms of degradation. In addition, these new tools have opened the door to the use of bacteria as protective agents.

This volume brings together recent research by conservation microbiologists working in diverse heritage environments and includes contributions on the effects of microbial biofilms and climate change on the biodeterioration process. It is hoped that this book will prove helpful to microbiologists, chemists and other scientists working in the field of conservation. It should also be useful to practising conservators and individuals in public policy concerned with the protection of our world's cultural heritage treasures.

We thank Archana Vasankathumar, Julianna Braun, and Heather E. Saunders for their assistance in preparing for this book.

Ralph Mitchell and Jennifer Clifford
Cambridge MA
July 2018

THE IMPACT OF THE ENVIRONMENT ON BIODETERIORATION AND PRESERVATION OF HERITAGE MATERIALS

THE IMPORTANCE OF MICROBIAL BIOFILMS IN THE DETERIORATION OF HERITAGE MATERIALS

Marc W. Mittelman

INTRODUCTION

Cultural heritage objects such as stonework, paintings, manuscripts, textiles and archaeological treasures are inextricably linked with life on their surfaces. Bacteria and fungi possess a number of adaptive mechanisms for responding to physicochemical factors that define their environment. These factors include nutrient availability, pH, Eh, temperature, organic and ionic content and the presence of antagonistic agents. Depending upon the types and numbers present, microorganisms can effect alterations in their physiology or physical state in response to the environment. Organic and inorganic acid production, heavy metal binding, transformation of xenobiotics and extracellular polysaccharide production are important adaptive tools for microorganisms in this regard. Microbial metabolic activities can also directly or indirectly cause damage to heritage materials associated with historic materials. In most ecosystems, these activities are dependent upon a microbe's ability to attach to surfaces.

Due to their size and net negative charge, microorganisms act as colloidal particles in that they attract in aggregate to a surface. A significant amount of research has focused on exploring the sort of intimate associations that exist between microorganisms and various surfaces. Microbial attachment to surfaces, surface colonization and extracellular polymeric substance (EPS) biosynthesis are integral to the development of a biofilm (Fig. 1). Biofilms nearly always comprise complex communities of bacteria and, sometimes, higher organisms.

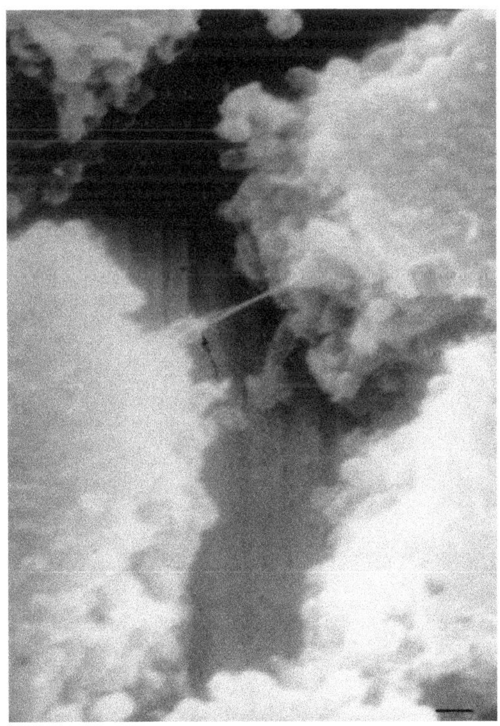

Figure 1 Bacterial biofilm, with regions of collapsed extracellular polymeric substance, on a metallic substrate (scale bar = 1 μm).

Life in a biofilm confers a number of advantages to the community members, including nutrient mobilization and protection from predation. Importantly, life in a biofilm also affords the community members protection from antagonistic agents such as antimicrobials and ultraviolet (UV) irradiation. Microbial biofilm interactions with natural and artificial surfaces are interesting and significant. Biofilms can lead to either the deterioration or protection of heritage items.

Understanding those factors that promote the transition of bacteria (and other microorganisms) from a planktonic to a sessile state with subsequent biofilm formation is essential to the development of effective biodeterioration monitoring and treatment programs. Biological fouling (biofouling) results from the deleterious effects of microorganisms including mechanical blockages, microbially influenced corrosion, product contamination, biodeterioration and/or discoloration (Characklis and Marshall 1990).

A large gap exists in understanding between the observed effects of microbial biofilms on substrata and the community structure, biomass constituents and metabolic activity of the biofilm consortia. For the most part, studies of biofilms have involved only single organisms or undefined consortia from natural environments (Dowling *et al.* 1991). Difficulties associated with the maintenance of diverse populations in continuous culture, combined with a lack of appropriate adhesion test systems, have limited the scope of biofilm studies under conditions which approximate those of the *in situ* environment. Studies of biofilm formation associated with non-aqueous environments, such as historic stone structures, are particularly difficult and time-consuming (McNamara and Mitchell 2005; Vázquez-Nion *et al.* 2017).

All surfaces are subjected to macro- and microenvironmental influences, some of which may directly result in aesthetic or structural damage. These influences include water availability, water chemistry (organic and inorganic constituents), temperature, the presence of volatile organic compounds (VOCs) and exposure to UV irradiation. The microbial community structure, biomass characteristics and *in situ* metabolic activity of biofilms are also affected by these same environmental influences. In many, but not all, instances, damage to heritage materials results from the interplay between the exposure environment and microorganisms present in biofilms. This contribution describes the development of biofilms on various substrata that are associated with cultural heritage objects, with a focus on bacteria, which are often the primary colonizers associated with fouling biofilms (Marshall *et al.* 1971; Maki *et al.* 1994; Bowen *et al.* 1998).

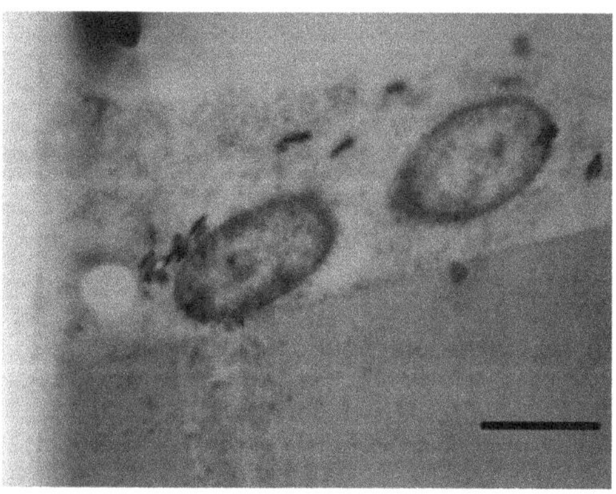

Figure 2 Transmission electron micrograph of bacteria adhering to a polymeric substrate; attachment mediated by extracellular polymeric substance (scale bar = 1 μm). (Courtesy of Dr. Gill Geesey.)

INITIAL EVENTS IN BIOFILM FORMATION

Adhesion may be defined as the discrete association between a bacterium and a surface. If a bacterium has adhered to a surface, energy is required to effect a separation (Mittelman 1999). A biofilm is comprised of these adherent cells, EPS produced by the attached organisms and entrained organic and inorganic material from the fluid or ambient air milieu. Attachment to surfaces is mediated and stabilized, in part, by EPS (Fig. 2). At the molecular level, bacteria and fungi adhere to surfaces by a combination of electrostatic and hydrophobic interactions. The nature of these interactions is dependent upon a combination of physicochemical factors, including pH, temperature, ionic strength, ligand density, dipole moment and charge density.

A number of mechanisms associated with the initial adhesion events in a wide range of environments have been described, such as chemotactic responses of bacteria and other microorganisms to nutrient gradients associated with surfaces, changes in fluid viscosity at the surface–fluid interface, specific lectin-receptor binding, non-specific hydrophobic interactions and responses by microorganisms to contact between substrata and flagella, pili or cilia (Lewin 1984; Ofek and Doyle 1994; Costerton *et al.* 1995; Mittelman 1997; Bhosle *et al.* 1998; Crawford *et al.* 2012). In non-aqueous environments such as those associated with heritage surfaces, electrostatic interactions likely are critical factors in the initial cell–substratum interaction. Non-specific advective forces, such as air movements, bring individual cells into contact with surfaces such as natural and engineered stones used for sculptures and monuments. The size and charge of bacteria and some fungal/microalgal spores are in the same range as colloidal solutes such as silica. Therefore, these microorganisms can exist as lyophobic or lyophilic solutions, association colloids, gels or as part of an emulsion, which contributes to adherence.

Microbial cells may also become more hydrophobic and adhesive through secretion of EPS, which can increase the rate of coaggregation of cells (Declerck 2010). Increases in coaggregation and ultimately biofilm formation provide additional protection against predation and impacts from other environmental stressors as well as enabling the recycling of low molecular weight organic compounds secreted or lyzed from neighboring cells (Álvarez *et al.* 2008). However, the role of coaggregation in the initial phases of biofilm development in heritage materials is poorly understood.

Biofilm formation and cell desorption is also controlled, in part, by quorum sensing, a bacterium–bacterium communication mechanism that is related to population density (Costerton *et al.* 2007; Jones *et al.* 2009). Inhibiting precursor molecules involved with this mechanism may prevent or disrupt bacterial biofilm formation but it remains unclear whether quorum sensing is an important factor for biofilm development and activity on heritage materials.

PHYSICOCHEMICAL VARIABLES

Naturally low-nutrient environments, such as those that typically exist on historic buildings, are often defined as containing less than 0.5 mg/L of dissolved organic carbon (DOC) (Amy and Morita 1983; Amy *et al.* 1983; Cavicchioli *et al.* 2002), or 0.1–50 nanomolar concentrations of low molecular weight organic compounds such as amino acids and sugars (Fuhrman 1987; Williams 2000). A significant portion of this DOC in low-nutrient environments is often considered recalcitrant and not bioavailable (Barber 1968). Similarly, cultural heritage material surfaces are generally limited in assimilable organic carbon concentrations. Nonetheless, microbial density and diversity in low-nutrient environments can be high, with up to 10^5 to 10^6 cells/mL (Cavicchioli *et al.* 2002; Hobbie and Hobbie 2013).

Exposed stone surfaces are typically nutrient deficient. Microorganisms have adopted a number of starvation-survival mechanisms, which include nutritional versatility, cellular morphological changes, viable but non-culturable (VBNC) state, formation of cysts and spores, and biofilm formation. Environmental stressors such as low DOC, temperature extremes, salinity, low-nutrient concentration and light extremes can trigger these mechanisms. Some of these stressors can be associated with environments experienced by heritage materials, in particular, stone structures. A single environmental stressor will likely trigger any one of the starvation-survival mechanisms available, while multiple stressors may favor a VBNC starvation-survival mechanism (Morita 1997: 529; Lleò *et al.* 2005). The starvation-survival mechanisms used are species-dependent and based on whether the microorganism can form cysts or spores (McDougald *et al.* 2006). Exposure to one stressor typically provokes

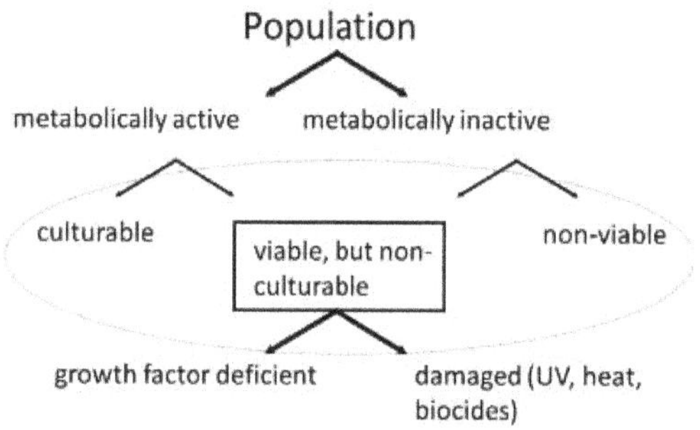

Figure 3 Viable but non-culturable (VBNC) populations in microbial communities.

a starvation-survival mechanism that may provide resistance against other, unrelated, environmental stressors (Hartke *et al.* 1998).

Microbial starvation-survival mechanisms in response to nutrient deficiency in the natural environment are variable and highly effective in reducing both respiration rates and microbial cell densities while enabling microbial cells to retain their essential metabolic activities. This may be of critical importance in biodeterioration processes on exposed historic artefacts, which often take place in low-nutrient environments. Non-spore-forming bacteria have been shown to maintain a VBNC state for greater than seven years in seawater (Moyer and Morita 1989a, b).

Mixotrophic environments are common in biofilms on archaeologic and monument stone surfaces. Mixotrophy is the ability to utilize a variety of different sources of energy and carbon, and mixotrophic organisms can adopt combinations of heterotrophy, phototrophy and phagotrophy to varying extents to obtain their required energy. Mixotrophy provides greater nutritional flexibility in low-nutrient environments. Such environments, common to stone surfaces, favor microorganisms able to utilize phototrophy, including Cyanobacteria and microalgae. Mixotrophy may be considered both a starvation-survival mechanism and a nutritionally versatile approach to avoid adopting other starvation-survival mechanisms discussed below.

In addition to the nutritional and morphological starvation-survival mechanisms, microbial cells are also able to dramatically alter their cellular metabolic rates and enter a VBNC state. Indeed, most bacteria can be VBNC, as shown in Figure 3. Xu *et al.* (1982) first identified the loss of bacterial culturability with maintenance of viability. VBNC has been described as as an organism's loss of cultureability on otherwise suitable media under some conditions, while maintaining metabolic activity; i.e. viability (McDougald *et al.* 1998). VBNC is characterized by a low respiration rate and low energy requirement. Under low-nutrient concentrations, microbial cells with an inherently slower growth rate are better adapted for these environments than their faster growth counterparts (Moyer and Morita 1989a, b). Microorganisms with high substrate affinities, which allow for rapid uptake of nutrients at low concentrations, are at a competitive selective advantage in an oligotrophic environment (Hirsch *et al.* 1978; Button 1991). Although these organisms will reach the enzyme saturation constant (V_{max}) at a lower concentration, in a nutrient deficient environment (e.g. the surface of a stone monument), this is unlikely to occur.

Some microbial cells increase budding rates and cell division creating a greater number of smaller microbial cells, a survival strategy of increasing the number of VBNC cells and hence increasing the chances of one cell surviving (Álvarez *et al.* 2008). Significant underestimates of viable or potential viable cells are associated with evaluations of bacteria in low-nutrient environments, such as historic stone surfaces (McNamara *et al.* 2003).

COLONIZATION AND BIOFILM FORMATION

The division of initially adhered microorganisms on surfaces is referred to as colonization. While assimilable organic carbon is an important limiting factor, for most heritage materials it is the availability of water that determines whether initially adhered cells will divide and eventually form a biofilm. Villa *et al.* (2016) have reviewed the development of subaerial biofilms (SABs) on historic stone surfaces as shown in Figure 4. They describe the role that surface irregularities

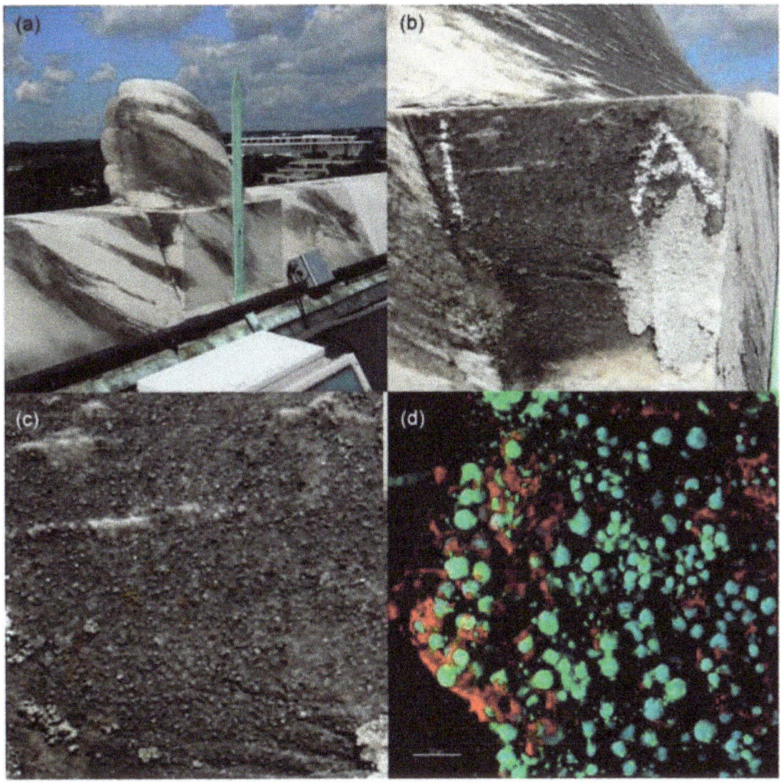

Figure 4 A subaerial biofilm (SAB) growing on the white marble of the Lincoln Memorial in Washington, DC: (a) antefix on the roof of the Lincoln Memorial; (b) and (c) close-up shots of a vertical SAB on the Lincoln Memorial; (d) confocal laser scanning microscopy image of a biofilm taken from this location. Microcolonies of photoautotrophic microbes are shown in blue, chemoheterotrophic microbes in green and extracellular polymeric substances in red. (Images used with permission of Villa *et al.* 2016.)

play in accumulating moisture, which is essential for EPS biosynthesis and eventual biofilm formation. Additionally, the hygroscopic nature of EPS enhances moisture retention.

In moist environments, biofilm formation proceeds through a four-step process involving development of an organic conditioning film on a substratum, primary adhesion of single or coaggregated microbial cells, colonization of the substratum and EPS elaboration. While less well understood, biofilm formation on seemingly dry surfaces proceeds in the same fashion. However, the paucity of liquid water greatly limits both the rate and extent (i.e. biomass) of biofilm development. While substratum mineralogical characteristics and organic constituents play a role, the ability of surfaces to sequester moisture seems to be the key determinant of fouling biofilm development. For example, the growth of cyanobacterial biofilms on different granite surfaces was primarily influenced by porosity and capillary water content (Vázquez-Nion *et al.* 2017). Surface age and weathering also influence EPS production; again, this is likely due to the ability of these 'damaged' surfaces to retain moisture through increased surface permeability (Coutinho *et al.* 2016).

EPS CHEMISTRY AND EFFECTS ON HERITAGE MATERIALS

EPS is primarily composed of polysaccharides, with lesser amounts of proteins, lipids and nucleic acids. Microbial EPS chemistry is complex, reflecting the diversity of biofilm

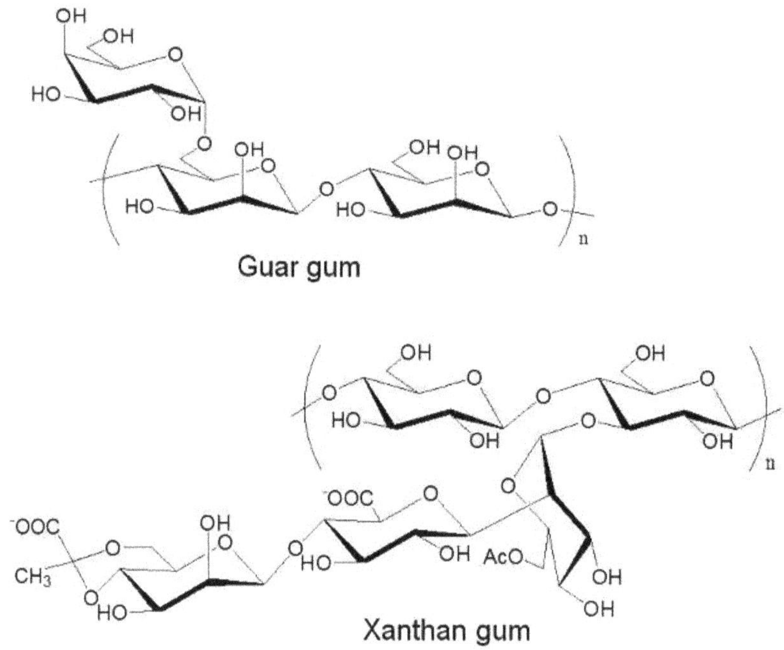

Figure 5 Structure of the uronic acid-containing polysaccharide from *Xanthomonas campestris*. Charged moieties available for binding to metals and other cations are circled.

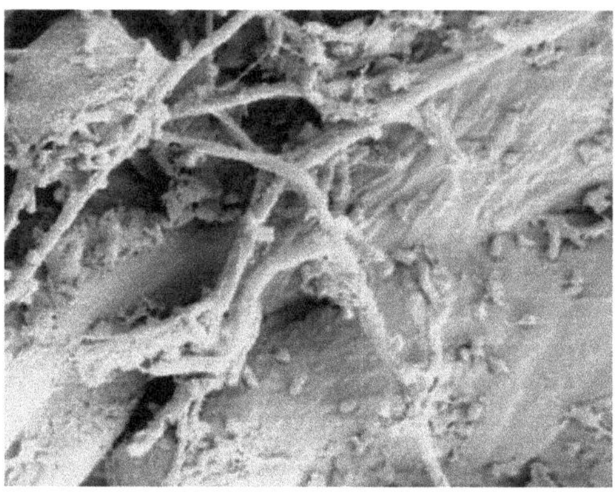

Figure 6 A mixed species biofilm on the surface of carbonate stone. Bacteria, fungi and biofilm exoploysaccharide are visible. (Courtesy of Dr. Christopher McNamara.)

communities and their various metabolic activities. Sutherland (1982) and others (Guo *et al.* 2008; Becker 2015; Schmid *et al.* 2015) reviewed the biosynthesis and chemical constituents of microbial EPS, primarily bacterial EPS. EPS may contain a high concentration of uronic acids, which are positively charged sugars present in the polysaccharide fraction (Fig. 5). These charged sugars are associated with a number of biochemical and biogeochemical reactions, all with relevance to heritage materials: these include cell-substratum binding, sequestration of heavy metals and other divalent cations, and nutrient mobilization.

Degradation of historic stone surfaces through freeze–thaw cycles in high latitude environments is mediated, in part, by endolithic biofilms and their associated EPS (McNamara and Mitchell 2005). Villa *et al.* (2016) reviewed biodeterioration mechanisms linked with SABs and their associated EPS, including alterations in water-vapor transport and particle separation during desiccation/hydration cycles. Water absorption by the biofilm EPS results in cycles of shrinkage and swelling, which can create fissures in archaeological and building stone (Warscheid and Braams 2000).

While metabolically produced acids and pigments can directly damage various surfaces, causing both losses in structural integrity and aesthetic changes (Fig. 6), EPS may also increase

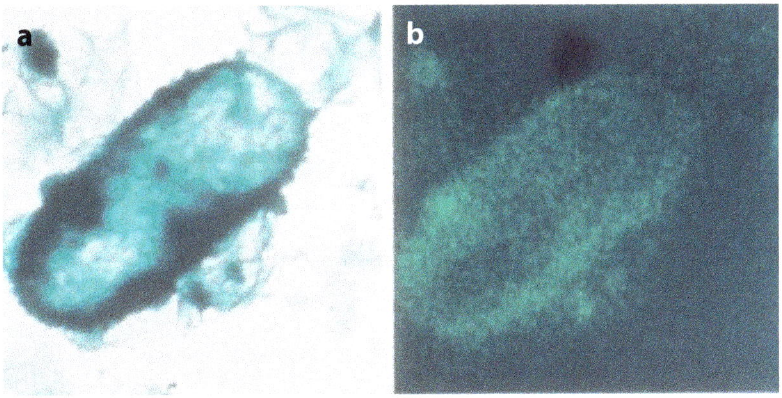

Figure 7 Scanning transmission electron micrograph (a) of *Acidovorax delafieldii*, with (b) energy-dispersive X-ray spectroscopic image of copper binding to cell surfaces.

the rate of calcium carbonate dissolution, resulting in deterioration (Perry *et al.* 2004). EPS binding of heavy metals can provide an adaptive advantage to biofilm organisms, effectively sequestering otherwise toxic metals from labile cellular components (Fig. 7) (Mittelman and Geesey 1985; Geesey *et al.* 1986). While most studies of metal binding have been associated with aquatic environments and/or corrosion activities, it is likely that similar reactions occur on oil paintings containing various metals as colorants.

BIOFILMS AND BIODETERIORATION PROCESSES

Ogawa *et al.* (2017) described a complex biofilm microbial community on the walls of an historic fortress in Niteroi, Brazil. The biofilm was comprised of Chloroflexi, Proteobacteria, diatoms, microalgae and archaea. Discolorations and underlying damage to the surface were observed. Hauer *et al.* (2015) discuss the role that Cyanobacteria and associated biofilms play in the weathering of rocks and stone surfaces. The authors reviewed 180 related publications describing over 700 species of Cyanobacteria on rock surfaces. They found that heterocystous and coccoid Cyanobacteria were most prevalent in terms of biomass, but that a wide diversity of species was present. The community diversity of biofilms associated with heritage materials may be lower than that associated with aquatic environments due to the oftentimes extreme environmental conditions on relatively dry substrata (Gorbushina and Broughton 2009). Such environments may select for an increased population of xerotolerant biofilm community members, in particular, fungi with the ability to develop biofilms in relatively low water activity (a_w) environments. Xerophilic fungi, including members of *Aspergillus*, *Paecilomyces*, *Chrysosporium*, *Penicillium* and *Cladosporium*, can grow and form biofilms in materials with an a_w of 0.70–0.85 (Pinzari and Montanari 2011), a significantly lower moisture content than would be associated with bacterial biofilm formation; i.e. >.85.

Discoloration, structural integrity losses and surface spalling of temples at Angkor Wat and elsewhere in Cambodia (Lan *et al.* 2010; Andre *et al.* 2012, 2014; Kusumi *et al.* 2013; Bartoli *et al.* 2014; Caneva *et al.* 2016) and Mayan ruins (McNamara *et al.* 2006; Ramirez *et al.* 2010) in Mexico have been associated with biofilm-mediated biodeterioration activities, as shown in Figures 8 and 9.

Oil paintings – both centuries old and those completed in the recent past – are subject to microbial contamination and fouling biofilm formation. A biofilm growing on a 16th-century painting is shown in Figure 10. Chromatic and paint structure alterations are associated with

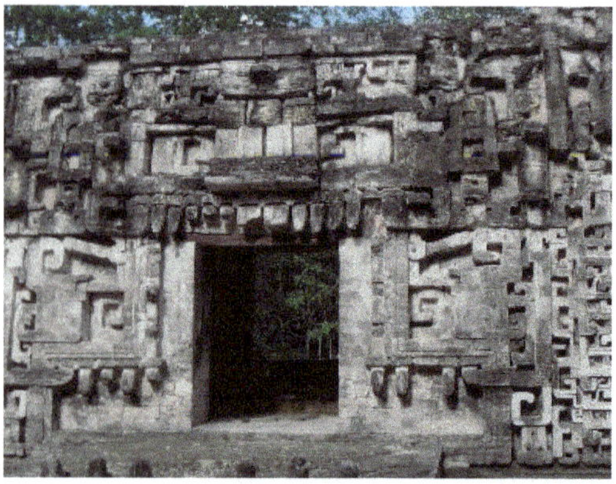

Figure 8 Example of mixed species biofilms associated with Mayan temple ruins, Compeche/Yucatan, Mexico. (Courtesy of Drs. Otto Morales and Jean Guezennec.)

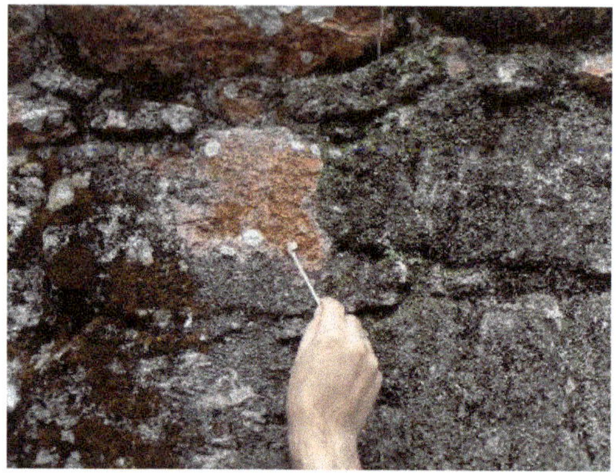

Figure 9 Biofilm sampling from stone surfaces of Mayan temple ruins, Compeche/Yucatan, Mexico. (Courtesy of Drs. Otto Morales and Jean Guezennec.)

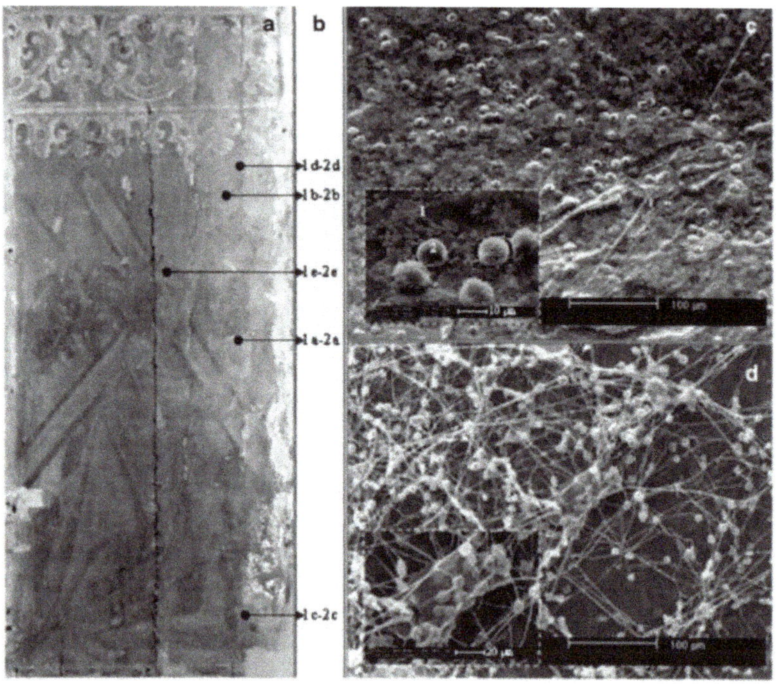

Figure 10 (a) A 16th-century painting on wood entitled *Cristo con la Cruz a Cuestas*; (b) two samples (1 and 2) from every sampling area (areas a–e) were obtained. Electron micrographs of surface material were observed from specimen 1b (c) and specimen 1e (d). The inset shows bacteria (1) and fungal spores (2) which are attached to the surface material. (Images used with permission of Santos *et al.* 2009.)

microbial biofilms (Salvador *et al.* 2016). In some cases, the presence of proteins such as ovalbumin, collagen and casein in the paint matrix serve as assimilable organic carbon sources (Salvador *et al.* 2017). The presence of atmospheric deposits on paintings can also influence the development of biodeteriorative biofilms. For example, volcanic ash deposits on the surfaces of mural paintings in Pompeii, Italy, were found to promote more extensive fungal biofilms relative to surfaces that were generally free of the deposits (Veneranda *et al.* 2017).

Lavin *et al.* (2016) described fungal and bacterial biofilm development on manuscripts that led to discoloration and cellulose degradation. The authors attributed the observed 'foxing' – the rusty-red staining on paper – to the production of EPS formation. *Scopulariopsis* spp., *Fusarium* spp. and *Bacillus* spp. were isolated from areas showing foxing and a pitting-type attack of the material. Mixed species bacterial/fungal biofilms have also been associated with photographs and maps. Degradation of cellulose and pigment and acid production were analogous with fungi colonizing documentary materials at museums in Cuba (Guiamet *et al.* 2011). Bacterial cellulolytic and proteolytic activity was also reported in the biofilms affecting these materials, where higher levels of bacteria than fungi were present.

Microbially influenced corrosion (MIC) of ferrous and nonferrous metals is a worldwide problem affecting culturally significant heritage items, examples of which are shown in Figures 11 and 12. Little *et al.* (2014 and references therein) reviewed mechanisms of iron MIC, including effects on cultural artefacts, such as the *Titanic* and other shipwrecks in deep ocean waters. Further, Rémazeilles *et al.* (2010) provided a case history of MIC of buried iron artefacts, including nails and iron associated with submerged wood. Despite the general toxicity to microbes associated with silver and copper, nonferrous artefacts made from these elements recovered from shallow soil or shallow seawater are also subject to MIC (McNeil and Little 1992). As with other biodeteriorative mechanisms, MIC is, fundamentally, modulated by the microbial biofilm. Water, an electrolyte and an electrochemical potential are all prerequisites for corrosion (Videla 2001), including corrosion initiated or propagated by microbial activity. Multispecies biofilms are nearly always involved, providing a means for various types of co-metabolic activities in a localized area. It is the localization of byproducts of microbial metabolism that may create an electrochemical potential that drives localized corrosion activities. *In situ* generation of corrosive hydrogen sulfide, hydrogen embrittlement, establishment of differential oxygen cells, acid production and localized oxygen production (e.g. from diatoms) are among the mechanisms associated with microbial biofilms and MIC (Dowling *et al.* 1991).

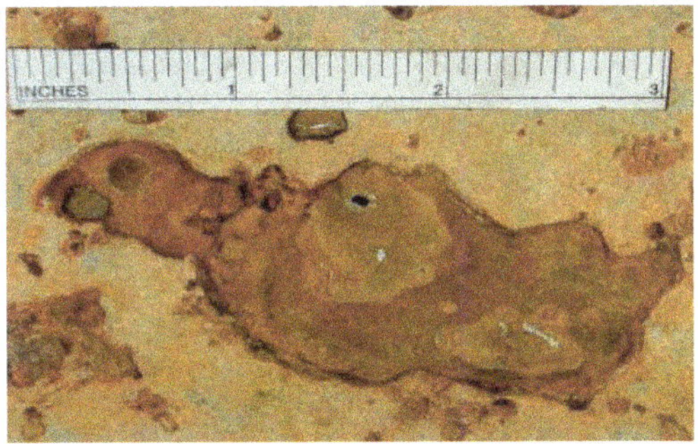

Figure 11 Microbially influenced corrosion of mild steel in a seawater environment.

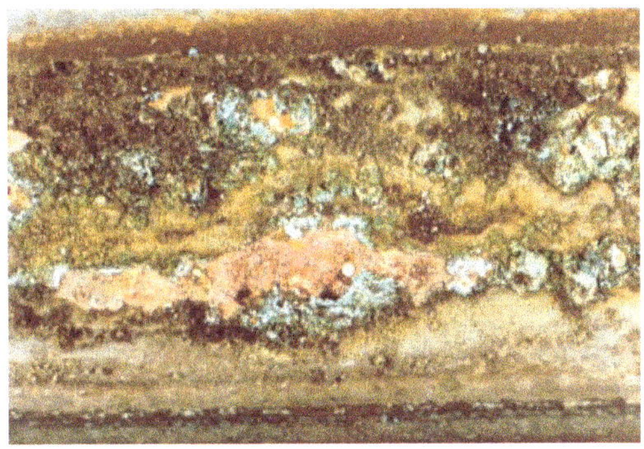

Figure 12 Microbially influenced corrosion of copper tube material in a freshwater environment.

While a variety of preventive and restorative treatments has been developed to control biofilms associated with the heritage materials described herein, none are without disadvantages to the treated materials, restoration personnel or the environment. Interestingly, some biofilms may be protective of some substrata. Calcium oxalate and other surface deposits mobilized by microbial biofilms may provide resistance to UV damage and wind/precipitation erosion (McNamara and Mitchell 2005). Conversely, some mineral-based films, including oxalates, produced by bacterial and microalgal biofilms have been associated with damage to mural paintings (Rosado *et al.* 2013) and other surfaces (Zammit *et al.* 2011). Some bacterial biofilms appear to be protective against MIC mediated by sulfate-reducing bacteria and other putatively corrosive microorganisms (Videla and Herrera 2009; Zarasvand and Rai 2014; Kip and van Veen 2015).

An argument has also been made for some cultural heritage materials as to 'whether the process of biodeterioration of cultural heritage can or should be stopped under all circumstances, or whether we have to accept it as a natural and an implicit consecution of its creation' (Sterflinger and Piñar 2013: 9637). Finally, climate change is likely to influence biofilm-mediated deterioration of heritage materials as a result of changing biofilm community structures subsequent to changing macro- and microenvironmental conditions (Viles and Cutler 2012).

REFERENCES

Aaronson, S. 1973. 'Digestion of phytoflagellates', in *Lysosmes in Biology and Medicine*, J.T. Dingle (ed.), 18–27. Amsterdam: North Holland Publishing.

Álvarez, B., López, M.M. and Biosca, E.G. 2008. 'Survival strategies and pathogenicity of *Ralstonia solanacearum* phylotype II subjected to prolonged starvation in environmental microcosms', *Microbiology* 154: 3590–98.

Amy, P.S. and Morita, R.Y. 1983. 'Starvation-survival patterns of 16 freshly isolated open-ocean bacteria', *Applied and Environmental Microbiology* 45(3): 1109–15.

Amy, P.S., Pauling, C. and Morita, R.Y. 1983. 'Starvation-survival processes of a marine vibrio', *Applied and Environmental Microbiology* 45(3): 1041–8.

Andre, M.F., Voldoire, O., Kousrel, E., Vautier, F., Phalip, B. and Peou, H. 2012. 'Contrasting weathering and climate regimes in forested and cleared sandstone temples of the Angkor region', *Earth Surface Processes and Landforms* 37(5): 519–32.

Andre, M.F., Vautier, F., Voldoire, O. and Roussel, E. 2014. 'Accelerated stone deterioration induced by forest clearance around the Angkor temples', *Science of the Total Environment* 493: 98–108.

Barber, R.T. 1968. 'Dissolved organic carbon from deep waters resists microbial oxidation', *Nature* 220: 274–5.

Bartoli, F., Municchia, A.C., Futagami, Y., Kashiwadani, H., Moon, K.H. and Caneva, G. 2014. 'Biological colonization patterns on the ruins of Angkor temples (Cambodia) in the biodeterioration vs bioprotection debate', *International Biodeterioration and Biodegradation* 96: 157–65.

Basson, A., Flemming, L.A. and Chenia, H.Y. 2008. 'Evaluation of adherence, hydrophobicity, aggregation, and biofilm development of *Flavobacterium johnsoniae*-like isolates', *Microbial Ecology* 55(1): 1–14.

Becker, A. 2015. 'Challenges and perspectives in combinatorial assembly of novel exopolysaccharide biosynthesis pathways', *Frontiers in Microbiology* 6: 687.

Bhosle, N., Suci, P.A., Baty, A.M., Weiner, R.M. and Geesey, G.G. 1998. 'Influence of divalent cations and pH on adsorption of a bacterial polysaccharide adhesion', *Journal of Colloid and Interface Science* 205(1): 89–96.

Bowen, K.L., Kaushik, N.K. and Gordon, A.M. 1998. 'Macroinvertebrate communities and biofilm chlorophyll on woody debris in two Canadian oligotrophic lakes', *Archiv für Hydrobiologie* 141(3): 257–81.

Busscher, H.J., Rinastiti, M., Siswomihardjo, W. and van der Mei, H.C. 2010. 'Biofilm formation on dental restorative and implant materials', *Journal of Dental Research* 89(7): 657–65.

Button, D.K. 1991. 'Biochemical basis for whole-cell uptake kinetics: specific affinity, oligotrophic capacity, and the meaning of the Michaelis constant', *Applied and Environmental Microbiology* 57(7): 2033–8.

Caneva, G., Bartoli, F., Savo, V., Futagami, Y. and Strona, G. 2016. 'Combining statistical tools and ecological assessments in the study of biodeterioration patterns of stone temples in Angkor (Cambodia)', *Scientific Reports* 6: article number 32601.

Cano, R.J. and Borucki, M.K. 1995. 'Revival and identification of bacterial spores in 25–40 million year old Dominican amber', *Science* 268: 1060–64.

Cavicchioli, R., Ostrowski, M., Fegatella, F., Goodchild, A. and Guixa-Boixercu, N. 2002. 'Life under nutrient limitation in oligotrophic marine environments: an eco/physiological perspective of *Sphingopyxis alaskensis* (formerly *Sphingomonas alaskensis*)', *Ecology* 45: 203–17.

Characklis, W.G. and Marshall, K.C. 1990. *Biofilms*. New York: John Wiley and Sons.

Costerton, J.W., Lewandowski, Z., Caldwell, D.E., Korber, D.R. and Lappin-Scott, H.M. 1995. 'Microbial biofilms', *Annual Review of Microbiology* 49: 711–45.

Costerton, J., Montanaro, L. and Arciola, C.R. 2007. 'Bacterial communications in implant infections: a target for an intelligence war', *International Journal of Artificial Organs* 30(9): 757–63.

Coutinho, M.L., Miller, A.Z., Rogerio-Candelara, M.A. *et al.* 2016. 'An integrated approach for assessing the bioreceptivity of glazed tiles to phototrophic microorganisms', *Biofouling* 32(3): 243–59.

Crawford, R.J., Webb, H.K., Truong, V.K., Hasan, J. and Ivanova, E.P. 2012. 'Surface topographical factors influencing bacterial attachment', *Advances in Colloid and Interface Science* 179: 142–9.

Crump, B.C., Ducklow, H.W. and Hobbie, J.E. 2013. 'Estuarine microbial food webs', in *Estuarine Biology*, J.W. Day, B.C. Crump, W.M. Kemp and A. Yàñez-Arancibia (eds), 263–84. Wiley-Blackwell.

Declerck, P. 2010. 'Biofilms: the environmental playground of *Legionella pneumophila*', *Environmental Microbiology* 12(3): 557–66.

Dowling, N.J.E., Mittelman, M.W. and White, D.C. 1991. 'The role of consortia in microbially influenced corrosion', in *Mixed Cultures in Biotechnology*, G. Zeikus and E.A. Johnson (eds), 341–72. New York: McGraw-Hill.

Fuhrman, J. 1987. 'Close coupling between release and uptake of dissolved free amino acids in seawater studied by an isotope dilution approach', *Marine Ecology Progress Series* 37: 45–52.

Gaylarde, C.C. and Gaylarde, P.M. 2005. 'A comparative study of the major microbial biomass of biofilms on exteriors of buildings in Europe and Latin America', *International Biodeterioration and Biodegradation* 55(2): 131–9.

Geesey, G.G., Mittelman, M.W., Iwaoka, T. and Griffiths, P.R. 1986. 'Role of bacterial exopolymers in the deterioration of metallic copper surfaces', *Materials Performance* 25: 37–40.

Gorbushina, A.A. and Broughton, W.J. 2009. 'Microbiology of the atmosphere–rock interface: how biological interactions and physical stresses modulate a sophisticated microbial ecosystem', *Annual Review of Microbiology* 63: 431–50.

Guiamet, P., Borrego, S., Lavin, P., Perdomo, I. and de Saravia, S.G. 2011. 'Biofouling and biodeterioration in materials stored at the Historical Archive of the Museum of La Plata, Argentine and at the National Archive of the Republic of Cuba', *Colloids and Surfaces B: Biointerfaces* 85(2): 229–34.

Guo, H., Yi, W., Song, J.K. and Wang, P.G. 2008. 'Current understanding on biosynthesis of microbial polysaccharides', *Current Topics in Medicinal Chemistry* 8(2): 141–51.

Hartke, A., Giard, J.C., Laplace, J.M. and Auffray, Y. 1998. 'Survival of *Enterococcus faecalis* in an oligotrophic microcosm: changes in morphology, development of general stress resistance, and analysis of protein synthesis', *Applied and Environmental Microbiology* 64(11): 4238–45.

Hauer, T., Muhlsteinova, R., Bohunicka, M., Kastovsky, J. and Mares, J. 2015. 'Diversity of Cyanobacteria on rock surfaces', *Biodiversity and Conservation* 24(4): 759–79.

Hirsch, P.M., Bernhard, Y.S., Cohen, S.S., Ensign, J.C., Jannasch, H.W., Koch, A.L., Marshall, K.C., Matin, A., Poindexter, J.S., Rittenberg, S.C., Smith, D.S. and Veldkamp, H. 1978. 'Life under conditions of low nutrient concentrations'. Group report in Strategies of Microbiological Life in Extreme Environments, Dahlem Workshop, Berlin.

Hobbie, J.E. and Hobbie, E.A. 2013. 'Microbes in nature are limited by carbon and energy: the starving-survival lifestyle in soil and consequences for estimating microbial rates', *Frontiers in Microbiology* 4: 1–11.

Horneck, G., Bücker, H. and Reitz, G. 1994. 'Long-term survival of bacterial spores in space', *Advances in Space Research* 14(19): 41–5.

Jones, S.M., Dang, T.T. and Martinuzzi, R. 2009. 'Use of quorum sensing antagonists to deter the formation of crystalline *Proteus mirabilis* biofilms', *International Journal of Antimicrobial Agents* 34(4): 360–64.

Kip, N. and van Veen, J.A. 2015. 'The dual role of microbes in corrosion', *ISME Journal* 9(3): 542–51.

Kusumi, A., Li, X.S., Osuga, Y., Kawashima, A., Gu, J.D., Nasu, M. and Katayama, Y. 2013. 'Bacterial communities in pigmented biofilms formed on the sandstone bas-relief walls of the Bayon Temple, Angkor Thom, Cambodia', *Microbes and Environments* 28(4): 422–31.

Lan, W.S., Li, H., Wang, W.D., Katayama, Y. and Gu, J.D. 2010. 'Microbial community analysis of fresh and old microbial biofilms on Bayon Temple sandstone of Angkor Thom, Cambodia', *Microbial Ecology* 60(1): 105–15.

Lavin, P., de Saravia, S.G. and Guiamet, P. 2016. '*Scopulariopsis* sp. and *Fusarium* sp. in the documentary heritage: evaluation of their biodeterioration ability and antifungal effect of two essential oils', *Microbial Ecology* 71(3): 628–33.

Lewin, R. 1984. 'Microbial adhesion is a sticky problem', *Science* 224: 375–7.

Little, B.J., Gerke, T.L. and Lee, J.S. 2014. 'Mini-review: the morphology, mineralogy and microbiology of accumulated iron corrosion products', *Biofouling* 30(8): 941–8.

Lleò, M.D.M., Bonato, B., Benedetti, D. and Canepari, P. 2005. 'Survival of enterococcal species in aquatic environments', *FEMS Microbiology Ecology* 54: 189–96.

Maki, J.S., Yule, A.B., Rittschof, D. and Mitchell, R. 1994. 'The effect of bacterial films on the temporary adhesion and permanent fixation of cypris larvae, *Balanus amphitrite* Darwin', *Biofouling* 8: 121–31.

Marshall, K.C., Stout, R. and Mitchell, R. 1971. 'Mechanisms of the initial events in the sorption of marine bacteria to surfaces', *Journal of General Microbiology* 68: 337–48.

McDougald, D.M., Rice, S.A., Weichart, D. and Kjelleberg, S. 1998. 'Nonculturability: adaptation or debilitation', *FEMS Microbiology Ecology* 25(1): 1–9.

McNamara, C.J. and Mitchell, R. 2005. 'Microbial deterioration of historic stone', *Frontiers in Ecology and the Environment* 3(8): 445–51.

McNamara, C.J., Lemke, M.J. and Leff, L.G. 2003. 'Underestimation of bacterial numbers in starvation-survival mode using the nucleic acid stain DAPI', *Archiv für Hydrobiologie* 157(3): 309–19.

McNamara, C.J., Perry, T.D., Bearce, K.A., Hernandez-Duque, G. and Mitchell, R. 2006. 'Epilithic and endolithic bacterial communities in limestone from a Maya archaeological site', *Microbial Ecology* 51(1): 51–64.

McNeil, M.B. and Little, B.J. 1992. 'Corrosion mechanisms for copper and silver objects in near-surface environments', *Journal of the American Institute for Conservation* 31(3): 355–66.

Mittelman, M.W. 1997. 'Adhesion to biomaterials', in *The Molecular and Ecological Diversity of Bacterial Adhesion*, M. Fletcher (ed.), 89–127. New York: Wiley.

Mittelman, M.W. 1999. 'Recovery and characterization of biofilm bacteria associated with medical devices', *Biofilms* 310: 534–51.

Mittelman, M.W. and Geesey, G.G. 1985. 'Copper-binding characteristics of exopolymers from a freshwater-sediment bacterium', *Applied and Environmental Microbiology* 49: 846–51.

Morita, R.Y. 1997. *Bacteria in Oligotrophic Environments: Starvation-Survival Lifestyle.* London: Chapman and Hall.

Moyer, C.L. and Morita, R.Y. 1989a. 'Effect of growth-rate and starvation-survival on the viability and stability of a psychrophilic marine bacterium', *Applied and Environmental Microbiology* 55(5): 1122–7.

Moyer, C.L. and Morita, R.Y. 1989b. 'Effect of growth-rate and starvation-survival on cellular DNA, RNA, and protein of a psychrophilic marine bacterium', *Applied and Environmental Microbiology* 55(10): 2710–16.

Ofek, I. and Doyle, R.J. (eds) 1994. *Bacterial Adhesion to Cells and Tissues.* New York: Chapman & Hall.

Ogawa, A., Celikkol-Aydin, S., Gaylarde, C.C., Neto, J.A.B. and Beech, I. 2017. 'Microbial communities on painted wet and dry external surfaces of a historic fortress in Niteroi, Brazil', *International Biodeterioration and Biodegradation* 123: 164–73.

Perry, T.D., Duckworth, O.W., McNamara, C.J., Martin, S.T. and Mitchell, R. 2004. 'Effects of the biologically produced polymer alginic acid on macroscopic and microscopic calcite dissolution rates', *Environmental Science & Technology* 38(11): 3040–46.

Pinzari, F. and Montanari, M. 2011. 'Mould growth on library materials stored in compactus-type shelving units', in *Sick Building Syndrome in Public Buildings and Workplaces*, S.A. Abdul-Wahab (ed.), 193–206. Berlin: Springer-Verlag.

Ramirez, M., Hernandez-Marine, M., Novelo, E. and Roldan, M. 2010. 'Cyanobacteria-containing biofilms from a Mayan monument in Palenque, Mexico', *Biofouling* 26(4): 399–409.

Rémazeilles, C., Saheb, M., Neff, D., Guilminot, E., Tran, K., Bourdoiseau, J.A., Sabot, R., Jeannin, M., Matthiesen, H., Dillmann, P. and Refait, P. 2010. 'Microbiologically influenced corrosion of archaeological artefacts: characterisation of iron(II) sulfides by Raman spectroscopy', *Journal of Raman Spectroscopy* 41(11): 1425–33.

Ronimus, R.S., Rueckert, A. and Morgan, H.W. 2006. 'Survival of thermophilic spore-forming bacteria in a 90+ year old milk powder from Ernest Shackelton's Cape Royds Hut in Antarctica', *Journal of Dairy Research* 73: 235–43.

Rosado, T., Gil, M., Mirão, J., Candeias, A. and Caldeira, A.T. 2013. 'Oxalate biofilm formation in mural paintings due to microorganisms: a comprehensive study', *International Biodeterioration and Biodegradation* 85: 1–7.

Rosenberg, M. and Kjelleberg, S. 1987. 'Hydrophobic interactions: role in bacterial adhesion', in *Advances in Microbial Ecology*, K.C. Marshall (ed.), 353–93. New York: Plenum Press.

Salvador, C., Silva, M., Rosado, T., Vaz Friere, R., Bordalo, R., Candeias, A. and Caldeira, A.T. 2016. 'Biodeterioration of easel paintings: development of new mitigation strategies', *Conservar Património* 23: 119–24.

Salvador, C., Bordalo, R., Silva, M., Rosado, T., Candeias, A. and Caldeira, A.T. 2017. 'On the conservation of easel paintings: evaluation of microbial contamination and artists materials', *Applied Physics A* 123(1): 80.

Schmid, J., Sieber, V. and Rehm, B. 2015. 'Bacterial exopolysaccharides: biosynthesis pathways and engineering strategies', *Frontiers in Microbiology* 6: 496.

Sterflinger, K. and Piñar, G. 2013. 'Microbial deterioration of cultural heritage and works of art – tilting at windmills?', *Applied Microbiology and Biotechnology* 97(22): 9637–46.

Sutherland, I.W. 1982. 'Biosynthesis of microbial exopolysaccharides', *Advances in Microbial Physiology* 23: 79–150.

Vázquez-Nion, D., Silva, B. and Prieto, B. 2017. 'Influence of the properties of granitic rocks on their bioreceptivity to subaerial phototrophic biofilms', *Science of the Total Environment* 610: 44–54.

Veneranda, M., Prieto-Taboada, N., de Vallejuelo, S.F.O., Maguregui, M., Morillas, H., Marcaida, I., Castro, K., Madariaga, J.M. and Osanna, M. 2017. 'Biodeterioration of Pompeian mural paintings: fungal colonization favoured by the presence of volcanic material residues', *Environmental Science and Pollution Research* 24(24): 19599–608.

Videla, H.A. 2001. 'Microbially induced corrosion: an updated overview (reprinted)', *International Biodeterioration and Biodegradation* 48(1–4): 176–201.

Videla, H.A. and Herrera, L.K. 2009. 'Understanding microbial inhibition of corrosion: a comprehensive overview', *International Biodeterioration and Biodegradation* 63(7): 896–900.

Viles, H.A. and Cutler, N.A. 2012. 'Global environmental change and the biology of heritage structures', *Global Change Biology* 18(8): 2406–18.

Villa, F., Stewart, P.S., Klapper, I., Jacob, J.M. and Cappitelli, F. 2016. 'Subaerial biofilms on outdoor stone monuments: changing the perspective toward an ecological framework', *Bioscience* 66(4): 285–94.

Warscheid, T. and Braams, J. 2000. 'Biodeterioration of stone: a review', *International Biodeterioration and Biodegradation* 46(4): 343–68.

Whitehead, K.A. and Verran, J. 2007. 'The effect of surface properties and application method on the retention of *Pseudomonas aeruginosa* on uncoated and titanium-coated stainless steel', *International Biodeterioration and Biodegradation* 60(2): 74–80.

Williams, P.J.L. 2000. 'Heterotrophic bacteria and the dynamics of dissolved organic matter', in *Microbial Ecology of the Oceans*, D.L. Kirchman (ed.), 153–200. New York: Wiley-Liss.

Xu, H.S., Roberts, N., Singleton, F.L. and Attwell, R.W. 1982. 'Survival and viability of nonculturable *Escherichia coli* and *Vibrio cholera* in the estuarine and marine environment', *Microbial Ecology* 8(4): 313–23.

Zammit, G., Sanchez-Moral, S. and Albertano, P. 2011. 'Bacterially mediated mineralisation processes lead to biodeterioration of artworks in Maltese catacombs', *Science of the Total Environment* 409(14): 2773–82.

Zarasvand, K.A. and Rai, V.R. 2014. 'Microorganisms: induction and inhibition of corrosion in metals', *International Biodeterioration and Biodegradation* 87: 66–74.

AUTHOR'S ADDRESS

Marc W. Mittelman, Mittelman and Associates, LLC, 227 Summit Ave. W310, Brookline, MA 02446 USA (marc@mittelmanassociates.com)

THE EFFECTS OF CLIMATE CHANGE ON THE BIODETERIORATION OF HISTORIC MATERIALS

Peter Brimblecombe

INTRODUCTION

Climate is just one of a number of environmental factors that influence the rate of biodeterioration. In the case of heritage management, biological processes may be especially sensitive to climate change. Although this change often lead to small shifts in temperature, humidity or rainfall, biochemistry, organisms and ecology effectively amplify them. Additionally, slow seasonal cycles or more rapid diurnal variations can pose threats to heritage that can be mediated by biology. The damage caused by biodeterioration is not as dramatic as that which occurs during extreme events such as war, vandalism and natural disasters. Although biodeterioration may appear in just a few days, it can seem more akin to the subtle changes imposed by increased tourism or poor restoration. Dramatic impacts on heritage assets often occur as a single event such as immediate and catastrophic damage from a severe earthquake, flood or fire. In contrast, climate change effects on biodeterioration can lack immediate and obvious impact. This creates its own problems however: on the one hand climate change is long term, allowing time to consider a response, but on the other it may be so slow or subtle that it is ignored in terms of day-to-day planning.

Biodeterioration is most typically a problem for organic materials, but the colonization of building stone by lichens, for example, can also be deleterious – even though outdoor materials are typically durable, the effects of biological weathering may nevertheless be readily apparent. Compared with the indoor environment, the aggressive nature of the external environment can be recognized in terms of the effect of rainfall and surface water, direct solar irradiation and higher wind speeds. In addition, materials are at risk from exposure to air pollutants, although these can also be prevalent indoors, particularly when they derive from indoor heating or volatilize from construction materials (Leslie and Lunau 1992). (For further information on the long-term effects of air pollution in relation to heritage see Brimblecombe and Grossi 2009 and 2010.)

HISTORICAL CLIMATE CHANGES

Climate change has a long history and there is much evidence of the environmental pressures faced by past civilizations. The Bronze Age crisis of the Mediterranean was especially notable as an unstable period, with the widespread collapse of great civilizations. This is often interpreted in terms of environmental change (Kaniewski *et al.* 2013) and supported by pollen evidence that suggests a significant drying of the region. The Roman Empire depended on Egypt for much of its wheat (Evans 1981) and was affected by increasing desertification of North Africa. The expansion of human achievement in the Middle Ages is associated with the optimal climate resulting from warmer conditions that prevailed early in the second millennium (the Medieval Warm Period). The climate subsequently became colder with the onset of the Little Ice Age (*c.*1300–1870), with frequent frosts recorded across Europe. Although proxy evidence of the kind used to construct the changing climate globally, as shown in Figure 1, can give a very satisfying picture, such generalized overviews must be regarded with some caution as there is likely to be considerable regional variation.

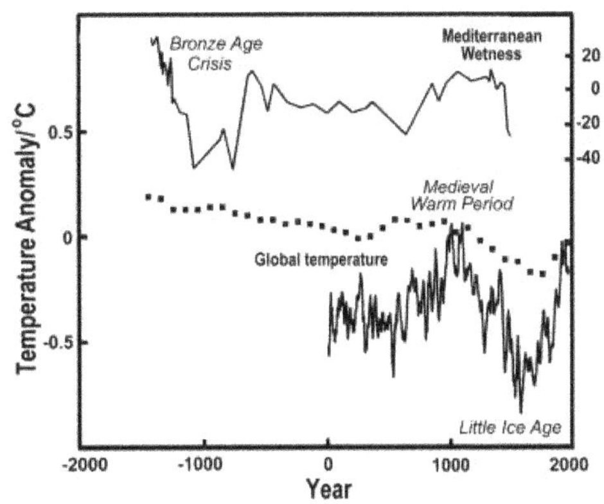

Figure 1 Long-term records of temperature and wetness largely derived from biological data such as pollen and tree rings. The estimate of wetness (dashed line) shown is from an analysis of pollen in the Mediterranean (Kaniewski *et al.* 2013). The long-term record of the global temperature anomaly (plotted as points) is taken from an assessment of many sets of proxy data that date back more than 11,000 years (Marcott *et al.* 2013) with 4000 years used here. The shorter record (line) is a 20-year running mean of the temperature anomaly (Moberg *et al.* 2005) taken mostly from tree rings.

HERITAGE CLIMATES

Climates can be regarded in terms of the vegetation they support, which is most apparent in the well-known climate classification system of Wladimir Peter Köppen developed at the end of the 19th century (Chen and Chen 2013). It divides climate into five major groups that provide clues to threats posed by biodeterioration:

1. Tropical/megathermal climates are characterized by constant high temperatures. Providing that conditions are humid, there can be significant biodeterioration and insects may be diverse and abundant.
2. Dry climates where precipitation is less than potential evaporation and transpiration, such that desiccation can often limit biodeterioration. Ground cover may be sparse.
3. Temperate/mesothermal climates where the average temperature is above 10 °C in the warmest months and the coldest months average between −3 °C and 18 °C. Warm and damp conditions lead to the potential for fungal attack and the warmer parts of the year may see increases in insect abundance.
4. Continental/microthermal climates where temperatures are above 10 °C in their warmest months and the coldest month average is below the freezing point. These typically occur in continental interiors and are not found in the Southern Hemisphere. Despite the harsh winters, insects that survive can later become abundant. Warmer times of the year may lead to rapid biodeterioration.
5. Polar climates where temperatures are below 10 °C all year round and have limited vegetation cover and biodeterioration.

This Köppen view of climate is described in terms of temperature and precipitation and the seasonality of these parameters. It refers principally to the relationship between climate and vegetation, but can also be generally relevant to ecological processes (Chen and Chen 2013). In the case of heritage materials, however, other factors such as humidity may be more pertinent than rainfall.

Despite the limitations, there are some important controls on biodegradation that have a good correlation to temperature and precipitation. Damage to wood by mould growth is often expressed in terms of the Scheffer Index (*Sch*)

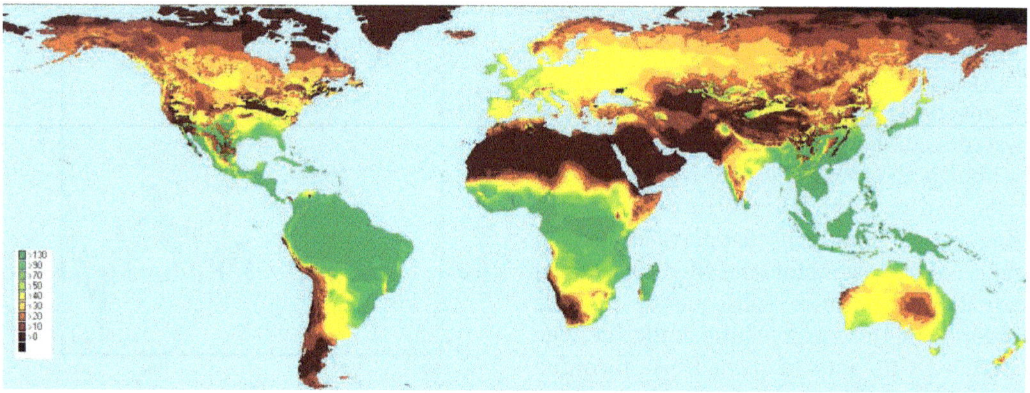

Figure 2 The global distribution of the Scheffer Index (*Sch*) for the fungal decay of wood. The colour index key represents *Sch* for that region.

(Scheffer 1971; Fernandez-Golfin *et al.* 2016) defined as:

$$Sch = \sum_{Jan}^{Dec} (D - 3)(T - 2)$$

where D is the number of wet days (defined at precipitation >0.25 mm) each month and T is the average monthly temperature, where T is greater than 2 °C. Although it is not always easy to acquire data for precipitation days >0.25 mm as these are more frequently only available for >1.0 mm, *Sch* can give a broad estimate of the relative risk for heritage materials (Fig. 2).

The development of new instruments in the early 17th century has allowed climatic changes in Europe to be tracked in considerable detail. One early, homogenized series is the Central England Temperature (CET) record, originally published in 1953 by Gordon Manley, which is updated regularly. It represents the climate of the Midlands region of England with monthly temperature values dating from 1659 and daily temperatures dating from 1772 (Manley 1974). There is a well-structured set of monthly precipitation data available for England and Wales that begins in 1766 (Simpson and Jones 2012). Many other long metereological datasets from early instrumental records are also available, providing considerable information on different regions. Rather than plot the CET, these have been used to derive output of some biological relevance such as the number of degree days each year above 10 °C (Fig. 3). Degree days allow an estimate of the total amount of heat above a threshold and are relevant to the development of vegetation and insects. For example, Figure 3 shows the estimated date of the first larva of the *Euonymus* caterpillar, a pest of trees and shrubs, in central England. This parameter has also been used to predict the likely presence of insects in heritage properties in England (Brimblecombe and Lankester 2012) and museums in Europe (Stengaard Hansen *et al.* 2012).

In many ways the simple classifications are not subtle enough to describe damage to heritage materials, which can be sensitive to a combination of factors. The simpler descriptions of meteorological observations embodied in the Köppen approach tend to neglect humidity (except in the guise of precipitation) and wind, which are important variables in damage to materials (Brimblecombe 2010a). Wind-driven rain can force water into porous building materials and create suitable conditions for mould growth, etc. Relative humidity (RH) can also present more favourable conditions for microbiological growth. As much sensitive material is stored within buildings, humidity then becomes a key variable. Nevertheless, airflow may be important 'in restricting mould growth because it can disrupt the stability of surface layers of air' (Scott 1994: 194). However, the role played by air movements in terms of mould spore deposition is not entirely clear. It is often recommended

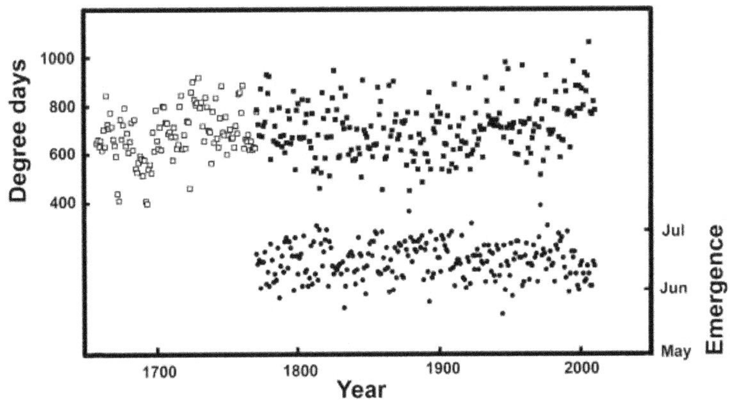

Figure 3 Degree days above 10 °C each year for central England since 1659 (left axis). The years lacking daily data (open squares) could only be estimated after tuning the data with the later daily values (data from http://www.metoffice.gov.uk/hadobs/hadcet/data/download.html). Filled circles represent the date of emergence (right axis) for the first larva of the *Euonymus* caterpillar estimated on the basis that this requires 126 degree days.

that air should be free to circulate behind fixed items of furniture in order to avoid the build-up of stagnant air with high RH.

These problems have led to a number of specialist approaches to climatology, such as building climatology, which aims to achieve a comfortable building climate together with an energy-saving structural design. The notion of a specific material climatology is reflected in the work by Brischke *et al.* (2008) on monitoring the 'material climate'. The Noah's Ark project (Sabbioni *et al.* 2010) on heritage and climate change quickly identified the need to develop heritage climatology (Brimblecombe 2010a). Heritage climatology recognizes the importance of a number of critical meteorological parameters, which are not always directly relevant to most traditional branches of meteorology. These often derive from combinations as shown in Figure 4, which illustrates the combination of meteorological parameters and the impact on heritage materials.

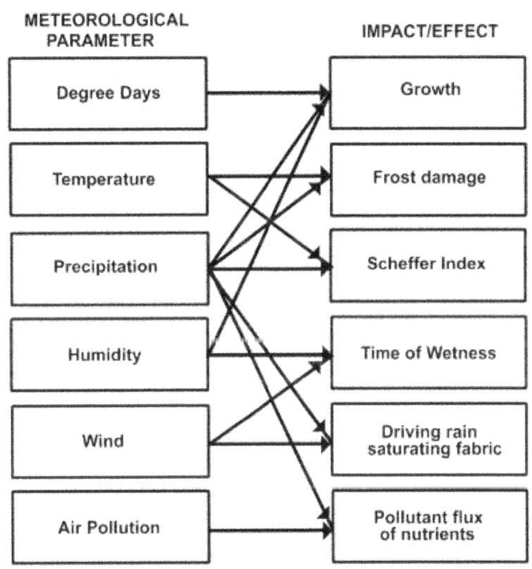

Figure 4 Combinations of meteorological parameters and some associated effects and impacts on heritage materials.

FUTURE CLIMATE

The issue of the 'greenhouse effect' is widely discussed, but climate change can affect far more than just temperature. However, temperature remains the best understood change and the increase is likely to be experienced across much of the globe. The report by the Intergovernmental Panel on Climate Change, *Climate Change 2013: Synthesis Report* is a reminder that there has been a century-long trend (1906–2005) indicative of an average global increase in land and ocean surface temperature of 0.85 [0.65 to 1.06] °C from 1880 to 2012. The 30-year period from 1983 to 2012 was probably the warmest in the Northern

Hemisphere for at least 800 years. The best estimates of the rise in global surface temperature by the end of the 21st century have been made using a range of scenarios from the stringent emission mitigation in RCP2.6 to one with very high greenhouse gas emissions as RCP8.5. These representative concentration pathways (RCPs) trace the likely future changes in greenhouse gases and allow the possible global temperature change to be estimated. RCP8.5 suggests an increase of 3.7 °C (range 2.6–4.8 °C), while RCP2.6 suggests 1.0°C (0.3–1.7 °C). Regional changes could be very different, however, most notably in high northern latitudes where the increase may be much closer to 10 °C.

Precipitation is more difficult to predict but increases are likely in high latitude regions with decreases in most mid-latitude and dry subtropical land regions. In areas affected by monsoons, precipitation will probably increase. Extreme precipitation events over most mid-latitude land and wet tropical masses as well as regions will probably intensify and there is likely to be increased variability in precipitation related to the El Niño/Southern Oscillation.

Over the next 100 years, therefore, it seems likely that there will be novel pressures on the natural and material environment that obviously threaten heritage assets. This will include changes to temperature, precipitation, extreme climate events, soil conditions, groundwater and sea levels that will affect our built heritage, the constituent materials and cultural landscapes. A number of projects such as Noah's Ark have mapped changes as a vulnerability atlas aimed at strategic management of future climate impacts (Sabbioni *et al.* 2010).

As previously mentioned, the changes are not equally spread across the globe. Frosts are likely to decrease in temperate climates such as the UK so it is possible that a wider range of insects will survive the winter. The impact of climate change is a process that will continue to be monitored and studied, but current evidence of its effects can already be seen. In high latitudes, where temperature changes are likely to be greatest with longer periods of temperatures above freezing, there would be a loss of permafrost. Thus, the middens of Viking Greenland, which record the Nordic settlements, are being disrupted (Brimblecombe 2010b). Similarly, there may be loss of archaeological information at more temperate sites. In the UK, dry summers cause a decrease in the water tables, exposing and damaging waterlogged wood, a process underway at Roman Alchester (Sauer 2005). As summers in the UK are expected to be drier in the future, a decrease in summer precipitation will lower water tables (see the paper by Charlotte Björdal on submerged archaeological wooden structures, in this volume) and more broadly cause the landscape to look very different.

Climate change is slow but it is likely to be accompanied by a greater frequency of extreme events. It is widely believed that violent windstorms and hurricanes will increase. Atlantic tropical cyclone (hurricane) activity, as measured by both frequency and the Power Dissipation Index (which combines storm intensity, duration and frequency) has also increased. The change in extreme weather is often of particular concern because of its impact on our lives (Easterling *et al.* 2000). It has also been argued that there will be a higher incidence of floods but this is less certain (Mudelsee *et al.* 2003) as they are influenced by drainage basins, which can reflect much broader environmental change.

As already stated, slow climatic changes make it possible to carefully plan preventive actions and responses to the protection of materials and structures of the historic built environment against any threat. Interest often focuses on the possibility that damage to heritage may be accelerated or worsened by climate change. However, it should be recognized that other changes (such as a reduction in frost weathering) might lead to less damage therefore any long-term strategy should focus on issues of the greatest threat.

PROPAGATING FUTURE INDOOR CLIMATES

Many estimates have been proposed relating to the changes in the external environment, both on a global and regional scale. However, fewer studies have been made to extrapolate from external climate change data the likely effects on internal environments. This is partially due to the perception that we have greater control of indoor environments – that it is just a matter of air conditioning. However, active control of indoor temperature and RH is not always appropriate within historic buildings. A number of studies of future indoor climate have been undertaken to establish energy requirements for the UK National Archives (Hong *et al.* 2012), National Trust properties (Blades *et al.* 2011), and across Europe (Leissner *et al.* 2015).

Others have determined future indoor environments more from a desire to assess the likely preservation issues. While temperature and RH seem the most obvious parameters, driving rain can penetrate walls and roofs, saturating material and dampening building fabric, increasing the risk of mould. Although interiors are protected from the climate, the less durable and organic materials are often more sensitive to damage – particularly those within European historic properties with limited levels of active climate control – making them more vulnerable.

Future internal conditions are generally assessed using a range of simulation techniques, which can estimate thermohygrometric parameters (Camuffo *et al.* 2014; Lankester and Brimblecombe 2012a). The approaches vary from simple transfer functions (Lankester and Brimblecombe 2012b; Kramer *et al.* 2013) to the use of sophisticated building simulation models (Leissner *et al.* 2015). These combine thermal building simulation with a hygrothermal simulation and can account for the type of use in the building, such as changing visitor flow or activities during special events. Simulations applied to buildings with active climate control such as heating, ventilation and air conditioning (HVAC) allow an understanding of any changes in the use or structure of rooms. The simple transfer functions generally only parameterize the difference between the outdoor and indoor temperature and RH.

However, even when complex models are employed, initialization to parameterize entire rooms or buildings can be so labour intensive that conditions are often calculated for a limited number of idealized rooms. For example, Lankester and Brimblecombe (2012a) modelled a simple room at seven different locations across Europe. Leissner *et al.* (2015) used a more detailed and thorough approach, creating more than 50,000 thematic maps in the Climate for Culture programme in which an automated model adopted 16 generic types of building interiors. All shared the same rectangular layout but had different sizes, window areas, construction and moisture buffering capacity.

These simulations suggest, not surprisingly, that room temperatures increase over time (Lankester and Brimblecombe 2012a). Leisner *et al.* (2015) studied indoor temperature ranges and suggested that in the future, Northern Europe (including Germany and Poland) will show a reduction in the indoor temperature range of up to 5 °C while Western and Southern Europe are likely to exhibit an increase of up to 4 °C. In the modelling of Amerongen Castle, a Dutch state monument built in the 17th century, it seemed likely that there would only be a small increase in the water content of air throughout the century (Leissner *et al.* 2015). Given the increase in temperature in the rooms, this agrees with Lankester and Brimblecombe (2012b), who argue for stable interior RH over the current century at Knole, a large historic house in southern England.

CHANGING THREAT OF BIODETERIORATION

The risks posed by climate change will not necessarily bring about new types of damage, nor will the threats be novel as they may already have been experienced in areas of warmer climates in the past. An understanding of the future risks will allow management strategies to take advantage of prior experiences as 'future indoor climates often already exist in other locations [and the] challenges have been managed in the past' (Lankester and Brimblecombe 2012a: 6).

Outdoors

The deterioration of outdoor materials, most notably building surfaces, has been the subject of many studies over the last 100 years, initially in response to the degradation of stone and metals caused by coal smoke (Brimblecombe and Grossi 2008). Sulfur dioxide concentrations led to a period of intense pollution and extremely high rates of stone recession and metal corrosion (Brimblecombe and Grossi 2009). Pollution has changed considerably with the move away from solid fuels such as coal and a predominance of petroleum-based fuel used for urban transport. Phytotoxic sulfur dioxide no longer dominates the urban atmosphere or reduces the biofilm on facades. The atmosphere now is richer in nitrogen compounds, which may enhance biological growth. In general terms there are concerns that climate change may reduce lichen diversity (Rubio-Salcedo *et al.* 2017; Ellis *et al.* 2007). Lichens are especially sensitive to sulfur and their prevalence in cities was much reduced in the past, although there is now evidence that they have been able to recolonize as the levels of sulfur dioxide have declined (Sérgio *et al.* 2016). The enhanced biofilm on urban facades has altered the appearance of buildings. Grossi and Brimblecombe (2008) studied the changing colour of a cathedral since its construction in 1910 to include future predictions to the year 2060. They suggest that warmer climate and altered water relations on the building surfaces could encourage the growth of organisms that contribute to biofilms.

The depth of moisture penetration in building stone was considered in studies by Smith *et al.* (2011) in Northern Ireland, where feedback associated with algal biofilms aiding moisture retention further facilitated salt penetration into stone. This deeper salt penetration might delay the onset of salt weathering, but stored salts within the stone will enhance future deterioration. Nevertheless, the greening has aesthetic impacts and the biofilm can trap dark coloured pollutant particles (Cutler *et al.* 2013). Further examples of bio-discoloration can be found in remote rural environments. For example, facades on the Mayan monuments in the Yucatan, which were once whitish, now have a pink coloration resulting from the presence of the algae, *Trentepohlia umbrina* (Gaylarde *et al.* 2006). Algal growth may result from changing humidity in central parts of the Yucatan.

Wood surfaces are also affected by the changing impacts of wind-driven rain, which allows droplets to penetrate deeper and saturate the wood. As more frequent storms are likely in the future, wood could remain wet for longer periods with the increased depth of wetting and changed seasonal cycles along with long-term trends in RH. Humidity cycles can cause substantial changes in the water content of wood, leading to stresses that form cracks. The altered seasonality of humidity may also affect the distribution of mould attack.

Climate scenarios suggest that soil moisture is likely to decline (Brimblecombe 2014) thereby exposing waterlogged wood, an important archaeological resource. The Historic Buildings and Monuments Commission for England suggests that the loss of waterlogged wood is likely to increase over the next 100 years as summers become hotter and drier (Jones 2010). Physical degradation is not the only problem – exposed wood will be subject to biological attack leading to loss of cell wall material. Anaerobic bacteria metabolizing in anoxic conditions cause deposition of iron sulfides in the cells of waterlogged wood and the subsequent oxidation can lead to the disintegration of lignin. Fungi, which require oxygen, will benefit from exposure and their growth will be encouraged by the significant amounts of water contained in the wood, leading to the loss of cell wall integrity, discoloration and disintegration.

Weather, especially temperature, has a strong impact on insect growth and survival, therefore logically a variation in climate is likely to result in an altered range. The shift in insect range has increasingly been modelled by software such as CLIMEX (Jung *et al.* 2016). However, the literature reveals surprisingly few documented examples of climate change causing an expansion of range (Battisti and Larsson 2015) and the potential for more frequent pest outbreaks (Hof and Svahlin 2016). The issue of climate change and pests has been widely examined, mainly because of its importance to future food security (Björkman and Niemelä 2015). Insects will bore into comparatively dry wood (8%

moisture content) and cause rapid disintegration (Jones 2010). Vodka beetles (*Attagenus smirnovi*) introduced into the UK in the 1970s, are now established in many museums and houses in the London area. This species does not live outdoors – adult insects fly only in the warmer indoor temperatures and are often attracted to light, increasing the potential for a spread of damaging infestations (Brimblecombe and Lankester 2012).

Purely climatic predictors should be viewed with caution as there are other potential controlling factors that may play a role in insect range such as differences in maximum daily temperatures, available food sources and geographical and ecological issues. Zhao *et al.* (2016) have suggested that in the case of six insect pests from China, landscape change may have a greater impact than climate change. In their study of probable changes in Swedish boreal forests, Hof and Svahlin (2016) report that warm and dry summers in the future may elevate stress levels in trees and enhance vulnerability to attack. They also note that the presence of one pest may attract another; for example, pheromones used by pioneer beetles attract other beetles.

In Japan, wood-boring beetles can be especially deleterious to temple complexes such as that at Nikkō, which is set within a forested landscape. In 2008, severe damage was evident in some wooden structural elements of Sanbutsudō, which appeared to be caused by *Priobium cylindricum*, a species of deathwatch beetle that is a rare anobiid in Japan (Komine *et al.* 2009). Although conservation efforts have overcome the immediate problems there are still concerns so the situation is being monitored (Komine *et al.* 2017). Long-term environmental change at Nikkō poses a range of threats (Brimblecombe and Hayashi 2018).

Indoor impacts

Future climate variations may cause physical damage. For example, higher temperatures might cause historic paper to degrade more rapidly (Lankester and Brimblecombe 2012a). The increasing release of formic or acetic acid at higher temperatures may further exacerbate the damage (Brimblecombe and Grossi 2012). The biggest threat to organic materials in internal environments is likely to be biological attack by fungi or bacteria, both of which can grow on a wide range of substrates. Although the prospect of climate warming is seen as a possible driver of insect numbers in the heritage environment, food and habitat availability are also important when considering a pest management practice.

It has been suggested that mould, which is commonly found in buildings, may become an increasing problem as the climate changes. Extreme amounts of precipitation, intense wind-driven rain or long periods of damp conditions may lead to some surfaces never fully drying out. A recent study of Blickling Hall in Norfolk, an important Jacobean building, suggests that the thermohygrometric conditions are suitable for the growth of mould (Aktas *et al.* 2017). It is not just a question of water – in some areas warmer monthly temperatures encourage rapid growth. A changing climate could increase both the available moisture and the indoor temperature. Lankester and Brimblecombe (2012b) explored the possible future changes in mould growth at Knole, where patches of mould can occasionally be observed on paintings in the Leicester Gallery. Modelling suggested that the potential for mould germination in this gallery is high whereas in the Cartoon Gallery it is lower due to the unfavourable temperature and humidity. However, this is likely to increase during the 21st century (Lankester and Brimblecombe 2012b). The maps created by Leissner *et al.* (2015) suggest that at latitudes greater than 50 degrees in Europe, the potential for damaging mould growth, particularly in Southern Sweden and Ireland, will increase towards the end of this century.

Warmer conditions have the potential to increase the number of insect pests in historic properties. Stengaard Hansen *et al.* (2012) determined the number of days over 20 °C in Scandinavian museums as a predictor of the effect of climate change on the activity of the vodka beetle. The potential for increased attack due to extra heat has been explored in the abovementioned Cartoon Gallery, which has suffered from damage caused by moths for hundreds of years (Brimblecombe and Lankester 2013). There is some support for the outcomes of such modelling exercises as there has been a higher prevalence of insects such as the webbing clothes moth (*Tineola bisselliella*) in historic properties over the last decade (Pinniger 2009; Brimblecombe and Brimblecombe 2015).

There is also evidence that novel and unusual insects may thrive in a warming climate, such as the rare *Priobium cylindricum*, which seems to have caused considerable damage at Nikkō (Harada *et al.* 2014). Brimblecombe *et al.* (2013) looked for evidence of change in insects caught using a decade of insect trapping data from properties managed by English Heritage. The evidence depended on the particular species: for example, the catch of booklice seemed to increase over the period 2000–2012, while the average catch of woolly bears (the larvae of *Anthrenus* spp.) suggests a decrease. A study of insects trapped in historic properties in London failed to reveal evidence that increases over time could be attributed to rising temperature (Brimblecombe and Brimblecombe 2015) but a wider range of changes such as habitat availability and food is likely to be an important factor in addition to environmental change.

A range of potential changes to other processes can also be affected by microbiology. Tarnowski *et al.* (2004) identified exopolymers that cause dust to adhere to textile fibres, which vacuuming fails to remove once the biofilm is established. The development of such films is controlled by humidity. At high RH, dust particles can very quickly become glued to fabrics by exopolymers or through calcification (Brimblecombe *et al.* 2009).

Biology and visitor effects

Although many threats in the 21st century will be attributable to direct shifts in climate, they are likely to be accompanied by broader changes in cultural landscapes. Changes to flora and fauna at heritage sites may result in new and exotic diseases in the gardens. New species will be incorporated and flowering periods will differ. Viewing the cherry blossom (*Hanami*) is an important traditional Japanese custom to celebrate the arrival of spring. Such celebrations are affected by climate change as the blooming dates in rural areas occur earlier (Allen *et al.* 2013). At the temples and shrines of Nikkō, the Yayoi Festival takes place between April 13 and 17 and floats decorated with cherry blossom parade through the site. The date is fixed, but now the the trees have finished flowering, so the floats often have to use artificial blossoms (Brimblecombe and Hayashi 2017). Viewing the autumnal colours of trees is another important custom, but the commencement of the autumnal tint to the Japanese maple has become progressively delayed – by more than 15 days over the last 50 years (Allen *et al.* 2013).

In Cornwall, frosts are now so rare that exotic plants thrive and in summer undesirable algae flourish in the warmer waters of rivers. New insect pests may move northward as seen with the appearance of the barkfly (*Atlantopsocus adustus*), normally found in Madeira, in the southwest of England (Alexander 2007). There is already discussion of new species that can potentially damage historic materials, but conversely, insects often live at temperatures that are close to the upper limits of their tolerance, so increases in temperature can be also be detrimental. In some cases, they form part of the heritage landscape such as the beautiful insects of considerable cultural importance that inhabit the Sun Moon Lake Butterfly Garden in Taiwan.

Extreme events and indoor damage

Although concerns mainly focus on the possibility of greater frequency or intensity in the future of extreme events such as hurricanes and typhoons, even if there is no increase, their tracks may shift. In August 2005, Hurricane Katrina caused flooding at Loyola University's library and the Earl K. Long library at the University of New Orleans, which heightened the risk of mould attack (Carlson 2005). Just three years later, during Hurricane Gustav, high winds removed loose ceiling tiles from the Earl K. Long library, a reminder that the threat continues. There are also worries about more frequent flash floods, potentially caused by changing climate, river hydrology and urban development. A flash flood swept down Manoa Valley inundating the library at the University of Hawaii in October 2004; water surged 3 m deep on the ground floor, depositing mud and enhancing the risk of mould growth (Davis 2007).

PREVENTIVE STRATEGIES AND RESPONSE

Not surprisingly, good housekeeping within museums and historic collections is key to the preservation of heritage materials as detailed in *The National Trust Manual of Housekeeping* (2011). This means the reduction of dirt and food accumulation as well as regular inspection of hidden nooks and crannies where mould, insects and vermin can flourish. Good management also requires careful inspection of incoming loan collections that may carry novel pests. Building maintenance is similarly a key issue – drainage across the walls, water leaks and broken windows need to be repaired quickly. In a changing climate, traditional guttering and pipes may be unable to cope with the increased flow of water. Enhanced visitor numbers also mean more dust, possible biological hazards, etc. – it is more difficult to manage properties as their popularity and dust deposition increases (Lloyd *et al.* 2007). Disaster management plans have acquired great importance especially as organizations traditionally responsible for civil emergencies often give cultural heritage protection only a low priority (Spennemann and Graham 2007). In Japan, however, there is strong public perception of the significance of cultural heritage and a desire to be involved in its recovery (Brimblecombe and Hayashi 2017).

Maintaining an environment appropriate to preventive conservation is vital. Generally, integrated pest management (IPM) is regarded as beneficial and relatively free from risk (Querner 2015). Some solutions suggest improving preservation by cleaning incoming air through elaborate air conditioning and filters. More direct approaches can be used close to objects, such as maintaining electric fields around books to prevent airborne spores from settling (Takikawa *et al.* 2014).

One approach aims to extend the maintenance survey to include a more formalized study in order to focus on potential future problems. This can be seen in the environmental monitoring at Blickling Hall by Aktas *et al.* (2017), who made a detailed assessment of moisture and seasonal changes in the indoor climate. With modern tools these changes can be predicted. Surveys of bacterial abundance (Tarnowski *et al.* 2004), mould prevalence (Smith 1997) and insect populations (Brimblecombe *et al.* 2013) can contribute to a strategy for managing biodeterioration. More elaborate monitoring can be both subtle and sophisticated. A number of studies have suggested that microbiological volatile organic compounds (VOCs) emitted during the decay of paper or mould growth may provide an early warning of potential problems. Mould infestations can also be detected (Micheluz *et al.* 2016). Furfural derived from acid hydrolysis and oxidation of cellulose is a marker of carbohydrate degradation (Gibson *et al.* 2012).

Deathwatch beetles (*Xestobium rufovillosum*) typically infest buildings where the timbers are decaying due to the presence of fungi (Belmain *et al.* 2002). The beetle orientation is affected by the VOCs from fungal decay, so detection of these compounds could be part of integrated pest management and used as attractants in traps. Modelling also represents an important means of assessing the future impacts on heritage. The Noah's Ark project (Sabbioni *et al.* 2009) and the Climate for Culture project (Leissner *et al.* 2015) have both sought to model future climate impacts on cultural heritage. Such approaches can be extended to flooding or storm frequency.

Although prevention is better than cure, well-structured plans for recovery after extreme events such as flooding are important. These may initally involve the freezing of collections damaged by water to prevent mould attack, as undertaken by the University of Hawaii library after the flood of 2004. In this instance, disaster recovery required taking action before mould could grow: over 60,000 maps and 70,000 aerial photographs were frozen within 72 hours (Davis 2007). Once mould outbreaks and insect infestation have already taken hold, more aggressive intervention may be required involving chemicals such as chlorine dioxide to eradicate mould (Southwell 2003) or fumigation with sulfuryl fluoride to remove wood-boring anobiids (Harada *et al.* 2014). Once building fabric has become saturated with water after flooding it should be dried out with care to minimize damage (Blades *et al.* 2008).

CONCLUSIONS

The exposure of heritage assets to climate and air pollution will vary in the future. Although there is much controversy surrounding climate predictions, there seems little doubt that the world will become warmer, and rainfall and storm patterns will change, even if we cannot be sure of the magnitude. As already stated, climate change occurs slowly, allowing time for careful long-term strategic planning, but all too often this lack of speed means that more immediate concerns take precedence and simple approaches, such as increased attention to maintenance, inspection and surveys to reduce the potential problems, are neglected.

REFERENCES

Aktas, Y.D., D'Ayala, D., Blades, N. and Calnan, C. 2017. 'An assessment of moisture induced damage in Blickling Hall in Norfolk, England, via environmental monitoring', *Heritage Science* 5(1): 5.

Alexander, K.N.A. 2007. '*Atlantopsocus adustus* (Hagen) (Psoc.: Psocidae) new to Britain from East Cornwall', *Entomologists Record* 119: 76.

Allen, J.M., Terres, M.A., Katsuki, T., Iwamoto, K., Kobori, H., Higuchi, H., Primack, R.B., Wilson, A.M., Gelfand, A. and Silander, J.A. 2013. 'Modeling daily flowering probabilities: expected impact of climate change on Japanese cherry phenology', *Global Change Biology* 20: 1251–63.

Battisti, A. and Larsson, S. 2015. 'Climate change and insect pest distribution range', in *Climate Change and Insect Pests*, C. Björkman and P. Niemelä (eds), 1–15. Oxford: CABI.

Belmain, S.R., Simmonds, M.S. and Blaney, W.M. 2002. 'Influence of odor from wood-decaying fungi on host selection behavior of deathwatch beetle, *Xestobium rufovillosum*', *Journal of Chemical Ecology* 28(4): 741–54.

Björkman, C. and Niemelä, P. 2015. *Climate Change and Insect Pests*. Oxford: CABI.

Blades, N., Cassar, M. and Biddulph, P. 2008. 'Optimizing drying strategies to reduce down times for actively-used flood damaged historic buildings', *Studies in Conservation* 53(suppl. 1): 104–8.

Blades, N., Poupard, S. and Barber, L. 2011. 'Analysing the energy consumption of conservation heating systems at the National Trust', *Journal of the Institute of Conservation* 34: 16–27.

Brimblecombe, P. 2010a. 'Heritage climatology', in *Climate Change and Cultural Heritage*, R.A. Lefevre and C. Sabbioni (eds), 54–7. Bari: Edipuglia.

Brimblecombe, P. 2010b 'Monitoring the future', in *Climate Change and Cultural Heritage*, R.A. Lefevre and C. Sabbioni (eds), 73–8. Bari: Edipuglia.

Brimblecombe, P. 2014. 'Refining climate change threats to heritage', *Journal of the Institute of Conservation* 37(2): 85–93.

Brimblecombe, P. and Brimblecombe, C.T. 2015. 'Trends in insect catch at historic properties', *Journal of Cultural Heritage* 16(2): 127–33.

Brimblecombe, P. and Grossi, C.M. 2008. 'Millennium-long recession of limestone facades in London', *Environmental Geology* 56(3–4): 463–71.

Brimblecombe, P. and Grossi, C.M. 2009. 'Millennium-long damage to building materials in London', *Science of the Total Environment* 407: 1354–61.

Brimblecombe, P. and Grossi, C.M. 2010. 'Potential damage to modern building materials from 21st century air pollution', *Scientific World Journal* 10: 116–25.

Brimblecombe, P. and Grossi, C.M. 2012. 'Carbonyl compounds indoors in a changing climate', *Chemistry Central Journal* 6: 21.

Brimblecombe, P. and Hayashi, M. 2017. 'Perception of the relationship between climate change and traditional wooden heritage in Japan', in *Public Archaeology and Climate Change*, D. Dawson, C. Nimura, E. Lopez-Romero and M.Y. Daire (eds), 288–302. Oxford: Oxbow Books.

Brimblecombe, P. and Hayashi, M. 2018. 'Pressures from long term environmental change at the shrines and temples of Nikkō', *Heritage Science* 6(1): 1–12.

Brimblecombe, P. and Lankester, P. 2012. 'Long-term changes in climate and insect damage in historic houses', *Studies in Conservation* 58(1): 13–22.

Brimblecombe, P., Thickett, D. and Yoon, Y.H. 2009. 'The cementation of coarse dust to indoor surfaces', *Journal of Cultural Heritage* 10(3): 410–14.

Brimblecombe, P., Brimblecombe, C.T., Thickett, D. and Lauder, D. 2013. 'Statistics of insect catch within historic properties', *Heritage Science* 1(1): 34.

Brischke, C., Rapp, A.O., Bayerbach, R., Morsing, N., Fynholm, P. and Welzbacher, C.R. 2008. 'Monitoring the "material climate" of wood to predict the potential for decay: results from *in situ* measurements on buildings', *Building and Environment* 43: 1575–82.

Camuffo, D., Bertolin, C., Bonazzi, A., Campana, F. and Merlo, C. 2014. 'Past, present and future effects of climate change on a wooden inlay bookcase cabinet: a new methodology inspired by the novel European Standard EN 15757:2010', *Journal of Cultural Heritage* 15(1): 26–35.

Carlson, S. 2005. 'Mold and muck threaten library collections', *Chronicle of Higher Education* 52(4): 0–A22.

Chen, D. and Chen, H.W. 2013. 'Using the Köppen classification to quantify climate variation and change: an example for 1901–2010', *Environmental Development* 6: 69–79.

Cutler, N.A., Viles, H.A., Ahmad, S., McCabe, S. and Smith, B.J. 2013. 'Algal "greening" and the conservation of stone heritage structures', *Science of the Total Environment* 442: 152–64.

Davis, L.A. 2007. 'Riding the surf: dealing with library disasters in island communities', *Public Library Quarterly* 25(3–4): 99–112.

Easterling, D.R., Meehl, G.A., Parmesan, C., Changnon, S.A., Karl, T.R. and Mearns, L.O. 2000. 'Climate extremes: observations, modeling impacts', *Science* 289: 2068–74.

Ellis, C.J., Coppins, B.J., Dawson, T.P. and Seaward, M.R.D. 2007. 'Response of British lichens to climate change scenarios: trends and uncertainties in the projected impact for contrasting biogeographic groups', *Biological Conservation* 140(3–4): 217–35.

Evans, J.K. 1981. 'Wheat production and its social consequences in the Roman world', *Classical Quarterly* 31(2): 428–42.

Fernandez-Golfin, J., Larrumbide, E., Ruano, A., Galvan, J. and Conde, M. 2016. 'Wood decay hazard in Spain using the Scheffer Index: proposal for an improvement', *European Journal of Wood and Wood Products* 74: 591.

Gaylarde, P., Englertb, G., Ortega-Morales, O. and Gaylarde, C.C. 2006. 'Lichen-like colonies of pure *Trentepohlia* on limestone monuments', *International Biodeterioration and Biodegradation* 58: 119–23.

Gibson, L.T., Ewlad-Ahmed, A., Knight, B., Horie, V., Mitchell, G. and Robertson, C.J. 2012. 'Measurement of volatile organic compounds emitted in libraries and archives: an inferential indicator of paper decay?', *Chemistry Central Journal* 6(1): 42.

Grossi, C.M. and Brimblecombe, P. 2008. 'Past and future colouring patterns of historic stone buildings', *Materiales de Construccion* 58(289–290): 143–60.

Harada, M., Kigawa, R., Komine, Y., Fujii, Y., Fujiwara, Y. and Kawanobe, W. 2010. 'Survey of records of restoration in relation with damage by wood-boring insects at Rinnohji Temple in Nikkō', *Science for Conservation* 48: 165–70 (in Japanese).

Harada, M., Kigawa, R., Komine, Y. and Fujii, Y. 2014. 'Report on fumigation process conducted at Rinno-ji Temple', *Hozonkagaku* 53: 215–24 (in Japanese).

Hof, A.R. and Svahlin, A. 2016. 'The potential effect of climate change on the geographical distribution of insect pest species in the Swedish boreal forest', *Scandinavian Journal of Forest Research* 31(1): 29–39.

Hong, S.H., Strlič, M., Ridley, I., Ntanos, K., Bell, N. and Cassar, M. 2012. 'Climate change mitigation strategies for mechanically controlled repositories: the case of the National Archives, Kew', *Atmospheric Environment* 49: 163–70.

Jones, D.M. 2010. *Waterlogged Wood*. Swindon: English Heritage.

Jung, J.M., Lee, W.H. and Jung, S 2016. 'Insect distribution in response to climate change based on a model: review of function and use of CLIMEX', *Entomological Research* 46: 223–35.

Kaniewski, D., Van Campo, E., Guiot, J., Le Burel, S., Otto, T. and Baeteman, C. 2013. 'Environmental roots of the Late Bronze Age crisis', *PLoS One* 8(8): e71004.

Komine, Y., Kigawa, R., Harada, M., Fujii, Y., Fujiwara, Y. and Kawanobe, W. 2009. 'Damage by a rare kind of anobiid, *Priobium cylindricum*, found during restoration work of Sanbutsu-do, Rinnohji Temple in Nikko', *Science for Conservation* 48: 207–13 (in Japanese).

Komine, Y., Harada, M., Saito, A., Saito, Y., Kigawa, R. and Fujii, Y. 2017. 'Report on a new insect monitoring method in wooden historic buildings in Nikkō', *Science for Conservation* 56: 77–88 (in Japanese).

Kramer, R.P., van Schijndel, A.W.M. and Schellen, H.L. 2013. 'Inverse modeling of simplified hygrothermal building models to predict and characterize indoor climates', *Building and Environment* 68: 87–99.

Lankester, P. and Brimblecombe, P. 2012a. 'Future thermohygrometric climate within historic houses', *Journal of Cultural Heritage* 12: 1–6.

Lankester, P. and Brimblecombe, P. 2012b. 'The impact of future climate on historic interiors', *Science of the Total Environment* 417: 248–54.

Leissner, J., Kilian, R., Kotova, L., Jacob, D., Mikolajewicz, U., Broström, T. and Antretter, F. 2015. 'Climate for culture: assessing the impact of climate change on the future indoor climate in historic buildings using simulations', *Heritage Science* 3(1): 38.

Leslie, G. and Lunau, F. 1992. *Indoor Air Pollution: Problems and Priorities*. Cambridge: Cambridge University Press.

Lloyd, H., Brimblecombe, P. and Lithgow, K. 2007. 'Economics of dust', *Studies in Conservation* 52(2): 135–46.

Manley, G. 1974. 'Central England temperatures: monthly means 1659 to 1973', *Quarterly Journal of the Royal Meteorological Society* 100: 389–405.

Marcott, S.A., Shakun, J.D., Clark, P.U. and Mix, A.C. 2013. 'A reconstruction of regional and global temperature for the past 11,300 years', *Science* 339(6124): 1198–201.

Micheluz, A., Manente, S., Rovea, M., Slanzi, D., Varese, G.C., Ravagnan, G. and Formenton, G. 2016. 'Detection of volatile metabolites of moulds isolated from a contaminated library', *Journal of Microbiological Methods* 128: 34–41.

Moberg, A., Sonechkin, D.M., Holmgren, K., Datsenko, N.M. and Karlén, W. 2005. 'Highly variable Northern Hemisphere temperatures reconstructed from low- and high-resolution proxy data', *Nature* 433: 613–17.

Mudelsee, M., Börngen, M., Tetzlaff, G. and Grünewald, U. 2003. 'No upward trends in the occurrence of extreme floods in central Europe', *Nature* 425: 166–9.

National Trust 2011. *The National Trust Manual of Housekeeping*. Swindon: National Trust.

Pinniger, D.B. 2009. 'Clothes moths numbers are surging', *Pest Magazine* 4: 17–19.

Querner, P. 2015. 'Insect pests and integrated pest management in museums, libraries and historic buildings', *Insects* 6(2): 595–607.

Rubio-Salcedo, M., Psomas, A., Prieto, M., Zimmermann, N.E. and Martínez, I. 2017. 'Case study of the implications of climate change for lichen diversity and distributions', *Biodiversity and Conservation* 26(5): 1121–41.

Sabbioni, C., Brimblecombe, P. and Cassar, M. 2010. *The Atlas of Climate Change Impact on European Cultural Heritage: Scientific Analysis and Management Strategies*, Anthem-European Union Series. London: Anthem Press.

Sauer, E. 2005. 'Alchester: in search of Vespasian', *Current Archaeology* 196: 168–76.

Scheffer, T.C. 1971. 'A climate index for estimating potential for decay in wood structures above ground', *Forest Products Journal* 21(10): 25–31.

Scott, G. 1994. 'Moisture, ventilation and mould growth', *Studies in Conservation* 39: 149–53.

Sérgio, C., Carvalho, P., Garcia, C.A., Almeida, E., Novais, V., Sim-Sim, M., Jordão, H. and Sousa, A.J. 2016. 'Floristic changes of epiphytic flora in the Metropolitan Lisbon area between 1980–1981 and 2010–2011 related to urban air quality', *Ecological Indicators* 67: 839–52.

Simpson, I.R. and Jones, P.D. 2012. 'Updated precipitation series for the UK derived from Met Office gridded data', *International Journal of Climatology* 32: 2271–82.

Smith, B.J., McCabe, S., McAllister, D., Adamson, C., Viles, H.A. and Curran, J.M. 2011. 'A commentary on climate change, stone decay dynamics and the "greening" of natural stone buildings: new perspectives on "deep wetting"', *Environmental Earth Sciences* 63(7): 1691–700.

Smith, L.K. 1997. 'A mold by any other name: one librarian's battle against a mold bloom', *College and Undergraduate Libraries* 4(2): 57–63.

Southwell, K.L. 2003. 'Chlorine dioxide: a treatment for mold in libraries', *Abbey Newsletter* 26(6): 98–9.

Spennemann, D.H.R. and Graham, K. 2007. 'The importance of heritage preservation in natural disaster situations', *International Journal of Risk Assessment and Management* 7(6–7): 993–1001.

Stengaard Hansen, L., Akerlund, M., Grontoft, T., Rhyl-Svendsen, M., Schmidt, A., Bergh, J. and Vagn Jensen, K. 2012. 'FuturePest status of an insect pest in museums, *Attagenus smirnovi*: distribution and food consumption in relation to climate change', *Journal of Cultural Heritage* 13(1): 22–7.

Takikawa, Y., Matsuda, Y., Nonomura, T., Kakutani, K., Kimbara, J., Osamura, K. and Toyoda, H. 2014. 'Electrostatic guarding of bookshelves

for mould-free preservation of valuable library books', *Aerobiologia* 30(4): 435–44.
Tarnowski, A., McNamara, C., Bearce, K. and Mitchell, R. 2004. 'Sticky microbes and dust on objects in historic houses', in *American Institute for the Conservation of Historic and Artistic Works (AIC) 32nd Annual Meeting, Portland, OR*, 9–14.
UKEA 2004. *Reconstructed River Flow Series from 1860s to Present*. Bristol: UK Environment Agency.
Zhao, Z., Sandhu, H.S., Ouyang, F. and Ge, F. 2016. 'Landscape changes have greater effects than climate changes on six insect pests in China', *Science China* 59: 627–33.

AUTHOR'S ADDRESS

Peter Brimblecombe, School of Energy and Environment, City University of Hong Kong, Tat Chee Avenue, Kowloon, Hong Kong SAR (pbrimble@cityu.edu.hk)

BIODETERIORATION AND PRESERVATION PROCESSES

MICROBIAL PROCESSES INVOLVED IN THE DETERIORATION OF PAPER AND PARCHMENT

Flavia Pinzari

HISTORY OF PAPER AND PARCHMENT BIODETERIORATION

The cultural role played by parchment and paper throughout human history is incalculable: manuscripts, ancient books and archival materials have been the most important repositories of our recorded history from the earliest times. Examples include the Dead Sea Scrolls, the Archimedes Palimpsest, the Magna Carta and Leonardo da Vinci's codices (Sterflinger and Pinzari 2012). An awareness of the historical information that ancient documents can still transmit has increased in recent decades with greater collaboration among researchers, scientists, conservators and heritage managers. The archaeological investigation of written heritage, especially ancient and medieval books, can uncover valuable clues leading to a better understanding of the historical, social and political contexts in which they circulated and were exchanged, and the people who created, read and owned them (Piñar *et al.* 2014b). This is one reason among many why it is crucial, even in the digital age, to preserve the materiality of archival and library heritage.

Books and documents are manufactured objects made chiefly of decomposable and oxidizable organic materials. From a chemical and microbiological perspective, it could be said that the transmission of texts through the ages represents a sort of 'miracle'. Indeed, texts have come down to us through many centuries, surviving wars and natural disasters and passing from one generation to the next, effectively amounting to mankind's triumph over environmental pressures. Documents made of parchment or paper inevitably undergo chemical degradation and biodeterioration processes, however, since organic materials serve as substrates for fungi and bacteria. With the passage of time, there is a high probability that saprophytic microorganisms will attack a book. The aim of conservators and those who study biodeterioration prevention strategies is to slow down, as much as possible, these natural processes (Cappitelli *et al.* 2010; Sequeira *et al.* 2014).

The biodeterioration of books is a problem that has been faced since ancient times. Marco Vitruvio Pollione (AD 25), in his *De architectura, Liber VI*, wrote that to delay deterioration and prevent damage caused by damp environments libraries should be built facing east in order to catch the morning sun. Nevertheless, several well-known public libraries ignored Vitruvius' suggested orientation, preferring to align their buildings in sympathy with their monumental context (Jason *et al.* 2013).

Two thousand years would pass before the role played by microorganisms in the biodeterioration of books was again discussed

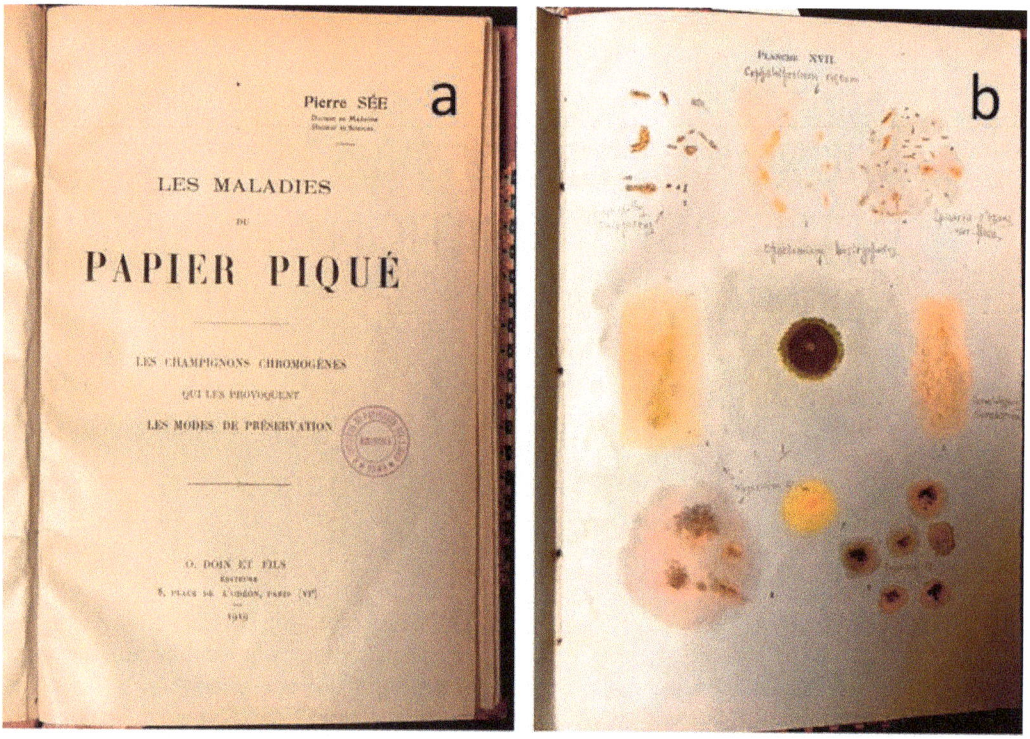

Figure 1 The first volume devoted entirely to paper biodeterioration was published by Pierre Sée in 1919, with the title *Les maladies du papier piqué*: (a) the title page and (b) one of the tables in colour accompanying the text and showing the appearance of fungal colonies of various species on paper. The annotations on this copy were made by Alfonso Gallo.

(Sibilia 1935; Gallo 1935). The study of fungi that cause blemishing and weakening of paper and parchment began in the 19th century. The first person to scientifically address the biodeterioration of paper was Christian Gottfried Ehrenberg in 1818 (Hueck 1968), although an awareness of the role played by microbes in the decay of materials existed long before this: Robert Hooke described the first microscopic observation of fruiting bodies of moulds in 1665 and Anton van Leeuwenhoek recorded bacteria and other microorganisms for the first time using a single lens microscope in 1676. In 1680, Athanasius Kircher was the first to note that 'putrid material is full of innumerable creeping animalcule' (quoted in Singer 1915).

Fungal species named on the basis of their association with paper date back to the early 1800s (e.g. *Chaetomium chartarum* named by Ehrenberg in 1818 and *Myxotrichum chartarum* named by Kunze in 1823) (Sterflinger and Pinzari 2012 and references therein). The first work devoted entirely to paper biodeterioration was produced by Pierre Sée who, in 1917, published his article *Sur les moisissures causant l'altération du papier*, and subsequently in 1919 the volume titled *Les maladies du papier piqué* (Fig. 1).

After Sée (1917), several other authors focused their studies on the microbial deterioration of books and paper. For example, Galippe (1919) published the first studies on chemical treatments to be applied to paper damaged by fungi. The Italian school for paper conservation was particularly active in the first decades of the 20th century, with some scientists, such as Alfonso Gallo, dedicating all their time to the study of both paper and parchment decay (Gallo 1935; Sibilia 1935; Verona 1939; Benveduti 1939; Bonaventura 1940; Bonaventura and Paganini 1940). In 1938, Gallo founded the Regio Istituto di Patologia del Libro (Gallo 1940, 2006) (now known as the Istituto Centrale per il Restauro e la Conservazione del Patrimonio Archivistico e Librario), an academic and scientific institution in Rome dedicated entirely to research in the area of library materials conservation.

Since then, the causes of biodeterioration in library heritage have continued to be extensively investigated by many scientists across the globe, and the main processes that cause library materials to lose their physical integrity and functionality are largely understood (Gallo 1992; Gallo *et al.* 1999; Zyska 1997; Nittérus 2000a; Florian 2002; Allsopp *et al.* 2004: 237). However, many questions remain unanswered and various forms of paper and parchment degradation still lack a comprehensive explanation (Arai 2000; Piñar *et al.* 2014a, 2015; Migliore *et al.* 2017), and the field of biological research in library materials biodeterioration remains open to investigation. Innovative technologies in the form of tools and methods using the latest scientific developments and biotechnologies provide great potential for further exploration. New techniques aimed at understanding decay mechanisms, and therefore the prevention of biodeterioration in materials, will improve our ability to conserve, restore and utilize library and archival heritage.

The costs associated with developing *ex novo* new technologies are very high: the cultural heritage sector, and even more so that of libraries and archives, cannot rely on dedicated research avenues. Advances in techniques are mostly concentrated in areas such as biomedical, pharmaceutical and agro-industry sciences. Technological innovations in biological research applied to the protection of books and documents depend to a large extent on the transfer of these new investigative and interventive techniques developed in other research areas.

Careful adaptation of the methods and experimental protocols developed in other contexts will permit the use of innovative techniques in the field of books and paper conservation, a sphere that is all too often economically uncompetitive. This has been the case over the last decade or so, with molecular techniques being applied in the study of paper and parchment only after wide application in other fields.

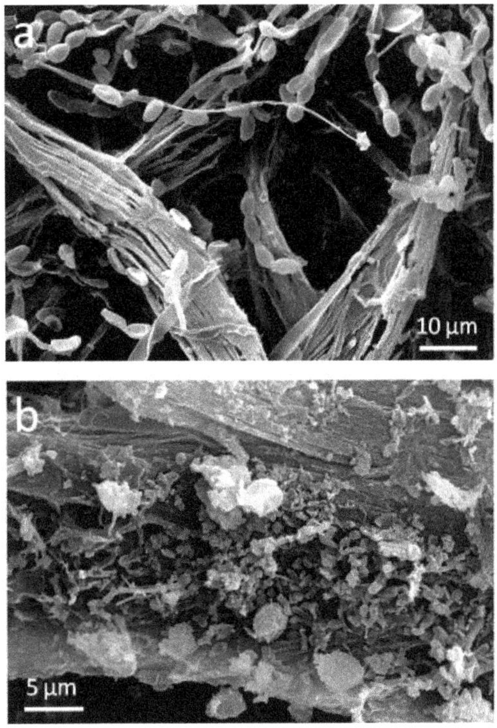

Figure 2 The structural destruction of fibres: (a) cellulose fibres that appear to be torn due to the enzymatic action of the fungus *Acrostalagmus* sp.; (b) parchment attacked by both bacteria and fungi, showing collagen fibres in a broken and depolymerized state.

The uniqueness and material heterogeneity of individual books and documents that must be investigated, or upon which action has to be taken, often demands the adaptation of ad hoc protocols and procedures because their side effects on cultural heritage can often be too invasive or their outputs not sufficiently sensitive. The need to carry out non-destructive analyses on paper and parchment by means of micro-invasive examinations and zero impact sampling often involves a huge effort being made simply to adjust procedures (Michaelsen *et al.* 2006; Pinzari *et al.* 2009, 2011).

MATERIALS AND DEGRADATION ORGANISMS

Books, documents and manuscripts are made predominantly of paper and/or parchment.

Covers or protective elements can contain wood, fabric, plastics or metals, but these generally represent only a small part of the overall object. Paper and parchment have both been used over the centuries for written materials. They were typically found and stored in similar environments and therefore share the same exposure to chemical, physical and biological deterioration factors (Mesquita *et al.* 2009; Nunes *et al.* 2013; Troiano *et al.* 2014). However, the two materials are very dissimilar in nature; they are attractive to specific microorganisms with distinct metabolisms and ecological behaviours and require different storage conditions (Fig. 2).

Parchment and paper are often combined and these volumes are subject to the same methods of preservation and aging events. The combination of these two materials can be seen in archival units that, unlike texts printed entirely on paper or written directly on parchment, are often composite objects. Archives frequently consist of multi-material collections in which parchment and paper are bound together to form a 'string', namely a bundle of documents joined together between covers by a long nail or a cord that enables a document to be removed or inserted. The leaves in the 'string' (often manuscripts) represent a documentary series that records information on past activities. Archival 'strings' are different from books in that they accommodate very different materials, usually generated over a range of time, and consist of documents made using different manufacturing methods. Protection from biodeterioration of such composite systems is even more challenging than it is for objects made of one material alone, as they provide substrates that offer a greater variety of nutrients for a broader range of saprophyte microorganisms.

Parchment

Most palimpsests, gospels and ancient texts known to modern scholars are made of parchment, a material that increased in popularity as a writing support in Western Europe from the 6th century onwards. Parchment is prepared from animal skins (Reed 1972, 1975) and although it is a far more durable material than paper or papyrus, it is still highly susceptible to degradation by microorganisms. Parchment is prepared with alkaline salts, such as lime and sodium chloride, and its chief constituent is collagen, an organic polymeric protein whose structural unit is a triple helix. Collagen helices are arranged to form fibrils that, at an upper hierarchical level, compose the final collagen fibre. During parchment manufacture, animal skin undergoes a series of processes resulting in a product composed of the dermal skin layer alone (Ryder 1969). Starting in the 4th century, dehairing was achieved by immersing the animal skins in a calcium hydroxide solution (Poole and Reed 1962). Calcium-based compounds were also used to render the surface fit for writing and to prevent inks from running. Lime, flour, egg whites and milk might be applied to the surface to make it smooth. Finished parchments also contain other compounds that derive from the skins (lipids, waxes, other peptides) and from manufacturing processes (salts and minerals, such as aluminium, silicon, magnesium, potassium, sulfur, etc.) (Piñar *et al.* 2014a, 2015). Salting (mainly with marine salt) was also commonly used in the initial phases of parchment manufacture to prevent hides from rotting.

The biodeterioration of parchment is caused by the activity of microorganisms that can metabolize proteins through the enzymatic hydrolysis of peptide bonds (Florian 2007). Microorganisms endowed with this ability are proteolytic bacteria and fungi that possess collagenases, enzymes capable of cleaving the collagen triple helix (Fig. 2b). Many microorganisms can spoil parchment components other than through the breakdown of collagen. Fungi, for example, often use as carbon sources the oils and waxes that are added during parchment manufacturing processes, further contributing to deterioration. The development of microbial agents in parchment is usually more intensive on the flesh side of the skin. In the areas most affected by microorganisms the parchment becomes rough, exhibits widespread staining, and a porous appearance that is sometimes perforated.

The studies conducted by Gallo and Strzelczyk (1971) confirmed that in the case of objects saturated with water, the most active bacteria on parchment are *Streptococcus*, *Micrococcus*, *Bacillus* and *Bacterium* genera, while in the drying phase, fungal species in genera such as *Aspergillus*, *Chaetomium*, *Cladosporium* and *Penicillium* can proliferate, since they have lipolytic and proteolytic attributes.

A previous study by Beöthy-Kozocsa *et al.* (1990) that examined about 34 codices from the 15th century showed that discoloration and structural deformations were ascribable to *Streptomyces fimicarius* strains. Kowalik (1980) reported some findings on the role played by anaerobic bacteria such as *Clostridium histolyticum*, *C. perfringens* and *C. capitovale* in parchment and collagen depolymerization, as well as the aerobic bacteria *Pseudomonas* sp., *Bacillus licheniformis* and *Achromobacter iophagus* which, according to the author, can be responsible for collagen biodegradation.

Polacheck (1989, 2005) studied the 'Aleppo' codex (which dates from the 10th century) and concluded that the discoloration and structural fragility of its parchment leaves were attributable to a fungal attack by an *Aspergillus* sp. Rebrikova and Dmitrieva (1996) conducted a series of investigations on some ancient manuscripts exhibiting obvious signs of damage such as discoloration and the microbiological destruction of collagen. The diagnosis showed the presence of numerous fungi from the Deuteromycetes and Basidiomycetes capable of proteolytic and collagenolytic activity. The authors stressed the fact that *in vitro* culturing of the microorganisms yielded negative results, while through the use of transmission electron microscopy (TEM) it was possible to identify different bacterial cells within the structure of the parchment. They attributed the discrepancy of these findings between the isolation technique and the TEM observations to the fact that the biodeterioration of parchment occurred at a different time, during which the bacteria, not the fungi, played an initial key role.

Petushkova and Koestler (1996) isolated several strains of *Bacillus subtilis* and *B. mesentericus* on a 16th-century codex. These species were characterized by proteolytic activity and proved capable of hydrolysing collagen, leading to the complete destruction of the parchment support. According to the authors, these organisms can be considered among the most important players in the decomposition of works of art supported on parchment. Similarly, Karbowska-Berent and Strzelczyk (2000) reported that *Streptomycetes* play a major role in the deterioration of many kinds of ancient parchment. These filamentous bacteria are known to produce many types of collagenases that are capable of hydrolysing collagen. As they are alkaliphiles they are therefore inclined to develop on parchment, whose pH is typically alkaline.

Among the alterations of biological origin that afflict parchment, there is one that has puzzled researchers for decades: the presence of purple spots. Using new and powerful techniques, scientists have discovered that these are associated with collagen fibre destruction, damage which has been described in the past. However, a common bacterial or fungal denominator has never been identified. The remarkable appearance and recurrence of the phenomenon in various places and on a wide range of objects inspired many scholars in the last decade of the 20th century to carry out a series of surveys on parchments that presented this kind of damage. One such survey isolated several microorganisms capable of producing violet pigments such as *Nocardia* spp., *Streptomyces* spp., *Aspergillus versicolor* (Vuill.) Tiraboschi, *Penicillium notatum* Westling and *Epicoccum* spp. (Gallo and Strzelczyk 1971). However, the current author's *in vitro* attempt at reproducing such blemishes was unsuccessful, and it proved impossible to generate such pigmented purple spots.

Using culture-independent methods and scanning electron microscopy (SEM), Piñar *et al.* (2014a) identified the fungal and bacterial communities present on a selection of parchments that exhibited the purple staining. The analyses detected the presence of both bacterial and fungal cells in the damaged areas, but the most interesting result was that the stains could be attributed to the presence of halophilic, halotolerant and proteolytic bacterial species which grew in the saline environment of the parchments (Larsen 1986). Members of the

Actinobacteria (mainly *Saccharopolyspora* spp.) and species of *Aspergillus* were detected in all investigated cases. In particular, *A. versicolor* appeared to be strongly associated with deteriorated parchment samples, suggesting a form of co-occurrence between this fungus and halophilic bacteria. The authors proposed a correspondence between purple spots on ancient parchments and that of so-called 'red heat', a phenomenon known to occur in protein-based products manufactured using marine salt.

Two recent articles confirmed the presence of *Saccharopolyspora* spp. on documents by means of DNA massive sequencing techniques (Teasdale *et al.* 2017; Migliore *et al.* 2017). Using a high-throughput DNA sequencing technique, Teasdale *et al.* (2017) obtained the microbial signature of the medieval York Gospel (York Minster Ms. Add. 1) and discovered the presence of *Saccharopolyspora* at high concentrations on the surface of different pages of the codex. Migliore *et al.* (2017), in a study of the same phenomenon observed on a parchment roll dated to AD 1244, hypothesized a two-phase model mechanism of parchment colonization where halophilic archaea, producing purple pigments, colonize parchment and are succeeded by Gammaproteobacteria with fungi as secondary colonizers. However, this hypothesis is based on a model generated by a single case study in which only three samples of damaged parchment and three undamaged samples were analysed. Moreover, in the molecular analysis carried out by the authors, fungi were not considered, even though they appeared in their succession model.

The possibility that the damage is the result of a co-occurrence of several species of microorganisms is, however, entirely plausible (Kraková *et al.* 2012). The subsequent colonization by filamentous fungi of parchment already affected by purple spots has been confirmed by two different studies in which the vitality of the fungal mycelium was verified. Pinzari *et al.* (2012a) quantified both adenosine triphosphate (ATP) and fungal β (1-4)-N-acetyl-D-glucosaminidase activity in damaged and undamaged samples, and concluded that the purple stains observed on the manuscript leaves were caused by bacteria that were no longer active and that some viable fungal mycelium was present, although not strictly associated with the stains. Piñar *et al.* (2014b) performed a molecular analysis on the Archimedes Palimpsest, also affected by purple spots, and described the presence of a homogeneous bacterial community consisting mainly of Proteobacteria and Actinobacteria. In the same study, quantification of the β-actin gene by real-time quantitative polymerase chain reaction analyses (qPCR) showed an increased fungal abundance on degraded areas in contrast to healthy areas of the parchment.

Paper

Paper has remained the most widely used writing support since the Middle Ages. For centuries, a growing number of books, documents and artistic drawings have been accumulating in libraries, archives and museums all over the world. In Italy alone, 262 million books are stored in state libraries and, within them, more than 50,000 artefacts predate the 16th century (Gallo 1992).

Paper is primarily composed of cellulose, the most abundant organic compound found in the terrestrial biosphere and the largest source of carbon on land. In plants, cellulose is one of the main elements of resistance to external mechanical and biological pressures. The degree of cellulose resistance to depolymerization mechanisms depends on the crystalline nature of its polymeric fibres. Cellulose is comprised of both crystalline and amorphous regions. It is a polysaccharide consisting of a linear chain of thousands of β-(1→4) linked D-glucose units. The multiple hydroxyl groups of the glucose from one chain form hydrogen bonds with oxygen atoms on the same or neighbouring chain, thereby keeping the chains firmly assembled in a side-by-side configuration and forming microfibrils with high tensile strength.

Cellulose is insoluble in water and most organic solvents, and can be broken down chemically into its glucose units by treatment with concentrated mineral acids at high temperature. Most of its properties depend on its degree of polymerization, i.e. the number of glucose units that make up one polymer molecule. Molecules with very small chain lengths resulting from the breakdown of cellulose are known as cellodextrins. In contrast to long-chain cellulose, cellodextrins are typically soluble in water and organic solvents.

Papermaking, which originated in China in AD ~105, is thought to have been invented by a high-ranking official named Cai Lun during the Han Dynasty. The evolution of paper as a writing support can be divided into two distinct phases: pre-19th century when paper was made by hand using cellulose obtained from cotton and linen rags and post-19th century, when mechanized papermaking using wood pulp as a raw material was introduced.

Cellulose in paper can be derived from various sources (cotton, linen, hardwood, softwood, hemp, etc.). In addition, paper usually contains sizing compounds, fillers, traces of metallic impurities (such as iron or copper) and other substances (e.g. inks, pigments, etc.) making it a complex and heterogeneous medium. Various paper constituents such as lignin, rosin or alum used in papers produced after 1850, and external factors such as water, atmospheric gases, light, pollutants and microorganisms, often act in a synergistic way, strongly promoting cellulose degradation.

The action of fungal and bacterial endoglucanase, exoglucanase and β-glucosidase enzymes (cellulases) can profoundly alter the structure of cellulose, degrading the surface of its fibres and gradually dismantling its architecture until only the crystalline parts are left intact. Crystalline cellulose is the most resistant to degradation by microorganisms, but many species do possess the metabolic tools to degrade it (Markham and Bazin 1991).

The action of microbial flora on the organic and inorganic structural elements of paper can inflict different kinds of damage and discoloration (Fig. 3). Damage can occur due to mechanical stress caused, for example, by the development of fungal structures between and inside cellulose fibres or degradation by enzymatic action. Moulds and bacteria can produce a wide range of enzymes and exudates that are able to change the structure and functionality of cellulose and other paper components, such as adhesives and mineral fillers (Pinzari *et al.* 2010). In particular, some filamentous fungi frequently associated with paper damage dissolve cellulose fibres through the action of their cellulolytic enzymes, and produce pigments and organic acids that cause paper discoloration and oxidation. Aged paper and chemical hydrolysis of cellulose chains can be attractive to bacterial and saprophytic fungal species, rendering already damaged paper even more vulnerable to microbial attack.

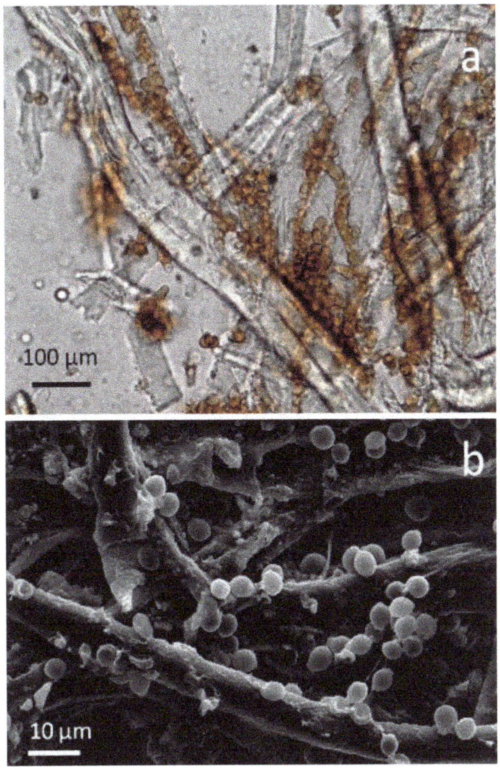

Figure 3 Fungi can damage paper fibres by growing between or within cellulose fibres, causing chromatic changes that depend on the pigmentation of cells, or producing expolymeric substances that remain attached to the fibres, consuming cellulose or other compounds present in paper such as glues, organic pigments and impurities. (a) Darkly pigmented mycelium of *Aureobasidium* sp. growing on cellulose fibres and developing within the lumen of cotton structures; (b) conidia of a fungus belonging to the *Microascus* genus that developed on paper, causing little or no damage to cellulose fibres, presumably by exploiting glues or sizing compounds as their main substrates.

The biodegradation of paper depends on several variables including the materials from which the cellulose is obtained, the manufacturing processes employed and the occurrence of other substances such as lignin or metallic compounds (Florian 2002). Pinzari *et al.* (2006) observed that a single paper characteristic could strongly influence the intensity and consequences of fungal action. For example, the activity of a fungal strain on papers containing fibres of the same origin, but of different size, can lead to the formation of profoundly different stains and alterations. Upon analysing the same fungal species and under the same growth conditions, fungal structures observed by SEM were present in different physiological states on different paper types, suggesting an important link between paper constituents and growth and sporulation ability of fungal species.

Available water is recognized as the chief limiting factor for the development of microorganisms on paper, followed by temperature and the presence of nutrients other than cellulose, such as nitrogen, potassium, magnesium, sulfur and phosphorus. At levels of relative humidity (RH) higher than 65%, and temperatures exceeding 20 °C, the moisture content of paper can reach 8–10%. Under these conditions, fungal and bacterial spores and resting structures can germinate and develop using paper as a growth medium (Florian 2002). In libraries and other archives, a RH level of ~55% is generally regarded as the threshold for fungal growth, therefore climate control is set at a lower value. However, poor ventilation and surface temperature dishomogeneity can create water condensation points, resulting in local microclimates with higher water availability than is found in other areas, thus encouraging the development of fungi on materials.

Water activity (a_w) is equal to the partial vapour pressure of water in a substance divided by the partial vapour pressure of pure water at the same temperature (Barbosa-Canovas *et al.* 2007). The lower a_w limit for the growth of microorganisms in organic substrates depends on the solutes that accumulate within cells that are exposed to external environments with reduced a_w. Intracellular accumulation of solutes such as amino acids and polyols counteract in fungi and some bacteria, resulting in the osmotic imbalance across the cell membrane at low a_w. In xerotolerant fungi, intracellular accumulation of glycerol, arabitol or mannitol may play a role in preventing the dehydration of aerial mycelia. These types of fungi are able to grow at much lower humidity levels than bacteria or archaea, with some xerophilic and xerotolerant species such as *Eurotium* sp., *Aspergillus* sp. and *Wallemia* sp. capable of growing at a_w levels of 0.65 (Sterflinger 2010; Sterflinger and Pinzari 2012).

Microbial activity on paper is possible only in the presence of unbound water in the substrate that is available to sustain microbial growth. Bound water is the fraction of water that forms a monomolecular layer covalently bonded to cellulose molecules in the material and has an a_w value lower than 0.2. Water weakly bonded to cellulose molecules and forming a multilayered film located in capillaries less than 30 nm in diameter has an a_w value lower than 0.65–0.70. Only the free water that is present in capillaries measuring more than 30 nm in diameter has an a_w value greater than 0.65–0.70, thereby allowing microbial growth and cellulose degradation processes.

Fungal spoilage of paper occurs more often than bacterial spoilage at a_w levels <0.85. This is not due to the fact that fungi cannot grow at a_w levels >0.85, but rather because bacteria are highly competitive at higher a_w levels, and therefore become a dominant type of microflora at a_w >0.85. In general, the optimum a_w for fungal activity coincides with that for mycelial growth (Sarrette *et al.* 1992). Most of the strongly cellulolytic fungi require 0.99–1.00 as optimum a_w to achieve maximum exocellulase activity (Beuchat 1983; Magan and Lacey 1984; Magan and Lynch 1986), which means that fungi are more destructive when there is abundant free water present in paper.

The microbial and fungal community structure on biodeteriorated paper is more complex than previously thought. Research based on molecular techniques which focuses on the investigation of fungal flora responsible for the biodeterioration of paper materials has shown

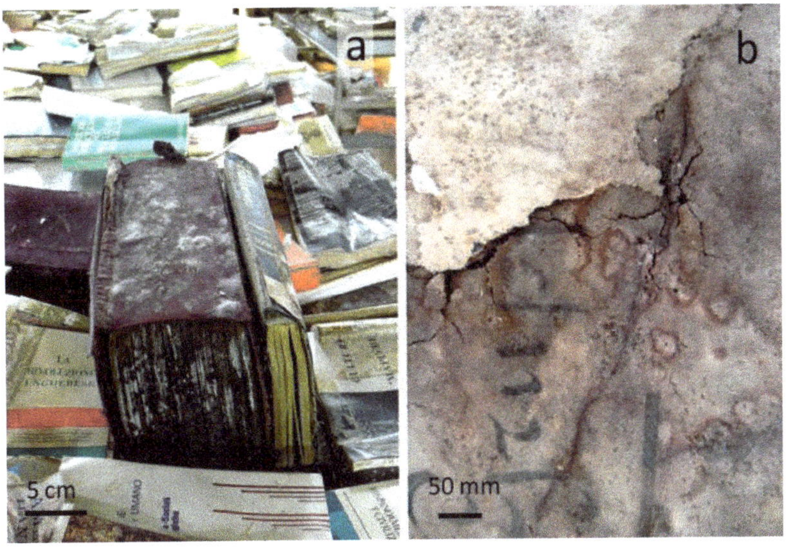

Figure 4 The moulds associated with 'water damage' consist of fungal species that require high water activity. These species are usually fast-growing organisms that emit strong odours, coloured stains and enzymes capable of destroying substrates (i.e. cellulolytic enzymes). (a) Books recovered after a flood but not dried quickly enough to prevent the formation of mould; (b) paper attacked by fungi and bacteria after a wetting event. The microorganisms caused extensive damage and the destruction of the physical resistance and functionality of the support. The paper appears brittle and marred by stains owing to fungal and bacterial pigments and the presence of fruiting bodies embedded in the cellulose (small dark spots, top left).

that the results obtained with culturing methods relate to only a few of the viable and culturable organisms (about 5%) among the full array of microorganisms actually present in materials (Di Bonaventura *et al.* 2003; Michaelsen *et al.* 2005, 2006, 2009, 2010; Rakotonirainy *et al.* 2007; Pangallo *et al.* 2009; Principi *et al.* 2011). Denaturing gradient gel electrophoresis (DGGE) fingerprints revealed a high biodiversity of both bacteria and fungi inhabiting biodeteriorated books, and DNA sequence analysis confirmed the coexistence of fungi and bacteria in paper samples. Many culture-independent studies on paper biodeterioration identified a large number of fungal species typically associated with dry environments, suggesting that indoor settings, and in particular paper-based objects, are favourable for growth and colonization of osmophilic fungal species. Additionally, many of the bacterial sequences amplified from books made of paper belong to phylotypes that exhibit cellulolytic activity, indicating that bacteria can colonize and grow on paper along with fungi. The co-occurrence of bacteria and fungi, as well as the role played by these relationships in substrate exploitation, is probably greatly underestimated (Michaelsen *et al.* 2005, 2006, 2009, 2010).

A different kind of paper biodeterioration is found in libraries and archives when water suddenly becomes widely available either from a naturally occurring flood or ruptured water pipe (Florian 2002). These situations are familiar to archivists and librarians who are well aware that when paper becomes sodden, mould grows profusely and odoriferous bacteria develop rapidly. These microbial attacks cause extensive damage including the loss of text legibility and compromise of the paper's strength and functionality (Barton and Wellheiser 1985; Dicus 2000). The moulds associated with 'water damage' (Nielsen 2002; Nielsen *et al.* 2004; Vereecken and Roels 2012) consist of fungal species that require a high a_w (Fig. 4). The relevant variables are the availability of water, temperature and the amount of time that the fungus is active before sodden books

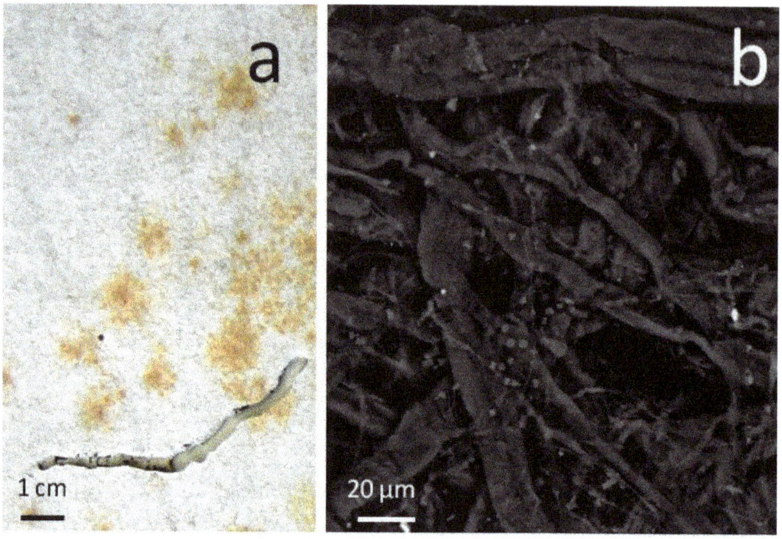

Figure 5 Chromatic alterations on paper are known as 'foxing spots'. These are often caused by the residues and substances released by tonophilic fungi on paper. (a) The brown-hued products and oxidative reactions on paper result in localized foxing spots; (b) dehydrated and collapsed fungal structures (hyphae and conidia) visualized from rust-coloured spots on paper.

and documents are dried out. The metabolic characteristics of the fungus itself are also a determining factor. These species are usually rapid-growing organisms that can emit strong odours (*Trichoderma* spp.), coloured stains (*Chaetomium* spp., *Monoascus* sp., *Epicoccum* spp.) and even toxic compounds (*Stachybotrys* spp.).

Fungal strains with markedly different physiologies can have a very similar effect on cellulose fibres and other constituents of paper, thereby resulting in unequivocal damage. The physiological state of the fungal mycelia, such as the degree of maturity of fungal structures, can exert different effects on the chemical-physical properties of paper. For example, senescent hyphae may release chemicals that young fungal mycelia do not produce, thus causing discoloration and cellulose degradation only under certain age-related conditions. This situation results in a conundrum which further complicates matters for researchers who need to be able to anticipate the effects of a degradative agent on a given paper support.

A fairly common scenario encountered by conservators is when the fungus is no longer present or living on the paper, even though permanent degradation has occurred, or oxidizing compounds have been released. One of the most notorious forms of chromatic alteration that spoils the appearance of ancient paper, foxing spots, often seems to be caused by residues and substances released by moribund fungi (Choi 2007) (Fig. 5). Arai (1987, 2000) believes that foxing results from melanoidins formed by the Maillard reaction that takes place between cello-oligosaccharides and amino-butyric acid and other amino acids produced by absolute tonophilic fungi growing on paper (Arai 1987). Arai *et al.* (1990) proposed the following process for the formation of foxing spots: absolute tonophilic fungi conidia and/or ascospores attach to paper, germinate and grow their hyphae, which then form colonies around dust particles. These fungi metabolize mainly malic acid, cello-oligosaccharides and aminobutyric acid in the colonies. These components conjointly undergo chemical reactions on the materials at a_w 0.80 and 20–35 °C and generate brown-hued products and oxidative reactions on paper that result in localized foxing spots.

Florian and Manning (2000) studied the role of fungi in foxing using light and electron microscopy and found that in some cases the initial cause of the spots was a group of spores of conidia that had been deposited on the surface of the paper prior to printing and had germinated *in situ* during the slow drying of the paper. Piñar *et al.* (2015) reported on an exceptional case of paper alteration due to the residue of past infections when Leonardo da Vinci's celebrated self-portrait, which is severely affected by foxing spots, was analysed using molecular and microscopic techniques. The study showed that the portrait was free from active fungal or bacterial infections, but was contaminated with airborne spores and fungal material that could have played a part in its marring. Dead structures of the fungus *Eurotium halophilicum* (Christensen *et al.* 1959) were identified in the foxing spots. Oxalates of fungal origin were also documented. These findings are consistent with the hypothesis that absolute tonophilic fungi germinate on paper, metabolizing mainly organic acids, oligosaccharides and proteic compounds. These components react chemically with paper at a low a_w, giving rise to brown-hued products and oxidative reactions that result in localized foxing spots.

BIODETERIORATION PHENOMENA RELATED TO MINERAL COMPOUNDS

Fungal activity in library materials can also result in visible effects on the mineral components of paper (Fig. 6). Fungi are effective mineral leaching agents, capable of translocating or sequestering leached metals in their hyphae and producing secondary minerals both within and outside their structures (Gadd 2010). Paper and parchment both contain, in varying amounts, inorganic compounds. In fact, salts originating from manufacturing processes such as sizing minerals or impurities, metals from inks made with iron and copper, and mineral compounds related to the use of pigments (i.e. orpiment, cinnabar, lapis lazuli, gold, silver, white lead, etc.) are found in most library materials. The presence of pigments and related minerals can create a patchy microenvironment in which the distribution and activity of potentially damaging microorganisms can vary and produce a broad range of discoloration and damage, which is difficult to analyse and manage.

For example, a micrometric-level fungal interaction with carbonate substrates results in a clear replacement of the original minerals with newly formed crystals (Pinzari *et al.* 2010). These biogenic minerals are produced in paper by fungal strains that are able to utilize calcium contained in $CaCO_3$ to produce calcium oxalate crystals that precipitate on cellulose fibres and fungal hyphae as encrustations and organized clusters (Pinzari *et al.* 2010; Sequeira *et al.* 2017a). Oxalate production is due to the metal-binding capacity of the organic acids generated by the fungi. These, in the presence of hydrogen ions, cause substrate mineral hydrolysis and new mineral formation through metal–acid complexation. The action of these acids is combined with the action of fungal respiratory CO_2, which generates carbonic acid in the growth environment. It can be hypothesized that the combined action of these acid groups produces the dissolution pattern observed on the calcium-containing materials, and that a subsequent spontaneous recrystallization of fungal oxalic acid with the Ca^{++} ions on chitin nucleation sites brings about the formation of the new crystals close to fungal structures. According to Pinzari *et al.* (2010), the removal of fungal mycelium from paper samples by washing does not alter the oxalate crystals, suggesting that they are insoluble and firmly attached to the cellulose fibres.

The effect of calcium oxalate precipitation among cellulose fibres, and the consequent diminution of the alkaline reserve in paper, can be the result of chemical and mechanical damage (Strlič *et al.* 2004 and references therein). The presence of poorly identified 'crystals' on ancient paper from manuscripts was described by some authors as being associated with biodeteriorated materials (Gallo and Pasquariello 1989; Montemartini Corte *et al.* 2003), suggesting that this phenomenon occurs naturally on library materials

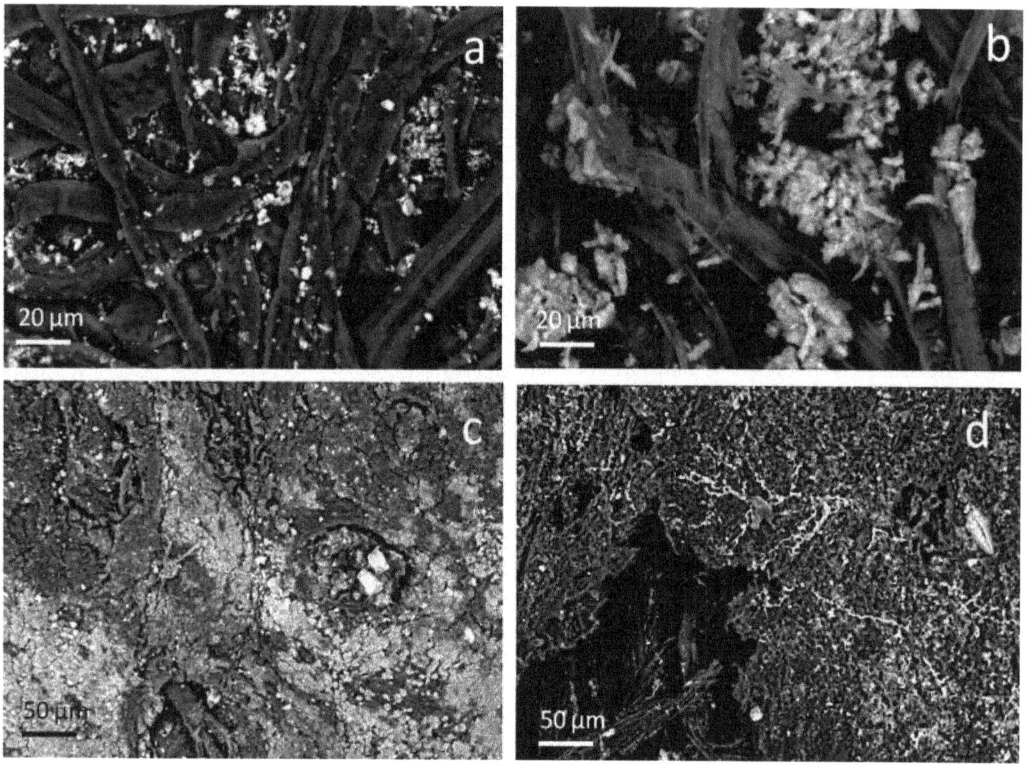

Figure 6 Fungi and bacteria are able to modify the substrates on which they grow by changing the chemistry of their immediate microenvironments, a process that involves precipitating and translocating microelements. As a result, at the micro scale, mineral weathering by microorganisms and precipitation of new minerals can contribute to the biodeterioration of both paper and parchment. (a) Modern paper originally containing calcium carbonate (before damage) and (b) paper destroyed by a fungus that used cellulose but also precipitated 'large' crystals of oxalates between the fibres (after fungal attack); (c) a parchment sample prior to biodeterioration; (d) the same parchment sample attacked by bacteria and fungi which used collagen and displaced salts that appeared to be newly crystallized into ramified, slender structures resembling filaments.

that are contaminated by fungi which produce organic acids.

Calcium oxalates can also be produced on parchment attacked by fungi. Pinzari *et al.* (2012a), reported on biodeterioration in an illuminated parchment, where a species of *Acremonium* exhibited the production of abundant biogenic crystals. The marring of the parchment was due to both collagenolytic activity and the conspicuous biotransformation of calcium-based minerals by fungi.

Tanney *et al.* (2015) described the case of a rare halophilic fungus (*Diploöspora rosea* Grove) growing on the preparative layer, composed of gypsum and calcium carbonate, of an illuminated parchment. The fungus had developed on gilded areas and green pigments used for illumination of the documents. The 'emerald green' layer was prepared using chromium, iron and arsenic, and the gold layer showed the presence of lead, presumably originating from lead white pigments (basic lead carbonate, $PbCO_3$). This suggests the capability of *D. rosea* to grow in the presence of toxic metals such as arsenic, chromium and lead, causing the exfoliation of large portions of the decorative gold illustrations.

Pinzari *et al.* (2012a) demonstrated salt displacement by microbial activity in ancient parchment using variable pressure scanning electron microscopy (VP-SEM) in combination with a backscattered electron (BSE) detector

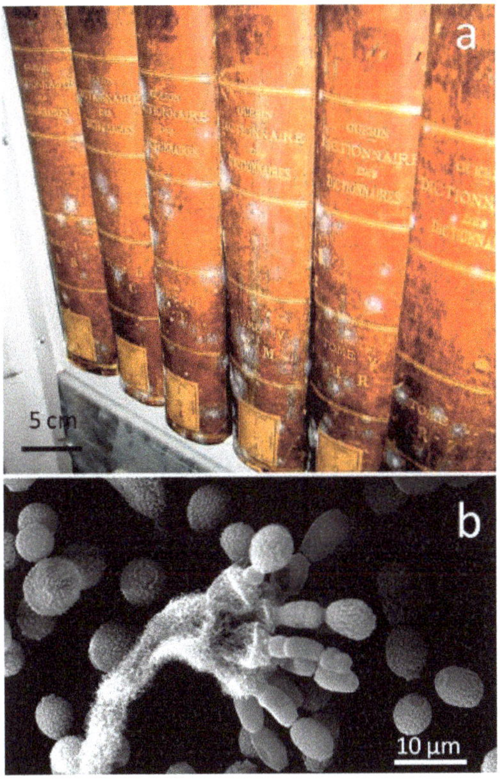

Figure 7 Monospecific contamination of books by the fungus *Eurotium halophilicum* and its anamorphous state *Aspergillus halophilicus*. Its detection can be greatly underestimated due to its very slow growth on commonly used media. *E. halophilicum* can develop at 50–60% RH and temperatures ranging from 20°C to 22°C. (a) Fungal colonies on book covers and (b) conidia and conidiophore of *A. halophilicus*.

which revealed the effect of microbial attack on the salt distribution pattern on the parchment surface. The biodeteriorated zones had reduced aluminium and silicon levels, and increased sulfur and phosphorus levels. The authors suggested that fungal and microbial species exerted a leaching effect on minerals (aluminium and silicon), and were simultaneously able to translocate and concentrate other salts, such as sulfates and phosphates. This hypothesis was supported by the SEM-BSE images that showed a different elemental distribution pattern before and after microbial damage had occurred. Calcium appeared to be dislocated in a different way in the parchment's blemished areas, changing from a scattered distribution trend to a pattern of ramified thin structures resembling fungal filamentous structures (Pinzari *et al.* 2012a).

The presence of metals in documents could also represent a deterrent for microbial development. The existence of unblemished areas, on an otherwise contaminated material, in proximity to decorative elements could be attributable to the biostatic effect of mineral compounds. Pinzari *et al.* (2012a) reported a case where a parchment document, bearing purple spots of microbial origin, showed the absence of microbial stains in ink-bearing zones due to the presence of copper sulfate and to the lower pH resulting from the use of iron-gall tannic inks.

Bicchieri and Pinzari (2016) reported a unique case study of a 17th-century manuscript in which the black iron-gall ink had faded into a yellowish-white shade. Raman and SEM analyses identified and documented a biogenic formation of jarosite on the surface of the ink. The crystals presented a typical rhombohedral form and appeared to be associated with a biofilm, composed of both fungal and bacterial structures. At a certain juncture the book had suffered a soaking event that left it covered with mud, which may have contained the microflora that caused the biogenesis of jarosite coinciding with the iron rich inked areas.

COLONIZATION OF MATERIALS AND ENVIRONMENTAL FACTORS

The microbial assemblages that can attack and destroy a book are akin to the communities of decomposers that, in natural environments, are responsible for nutrient cycling (Pinzari 2011). In a natural setting, there may be several organisms competing for a given substrate. Manmade repositories for books and documents represent a new niche for fungal and microbial species to colonize (Pinzari 2011). The storage of large amounts of cellulosic materials in a closed environment creates a simplified system, in contrast to a natural environment where the spread of a few species that develop with little or no competition would be expected. According to the niche theory, species' range is defined by physiological tolerances and dispersal barriers (Broennimann *et al.* 2007). The formation of a microbial community on a shelf in a library or on an individual book depends on the species that reaches the material's surface, the micro-environment (i.e. temperature, amount of moisture present and light). These can be single events such as a ruptured water pipe, an insect infestation or a more sustained 'immigration' phenomenon originating from external sources of fungal contamination (Borrego *et al.* 2012).

Some fungal and bacterial species, rarely encountered in nature, can be found in niches created by man, such as paper and parchment, which allow fungi to proliferate without competition (Bech-Andersen and Elborne 2004). These species are often considered 'invasive' and are able to shift from extreme natural environments to manmade niches with alarming ease. This is evidenced by *Saccharopolyspora* sp., a bacterium isolated from parchments manufactured over a long period of time and stored in distinct locations. Similarly, *D. rosea*, which was described 100 years ago, was not seen again until its identification as the main cause of pigments flaking away from an illuminated parchment. Yet another case of a fungal species creating major problems for books stored in libraries and repositories is the spread of *E. halophilicum* (Christensen *et al.* 1959), and its anamorph, *Aspergillus halophilicus*. This fungus is a xerophilic organism with a high tolerance to water stress. The minimum a_w required for the germination and growth of this species is 0.675. It has been isolated largely from dry foodstuffs or indoor dust in association with dust mites, but recently it has also been shown to cause biodegradation in books and paper (Michaelsen *et al.* 2010). Since the first reported case of this species as a book contaminant, it has been identified in many libraries and archives throughout Italy as the principal, if not the only, fungus thriving on volumes and archival material (Pinzari and Montanari 2011; Montanari *et al.* 2012; Micheluz *et al.* 2015; Polo *et al.* 2017). It has also been linked to the appearance of stains resembling foxing on materials (Piñar *et al.* 2015; Sclocchi *et al.* 2016).

E. halophilicum detection can be greatly underestimated due to its very slow growth on commonly used media. However, it is capable of prolific development in climatic conditions that are considered satisfactory for the conservation of library materials (RH values ranging from 50 to 60%, and temperatures from 20 to 22 °C) (Fig. 7). This fungus produces volatile organic compounds (VOCs) (Micheluz *et al.* 2016) and represents a potential health hazard for staff who handle the infected materials (Crook and Burton 2010).

In addition to health and safety issues, there are some interesting ecological considerations. A library or a single book can be compared to a tract of virgin land where colonizing organisms behave like pioneer species on bare soil. In the case of a single book, or multiple books stored in a closed environment, the colonization and biodegradation of constituent organic matter depend on the species concerned and the material's composition, since only a few organisms can exploit the bulk of the substrate. Not all the fungal and bacterial species that access stored materials germinate, colonize and are deleterious. Apparently, only a few pioneer organisms can initiate colonization and the conditions that promote their settlement and proliferation can often be found at the micro scale.

A new approach to paper and book microbiological diagnostics is now possible, thanks to molecular, biochemical and microscopic techniques. If correctly employed, these can reveal the mechanisms supporting the establishment of species assemblages on artefacts (Pinzari *et al.* 2009). A better understanding of the mechanisms concerned could be helpful in defining new control strategies and treatment procedures, thereby ensuring both human and environmentally friendly solutions to biodeterioration issues.

BIODETERIORATION MECHANISMS

In the past 15 years, the approach to the study of biodeterioration of cultural heritage has changed completely. Molecular methods and electron microscopy, especially when combined,

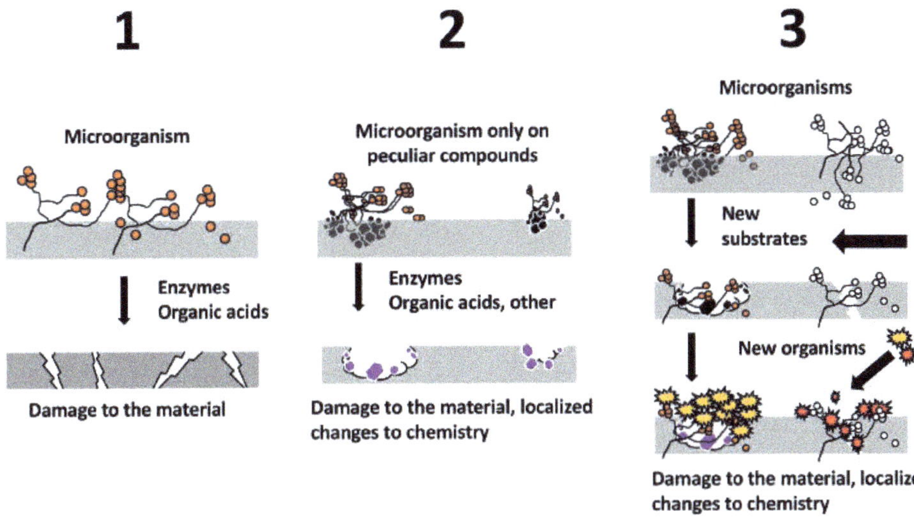

Figure 8 A diagram illustrating the three mechanisms of paper or parchment biodeterioration and showing increasing complexity in the resulting damage: (1) localized biodeterioration producing compounds that cause chemical damage; (2) localized compounds eliciting the establishment and growth of special organisms ('hot spots' containing organic or inorganic compounds that render the location suitable as a niche for a single species or an assemblage of species); (3) localized biodeterioration producing compounds that give rise to the establishment and growth of other damaging organisms. This can result in a true ecological succession, with several different species occurring on materials.

have revealed new mechanisms and microbial scenarios. Comparisons between diagnostic studies based solely on *in vitro* culture (Gallo *et al.* 1999; Gallo and Pasquariello 1989), culture-independent approaches and SEM have shown that among the various atmospheric biocontaminants that inhabit indoor air, only a few are harmful to paper and parchment (Michaelsen *et al.* 2006, 2009, 2010; Piñar *et al.* 2014a, b, 2015).

However, the level of detail on the presence of microbes made possible by new techniques may be useless if no information on the active microbial processes at play can be provided. For example, it is possible to obtain information on microbial communities from very small samples using metagenomics analysis. Additional information on the primary contributors and biodeterioration mechanisms using complementary techniques is required in order to gain insight into single species or metabolites associated with the ongoing decay processes. Despite the deployment of highly informative techniques, the identification of the agents responsible for any phenomenon observed can still be extremely complicated and may even require the use of multiple techniques suited to the particular object in question on a case-by-case basis. In fact, it is unacceptable to establish hard and fast rules in the area of diagnostics, since valuable paper or parchment heritage items are often unique objects that cannot be subject to standardization.

Based on the case studies examined so far, the complexity of damage suffered by paper and parchment can be summarized by three distinct types of biodeterioration mechanisms (Fig. 8):

1 Localized production of biodeterioration-inducing compounds that cause chemical damage. In this situation, microorganisms alter the chemistry of a material. Fast-growing cellulolytic fungi that attack paper after a waterlogging event can be included in this category. Such species produce cellulolytic enzymes and cause direct damage.
2 Localized compounds that elicit the establishment and growth of specific organisms. This occurs when materials offer 'hot spots' containing organic or inorganic compounds that render the location suitable as a niche

for single species or an assemblage of species, for example, when halophilic bacteria colonize parchment. The presence of salts in the material, more than the availability of collagen, is the selective factor that favours the establishment of such species and the onset of a degenerative process in parchment.

3 Localized biodeterioration in which compounds are produced that give rise to the establishment and growth of other damaging organisms. This can result in a true ecological succession, with different species occurring on materials, depending on the types of substances made available by the same degenerative phenomenon. This is a complex situation, where species coexist on materials, making it difficult to determine which species is the first colonizer and the type of damage caused. The co-occurrence of fungi and bacteria on both paper and parchment is a very common situation. The possibility that the damage to materials is due to the cooperation of multiple species is still underestimated and not well understood, although almost all studies using molecular techniques reveal the presence of active mixed communities on materials.

The type of mechanisms at play and the number of organisms involved can have very important consequences for the efficacy of treatments, and may have an impact on decision-making with regard to the chemicals and approaches to be used in conservation drives. There is still the need for *in vitro* studies aimed at creating models that can unravel the kind of interaction between materials and organisms, with the objective of developing advanced and better targeted solutions.

PREVENTION AND TREATMENT

Each of the biodeterioration mechanisms summarized previously results in damage to the constituent material and therefore requires a conservator's intervention. The type of biodeterioration phenomenon involved defines the kind of intervention required. Environmental control, drying, disinfection and cleaning are forms of intervention that conservators have long employed, but like diagnostic methods, they evolve in parallel with advances in technology, thereby affording the conservator greater control over outcomes (Cappitelli and Sorlini 2005; Sequeira *et al.* 2012; Boudalis *et al.* 2016). The key lies in understanding which intervention is actually necessary. Since no interventions are fail-safe and free from side effects, this is not as straightforward as it might seem (Gallo 1992; Nittérus *et al.* 2000b; Sequeira *et al.* 2012; Michaelsen *et al.* 2013; Boudalis *et al.* 2016).

The activity of fungi and bacteria that proliferate in damp paper or parchment depends on the presence of free water. In the case of an attack by microorganisms that can completely destroy paper or parchment after a waterlogging event, immediate action must be taken to dry the material as quickly as possible to arrest the harmful activity (Schmidt 1985; McCleary 1987; Florian 1990, 2002). After drying, when the microbial spots have formed and there are dead mycelium and presumably still viable spores present, conservators face a situation as summarized in the third type of biodeterioration mechanism whereby localized compounds elicit the establishment and growth of other species, which in turn feed on the products released by the pioneer organisms. In this circumstance it is necessary to carry out a deep cleaning operation to eliminate spores and microbial metabolites and eventually remove any stains. An efficient drying of materials coupled with a targeted cleaning operation can be much more effective and long lasting than any biocidal treatment.

In recent decades, a number of bulk drying (i.e. wind-drying) and freeze-drying treatments have been used and documented in cases where library materials have been damaged by floods or submerged under water (McCleary 1987; Hájek *et al.* 2011; Price 2013). In freeze-drying processes, the water contained in the material is frozen and then removed by sublimation, from a solid to a gaseous state, thereby avoiding the liquid phase. The treatment prevents the adhesion of paper sheets to one another and impedes the migration of inks or other soluble compounds. Lyophilization is capable of killing or suppressing fungal mycelium and, to a limited extent (in a way which varies according to the species), conidia (Michaelsen *et al.* 2013). Both wind-drying and freeze-drying halt fungal and bacterial growth but rarely kill the microorganisms, whose spores remain viable. Ascospore germination of the many fungal species can be stimulated by physical shocks such as heat and frost (Mazur 1968). This may explain the significantly higher growth of the cellulolytic fungus *Chaetomium globosum* subsequent to freeze-drying compared to a control strain (Troiano *et al.* 2012). Fungal ascospores and other resting structures allow the fungi to survive adverse conditions. In unfavourable environmental conditions, resting spores have low water content and their metabolism is inactive but reversible (Florian 1990; Sussman 1966). Fungal vegetative hyphae are relatively easy to control by monitoring the storage climate (i.e. temperature, RH and a_w) (Nittérus 2000a).

Procedures for incorporating adequate filtering systems that trap biological products and cleaning affected materials must be performed immediately after drying and freeze-drying treatments. These are safe and well-standardized drying procedures – considerably more challenging are the treatment of stains and the removal of microbial biomass embedded in materials.

Stains on books and documents may limit accessibility and legibility of the text (Fig. 9). Fungal and bacterial pigments can be classified into endopigments, which endow organisms with colour, and exopigments (mainly flavonoids, carotenoids, flavins, quinones, monascins, melanins, prodigiosins, monascins, violacein, indigo) that are secreted into the surrounding environment (Gill 2003; Fiacconi *et al.* 2012). The biological function of numerous fungal pigments is not yet understood: some may act as a filter against light radiation, while others may have antibiotic properties. Their production is influenced by environmental variables such as nutrient availability, the pH of the substrate, light and temperature (Gill 2003; Souza *et al.* 2016). Chemical and mechanical treatments aimed at removing microbiological stains and fungal material from paper and parchment may be harmful or ineffective (Szczepanowska and Lovett 1992; Nieto-Fernandez *et al.* 2003; Smith 2012). Enzymatic removal of fungal structures and soluble pigments may be preferable alternative treatments (Tavzes *et al.* 2013; El Monssef *et al.* 2016).

Enzymes have also been used in combination with gellan gum, which can bind the enzyme and absorb water-soluble elements through capillary action, thereby supporting enzyme activity (Fiacconi *et al.* 2012; Mazzuca *et al.* 2014). It is worth noting that results from the use of many mechanical, chemical and biological conservation measures and restoration treatments described in the literature are often contradictory, have undergone testing in oversimplified conditions or verified on too few types of materials to be considered safe and/or effective (Ďurovič and Zelinger 1993; Boudalis *et al.* 1996; Zou *et al.* 1996).

When microorganisms are active and cannot be removed from materials because they have penetrated the paper or parchment structure, or when too much time is needed for effective drying and risks causing serious damage to precious documents, it becomes necessary to resort to a biocidal treatment. Some very accomplished and comprehensive reviews on biocidal methods to be applied to library materials have been published including Nittérus (2000b), Allsopp *et al.* (2004: 237) and Sequeira *et al.* (2012).

Currently, more stringent regulations are in place, especially covering the use of chemical products, and the appropriate biocide choice is governed by the European Biocidal Products Directive (Sequeira *et al.* 2012). Topical treatments with chemicals are still the most used strategies employed by conservators (Sequeira *et al.* 2014), although their efficacy depends on the sensitivity of the fungal species concerned and the ability of the relevant material to withstand treatment. Chemical disinfection methods are based on the use of toxic chemicals that denature microbial proteins, block and oxidize thiol and amino groups, affect the function of cell membranes or interfere with the primary metabolism of microorganisms

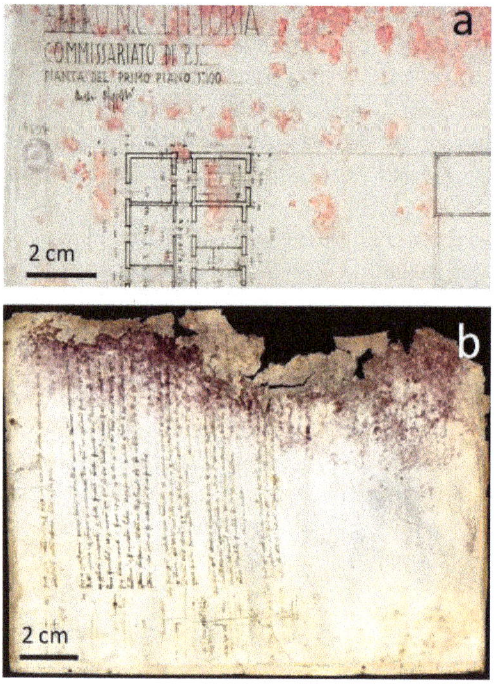

Figure 9 Stains on paper and parchment due to pigments of microbiological origin: (a) purple spots associated with collagen fibre destruction in ancient parchment (14th century) known to occur in protein-based products manufactured using marine salt, and believed to be due to bacteria producing red-purple pigments; (b) tracing paper dating from the first decades of the 20th century damaged by water and affected by the spread of *Chaetomium* sp., a fungal species that produces anthraquinone pigments.

(Sequeira *et al.* 2012; Ďurovič *et al.* 2014). Some biocides used for restoration purposes include formaldehyde releasers, quaternary ammonium compounds, parabens and isothiazolinone, 70% ethanol (Sequeira *et al.* 2017a), and essential oils (which produce very contradictory results and possible severe side effects) (Rakotonirainy and Lavédrine 2005; Borrego *et al.* 2012; Stupar *et al.* 2014). Constant evaluation of the efficacy and side effects of 'new' biocidal compounds and formulations is required in order to establish their impact on microorganisms and materials over time, with the possible inclusion of simulated aging processes (Zou *et al.* 1996). Gamma irradiation is the most commonly employed physics-based method used to kill microorganisms on archival materials (Ďurovič *et al.* 2014). Various research groups are assessing new approaches based on nanotechnologies (nanoparticles, plasma or laser treatments) (Gutarowska *et al.* 2012; Pietrzak *et al.* 2017).

The effects of treatments on the integrity of materials should also be evaluated in the light of new potential applications and future uses of cultural heritage. Recent developments in paleogenetics and the archaeology of manuscripts (Teasdale *et al.* 2017; Piñar *et al.* 2014b) involving non-invasive sampling techniques coupled with the analysis of total DNA allowed documentation of a rich palimpsest of biological information which has accumulated over thousands of years. Even after centuries of microbial or fungal cell death, DNA remains in the materials of biological origin. The source of skins from which parchment was made or traces of human microbiome in the most consulted pages can be investigated if the nucleic acids contained in a manuscript remain sufficiently intact and extractable.

Michaelsen *et al.* (2013) evaluated three different antifungal conservation treatments: freeze-drying, gamma irradiation and ethylene oxide (EtO) fumigation. These were compared and monitored to assess their short- and long-term effectiveness to inhibit fungal growth. Both gamma irradiation and EtO fumigation demonstrated total efficacy as disinfecting agents when evaluated by means of culturing techniques. Upon evaluation with molecular analyses, gamma ray exposure resulted in a short-term reduction in DNA recovery and DNA fragmentation; the latter phenomenon was also observed to a lesser extent in samples treated by freeze-drying. Only treatment with EtO eliminated DNA and RNA recovery, due to the structural changes caused to nucleic acids by this gas (Michaelsen *et al.* 2013). The fact that no DNA or RNA could be extracted using routinely applied methods from paper samples fumigated with EtO is a cause for concern. Museums and libraries have used EtO fumigation extensively in the past to treat insect or mould infestations, but often it is not traceable or clearly mentioned in the reports that accompany treated objects. Consequently, missed diagnoses could emerge from subsequent molecular studies. Consolidated techniques should be questioned from time to time, perhaps by applying new surveying techniques.

The previously described case of the spread of the fungus *E. halophilicum* in Italian libraries is an example that requires a revision of orthodox thinking (Montanari *et al.* 2012). This fungus was associated with books and paper biodeterioration, especially in specific micro-niches within museums, libraries or archives, which can develop when the overall environmental conditions are also ideal for the conservation of books. These niches are characterized by poor ventilation and the presence of a water vapour gradient that can lead to condensation forming on certain materials following a sudden drop in temperature or due to night–day thermohygrometric cycles. The climatic conditions considered suitable for the conservation of library materials, which are published as international standards, appear to be totally unsuitable for the prevention of some biodeteriorating fungi.

To conclude, it can be stated that prevention of biodeterioration in books and documents is possible to achieve, but organic materials will always remain vulnerable to natural decay processes. Digitization can save textual content, but the historical role of, and the information contained in, a book's material components are equally important. In parallel with digitization, material study campaigns and a continual verification and updating of conservation and prevention techniques should be promoted if we wish to avoid losing the scientific value of the physical objects over time.

ACKNOWLEDGEMENTS

I dedicate this article to my former colleagues at the Istituto Centrale per il Restauro e la Conservazione del Patrimonio Archivistico e Librario (ICRCPAL) in Rome, with whom I spent 12 highly rewarding years. I would like to thank Mark Livesey for his helpful suggestions during the preparation of the English text.

REFERENCES

Allsopp, D., Seral, K. and Gaylarde, C.C. 2004. *Introduction to Biodeterioration*. Cambridge: Cambridge University Press.

Arai, H. 1987. 'Microbiological studies on the conservation of paper and related cultural properties. Part 5: Physiological and morphological characteristics of fungi isolated from foxing, formation mechanisms and countermeasures', *Science for Conservation* 26: 43–52.

Arai, H. 2000. 'Foxing caused by fungi: twenty-five years of study', *International Biodeterioration and Biodegradation* 46: 181–8.

Arai, H., Matsumura, N. and Murakita, H. 1990. 'Microbiological studies on the conservation of paper and related cultural properties. Part 9: Induction of artificial foxing', *Science for Conservation* 29: 25–34.

Barbosa-Canovas, G., Fontana, A., Schmidt, S. and Labuza, T.P. 2007. *Water Activity in Foods: Fundamentals and Applications*. Hoboken NJ: FT Blackwell Press.

Barton, J.P. and Wellheiser, J.G. 1985. *An Ounce of Prevention: A Handbook on Disaster Contingency Planning for Archives, Libraries and Record Centres*. Toronto: Toronto Area Archivists Group Education Foundation.

Bech-Andersen, J. and Elborne, S.A. 2004. 'The true dry rot fungus (*Serpula lacrymans*) from nature to houses', in *European Research on Cultural Heritage: State-of-the-Art Studies*, vol. 2, M. Drádacký (ed.), 445–8. Prague.

Benveduti, P. 1939. 'Diagnosi precoce delle infezioni microbiche della carta', *Bollettino dell'Istituto. di Patologia del Libro* 3: 153–4.

Beothy-Kozocsa, I., Sipos-Richter, T. and Szlabey, G. 1990. 'Parchment codex restoration using parchment and cellulose fibre pulp', *Restaurator* 11: 95–109.

Beuchat, L.R. 1983. 'Influence of water activity on growth, metabolic activities and survival of yeasts and moulds', *Journal of Food Protection* 46: 135–41.

Bicchieri, M. and Pinzari, F. 2016. 'Discoveries and oddities in library materials', *Microchemical Journal* 124: 568–77.

Bonaventura, G. 1940. 'Diagnosi precoce di alcune alterazioni della carta'. Comunicazione al VI Convegno Nazionale dell'Associazione Italiana per le Biblioteche. Napoli, maggio, 1940, *Accademie e Biblioteche d'Italia* 5–6: 389–92.

Bonaventura, G. and Paganini, M.L. 1940. 'Primo contributo alla conoscenza della microflora del materiale bibliografico: microrganismi isolati da manoscritti cartacei yemeniti', *Bollettino dell'Istituto di Patologia del Libro* 2(1): 1–6.

Borrego, S., Lavin, P., Perdomo, I., Gómez de Saravia, S. and Guiamet, P. 2012. 'Determination of indoor air quality in archives and biodeterioration of the documentary heritage', *ISRN Microbiology*: article ID 680598, doi:10.5402/2012/680598.

Borrego, S., Valdés, O., Vivar, I., Lavin, P., Guiamet, P., Battistoni, P., de Saravia, S.G. and Borges, P. 2012. 'Essential oils of plants as biocides against microorganisms isolated from Cuban and Argentine documentary heritage', *ISRN Microbiology* article ID 826786, doi:10.5402/2012/826786.

Boudalis, G., Ciechanska, M., Engel, P., Ion, R., Kecskeméti, I., Moussakova, E., Pinzari, F., Schirò, J. and Vodopivec, J. (eds) 2016. *Mould on Books and Graphic Art.* Horn: Berger Verlag.

Broennimann, O., Treier, U.A., Müller-Schärer, H., Thuiller, W., Peterson, A.T. and Guisan, A. 2007. 'Evidence of climatic niche shift during biological invasion', *Ecology Letters* 10(8): 701–9.

Cappitelli, F. and Sorlini, C. 2005. 'From papyrus to compact disc: the microbial deterioration of documentary heritage', *Critical Reviews in Microbiology* 31: 1–10.

Cappitelli, F., Pasquariello, G., Tarsitani, G. and Sorlini, C. 2010. 'Scripta manent? Assessing microbial risk to paper heritage', *Trends in Microbiology* 18: 538–42.

Choi, S. 2007. 'Foxing on paper: a literature review', *Journal of the American Institute for Conservation* 46(2): 137–52.

Christensen, C., Papavizas, G.C. and Benjamin, C.R. 1959. 'A new halophilic species of *Eurotium*', *Mycologia* 51(5): 636–40.

Crook, B. and Burton, N.C. 2010. 'Indoor moulds, sick building syndrome and building related illness', *Fungal Biology Reviews* 24(3–4): 106–13.

Di Bonaventura, M.P., De Salle, R. and Bonacum, J. 2003. 'Tiffany's drawings, fungal spots and phylogenetic trees', in *Molecular Biology and Cultural Heritage*, C. Saiz-Jimenez (ed.), 131–5. Madrid: A.A. Balkema Publishers.

Dicus, D.H. 2000. 'One response to a collection wide mold outbreak: how bad can it be, how good can it get?', *Journal of the American Institute for Conservation* 39: 85–105.

Ďurovič, M. and Zelinger, J. 1993. 'Chemical processes in the bleaching of paper in library and archival collections', *Restaurator* 14: 78–101.

Ďurovič, M., Boumová, K., Drábková, K. and Bacílková, B. 2014. 'Effects of γ-radiation on micromycetes and paper support', paper presented at the *Men and Books: From Microorganisms to Megaorganisms Conference, St. Pölten, Rakousko, 28 April 2014.*

El Monssef, R.A.A., Hassan, E.A. and Ramadan, E.M. 2016. 'Production of laccase enzyme for their potential application to decolorize fungal pigments on aging paper and parchment', *Annals of Agricultural Sciences* 61(1): 145–54.

Fiacconi, F., Colaizzi, P., Ruggiero, D., Botti, L. and Pinzari, F. 2012. 'Colori indesiderati: prove di rimozione enzimatica di macchie di origine fungina dalla carta', in *Prima, durante ... invece del restauro CESMAR7*, C. Lodi and C. Sburlino (eds), 334–41, Parma, 16–17 November 2012.

Florian, M.L.E. 1990. 'The effects of freezing and freeze-drying on natural history specimens', *Collection Forum* 6: 45–52.

Florian, M.L.E. 2002. *Fungal Facts.* London: Archetype Publications.

Florian, M.L.E. 2007. *Protein Facts: Fibrous Proteins in Cultural and Natural History Artifacts.* London: Archetype Publications.

Florian, M.L.E. and Manning, L. 2000. 'SEM analysis of irregular fungal spots in an 1854 book: population dynamics and species identification', *International Biodeterioration and Biodegradation* 46: 205–20.

Gadd, G.M. 2010. 'Metals, minerals and microbes: geomicrobiology and bioremediation', *Microbiology* 156: 609–43.

Galippe, M.V. 1919. 'Des microrganisme vivants dans le papier, leur resistence à l'action de la chaleur et à celle du temps', *Comptes rendus de l'Académie des Sciences* v.CLXIX.

Gallo, A. 1935. *Le malattie del libro: le cure ed i restauri.* Milan: Mondadori.

Gallo, A. 1940. *Il Regio Istituto di Patologia del Libro.* Rome: Palombi.

Gallo, A. 2006. *Le malattie del libro: le cure ed i restauri.* Bibl. storia del libro e della stampa. C. Di Carlo (ed.). Forni Collana.

Gallo, F. 1992. *Il biodeterioramento di libri e documenti.* Centro di Studi per la conservazione della carta. Rome: ICCROM.

Gallo, F. and Pasquariello, G. 1989. 'Foxing, ipotesi sull'origine biologica', *Bollettino dell'Istituto di Patologia del Libro* 43: 136–76.

Gallo, F. and Strzelczyk, A. 1971. 'Indagine preliminare sulle alterazioni microbiche della pergamena', *Bollettino dell'Istituto di Patologia del Libro* 30(1/2): 71–87.

Gallo, F., Valenti, P., Colaizzi, P., Pasquariello, G., Scorrano, M., Sclocchi, M-C., Maggi, O. and Persiani, A.M. 1999. 'Research on the viability of fungal spores in relation to different microclimates and different materials', in *International Conference on Conservation and Restoration of Archival and Library Materials*, C. Federici and P.F. Munafò (eds), 213–30. Palermo: Palumbo.

Gill, M. 2003. 'Pigments of fungi (macromycetes)', *Natural Product Reports* 20: 615–39.

Gutarowska, B., Rembisz, D., Zduniak, K., Skóra, J., Szynkowska, M., Gliścińska, E. and Koziróg, A. 2012. 'Optimization and application of the misting method with silver nanoparticles for disinfection of the historical objects', *International Biodeterioration and Biodegradation* 75: 167–75.

Hájek, M., Ďurovič, M., Paulusová, H. and Weberová, L. 2011. 'Simultaneous microwave drying and disinfection of flooded books', *Restaurator* 32(1): 1–12.

Hueck, H.J. 1968. 'The biodeterioration of materials: an appraisal', in *Biodeterioration of Materials Microbiological and Allied Aspects*, A.H. Walters and J.J. Elphic (eds), 6–12. London: Springer.

ISO/DIS 11799, 2003. *Information and Documentation. Document Storage Requirements for Archive and Library Materials.*

Jason, K., Katerina, O. and Greg, W. 2013. *Ancient Libraries*. Cambridge: University Press.

Karbowska-Berent, J. and Strzelczyk, A. 2000. 'The role of streptomycetes in the biodeterioration of historic parchment', *Acta Mikrobiologica Polonica* 49(2): 177–8.

Kowalik, R. 1980. 'Microbiodeterioration of library materials. Part 2: Microbiodecomposition of basic organic library materials', *Restaurator* 4: 200–208.

Kraková, L., Chovanová, K., Selim, S.A., Šimonovičová, A., Puškárová, A., Maková, A. and Pangallo, D. 2012. 'A multiphasic approach for investigation of the microbial diversity and its biodegradative abilities in historical paper and parchment documents', *International Biodeterioration and Biodegradation* 70: 117–25.

Larsen, H. 1986. 'Halophilic and halotolerant microorganisms: an overview and historical perspective', *FEMS Microbiology Reviews* 39: 3–7.

Magan, N. and Lacey, J. 1984. 'Effect of water activity, temperature and substrate interactions between field and storage fungi', *Transactions of the British Mycological Society* 82: 83–93.

Magan, N. and Lynch, J.M. 1986. 'Water potential, growth and cellulolysis of fungi involved in decomposition of cereal residues', *Journal of General Microbiology* 132: 1181–7.

Markham, P. and Bazin, M.J. 1991. 'Decomposition of cellulose by fungi', in *Handbook of Applied Mycology. Volume 1: Soil and Plants. Part III Fungal Degradation Processes*, D.K. Arora, B. Rai, K.G. Mukerji and G.R. Knudsen (eds), 379–89. New York: Marcel Dekker.

Mazur, P. 1968. 'Survival of fungi after freezing and desiccation', in *The Fungi: An Advanced Treatise*, vol. Ill, G.C. Ainsworth and A.S. Sussman (eds), 325–94. London and New York: Academic Press.

Mazzuca, C., Micheli, L., Carbone, M., Basoli, F., Cervelli, E., Iannuccelli, S., Sotgiu, S. and Palleschi, A. 2014. 'Gellan hydrogel as a powerful tool in paper cleaning process: a detailed study', *Journal of Colloid and Interface Science* 416: 205–11.

McCleary, J.P. 1987. *Vacuum Freeze-Drying, a Method Used to Salvage Water-Damaged Archival and Library Materials: A RAMP Study with Guidelines*, PGI-87/WS/7. Paris: UNESCO.

Mesquita, N., Portugal, A., Videira, S., Rodríguez-Echeverría, S., Bandeira, A.M.L., Santos, M.J.A. and Freitas, H. 2009. 'Fungal diversity in ancient documents: a case study on the archive of the University of Coimbra', *International Biodeterioration and Biodegradation* 63: 626–9.

Michaelsen, A., Pinzari, F., Ripka, K., Lubitz, W. and Piñar, G. 2005. 'Application of molecular techniques for identification of fungal communities colonising paper material', in *Proceedings of the 13th International Biodeterioration and Biodegradation Symposium (IBBS-13) 4–9 September 2005, Madrid.*

Michaelsen, A., Pinzari, F., Ripka, K., Lubitz, W. and Piñar, G. 2006. 'Application of molecular techniques for identification of fungal communities colonising paper material', *International Biodeterioration and Biodegradation* 58: 133–41.

Michaelsen, A., Piñar, G., Montanari, M. and Pinzari, F. 2009. 'Biodeterioration and restoration of a 16th-century book using a combination of conventional and molecular techniques: a case study', *International Biodeterioration and Biodegradation* 63: 161–8.

Michaelsen, A., Piñar, G. and Pinzari, F. 2010. 'Molecular and microscopical investigation of the microflora inhabiting a deteriorated Italian manuscript dated from the 13th century', *Microbial Ecology* 60: 69–80.

Michaelsen, A., Pinzari, F., Barbabietola, N. and Piñar, G. 2013. 'Monitoring the effects of different conservation treatments on paper-infecting fungi', *International Biodeterioration and Biodegradation* 84: 333–41.

Micheluz, A., Manente, S., Tigini, V., Prigione, V., Pinzari, F., Ravagnan, G. and Varese, G.C. 2015. 'The extreme environment of a library: xerophilic

fungi inhabiting indoor niches', *International Biodeterioration and Biodegradation* 99: 1–7.
Micheluz, A., Manente, S., Rovea, M., Slanzi, D., Varese, G.C., Ravagnan, G. and Formenton, G. 2016. 'Detection of volatile metabolites of moulds isolated from a contaminated library', *Journal of Microbiological Methods* 128: 34–41.
Migliore, L., Thaller, M.C., Vendittozzi, G., Mejia, A.Y., Mercuri, F., Orlanducci, S. and Rubechini, A. 2017. 'Purple spot damage dynamics investigated by an integrated approach on a 1244 AD parchment roll from the Secret Vatican Archive', *Scientific Reports* 7: 1–12.
Montanari, M., Melloni, V., Pinzari, F. and Innocenti, G. 2012. 'Fungal biodeterioration of historical library materials stored in compactus movable shelves', *International Biodeterioration and Biodegradation* 75: 83–8.
Montemartini Corte, A., Ferroni, A. and Salvo, V.S. 2003. 'Isolation of fungal species from test samples and maps damaged by foxing, and correlation between these species and the environment', *International Biodeterioration and Biodegradation* 51: 167–73.
Nielsen, K.F. 2002. *Mould Growth on Building Materials: Secondary Metabolites, Mycotoxins and Biomarkers*, PhD dissertation, BioCentrum-DTU, Technical University of Denmark.
Nielsen, K.F., Holm, G., Uttrup, L.P. and Nielsen, P.A. 2004. 'Mould growth on building materials under low water activities: influence of humidity and temperature on fungal growth and secondary metabolism', *International Biodeterioration and Biodegradation* 54: 325–36.
Nieto-Fernandez, F.E., Centeno, S.A., Wypyski, M.T., Di Bonaventura, M.P. and Baldwin, A.M. 2003. 'Enzymatic approach to removal of fungal spots from drawing on paper', in *Art, Biology, and Conservation: Biodeterioration of Works of Art*, R.J. Koestler, V.H. Koestler, A.E. Charola and F.E. Nieto-Fernandez (eds), 110–27. New York: The Metropolitan Museum of Art.
Nittérus, M. 2000a. 'Fungi in archives and libraries: a literary survey', *Restaurator* 21: 25–40.
Nittérus, M. 2000b. 'Ethanol as fungal sanitizer in paper conservation', *Restaurator* 21: 101–15.
Nunes, I., Mesquita, N., Verde, S.C., Bandeira, A.M.L., Carolino, M.M., Portugal, A. and Botelho, M.L. 2013. 'Characterization of an airborne microbial community: a case study in the archive of the University of Coimbra, Portugal', *International Biodeterioration and Biodegradation* 79: 36–41.
Pangallo, D., Chovanová, K., Šimonovičová, A. and Ferianc, P. 2009. 'Investigation of microbica community isolated from indoor artworks and their environment: identification, biodegradative abilities, and DNA typing', *Canadian Journal of Microbiology* 55: 277–87.
Petushkova, J.P. and Koestler, R.J. 1996. 'Biodeterioration studies on parchment and leather attacked by bacteria in the Commonwealth of Socialist States', in *International Conference on Conservation and Restoration of Archival and Library Materials*, C. Federici and P.F. Munafò (eds), 195–211. Palermo: Palumbo.
Pietrzak, K., Otlewska, A., Danielewicz, D., Dybka, K., Pangallo, D., Kraková, l., Puškárová, A., Bučková, M., Scholtz, V., Ďurovič, M., Surma-Slusarska, B., Demnerová, K. and Gutarowska, B. 2017. 'Disinfection of archival documents using thyme essential oil, silver nanoparticles misting and low temperature plasma', *Journal of Cultural Heritage* 24: 69–77.
Piñar, G., Sterflinger, K. and Pinzari, F. 2014a. 'Unmasking the measles-like parchment discoloration: molecular and micro-analytical approach', *Environmental Microbiology* 17: 427–43.
Piñar, G., Sterflinger, K., Ettenauer, J., Quandt, A. and Pinzari, F. 2014b. 'A combined approach to assess the microbial contamination of the Archimedes Palimpsest', *Microbial Ecology* 69: 1–17.
Piñar, G., Tafer, H., Sterflinger, K. and Pinzari, F. 2015. 'Amid the possible causes of a very famous foxing: molecular and microscopic insight into Leonardo da Vinci's self-portrait', *Environmental Microbiology Reports* 7: 849–59.
Pinzari, F. 2011. 'Microbial ecology of indoor environments: the ecological and applied aspects of microbial contamination in archives, libraries and conservation environments', in *Sick Building Syndrome in Public Buildings and Workplaces*, S.A. Abdul-Wahab (ed.), 193–206. Berlin and Heidelberg: Springer.
Pinzari, F. and Montanari, M. 2011. 'Mould growth on library materials stored in compactus-type shelving units', in *Sick Building Syndrome in Public Buildings and Workplaces*, S.A. Abdul-Wahab (ed.), 193–206. Berlin and Heidelberg: Springer.
Pinzari, F., Pasquariello, G. and De Mico, A. 2006. 'Biodeterioration of paper: an SEM study of fungal spoilage reproduced under controlled conditions', *Macromolecular Symposia* 238: 57–66.
Pinzari, F., Montanari, M., Michaelsen, A. and Piñar, G. 2009. 'Analytical protocols for the assessment

of biological damage in historical documents', *Coalition* 19: 6–13.

Pinzari, F., Zotti, M., De Mico, A. and Calvini, P. 2010. 'Biodegradation of inorganic components in paper documents: formation of calcium oxalate crystals as a consequence of *Aspergillus terreus* Thom growth', *International Biodeterioration and Biodegradation* 64: 499–505.

Pinzari, F., Troiano, F., Piñar, G., Sterflinger, K. and Montanari, M. 2011. 'The contribution of microbiological research in the field of book, paper and parchment conservation', in *Approaches to Book and Paper Conservation-Restoration*, P. Engel, J. Schirò, R. Larsen, E. Moussakova and I. Kecskeméti (eds), 575–94. Horn and Vienna: New Verlag Berger.

Pinzari, F., Cialei, V. and Piñar, G. 2012a. 'A case study of ancient parchment biodeterioration using variable pressure and high vacuum scanning electron microscopy', in *Historical Technology, Materials and Conservation: SEM and Microanalysis*, N. Meeks, C. Cartwright, A. Meek and A. Mongiatti (eds.) 93–9. London: Archetype Publications.

Pinzari, F., Colaizzi, P., Maggi, O., Persiani, A.M., Schütz, R. and Rabin, I. 2012b. 'Fungal bioleaching of mineral components in a twentieth-century illuminated parchment', *Analytical Bioanalytical Chemistry* 402: 1541–50.

Polacheck, I. 1989. 'Damage to an ancient parchment document by *Aspergillus*', *Mycopathologia* 106: 89–93.

Polacheck, I. 2005. 'Fungi and cultural heritage: bio and material cultures at Qumran', in *Papers from a COST Action G8 Working Group Meeting*, J. Gunneweg, C. Greenblatt and A. Adriaens (eds), Jerusalem, 22–23 May 2005. Jerusalem: COST.

Polo, A., Cappitelli, F., Villa, F. and Pinzari, F. 2017. 'Biological invasion in the indoor environment: the spread of *Eurotium halophilicum* on library materials', *International Biodeterioration and Biodegradation* 118: 34–44.

Poole, J.B. and Reed, R. 1962. 'The preparation of leather and parchment by the Dead Sea Scrolls community', *Technology and Culture* 3(1): 1–26.

Price, L.O. 2013. *Managing a Mold Invasion: Guidelines for Disaster Response*. CCAHA Technical Series No. 1. Philadelphia: Conservation, Center for Art and Historic Artifacts.

Principi, P., Villa, F., Sorlini, C. and Cappitelli, F. 2011. 'Molecular studies of microbial community structure on stained pages of Leonardo da Vinci's Atlantic Codex', *Microbial Ecology* 61: 214222.

Rakotonirainy, M.S. and Lavédrine, B. 2005. 'Screening for antifungal activity of essential oils and related compounds to control the biocontamination in libraries and archives storage areas', *International Biodeterioration and Biodegradation* 55(2): 141–7.

Rakotonirainy, M.S., Heude, E. and Lavédrine, B. 2007. 'Isolation and attempts of biomolecular characterization of fungal strains associated to foxing on a 19th century book', *Journal of Cultural Heritage* 8: 126–33.

Rebrikova, N.L. and Dmitrieva, M.B. 1996. 'Experimental investigation of parchment manuscripts with traces of microbiological damage', in *International Conference on Conservation and Restoration of Archival and Library Materials*, C. Federici and P.F. Munafò (eds), 275–83. Palermo: Palumbo.

Reed, R. 1972. *Ancient Skins Parchments and Leathers. Department of Food and Leather Science*. London and New York: University of Leeds, Seminar Press.

Reed, R. 1975. *The Nature and Making of Parchment*. Leeds: Elmete Press.

Ryder, M.L. 1969. 'Remains derived from skin', in *Science in Archaeology*, D.R. Brothwell and E.S. Higgs (eds), 539–54. London: Thames & Hudson.

Sarrette, D.M., Nout, M.J.R., Gervais, P.F. and Rombouts, M. 1992. 'Effect of water activity on production and activity of *Rhizopus oligosporus* polysaccharidases', *Applied Microbiology and Biotechnology* 37(4): 420–25.

Schmidt, J.D. 1985. 'Freeze drying of historical cultural properties', *Technology and Conservation* 9(1): 20–26.

Sclocchi, M.C., Kraková, L., Pinzari, F., Colaizzi, P., Bicchieri, M. and Šaková, N. 2016. 'Microbial life and death in a foxing stain: a suggested mechanism of photographic prints defacement', *Microbial Ecology* 73: 1–12.

Sée, P. 1917. *Sur les moisissures causant l'altération du papier*. T.CLXIV, 29. Paris: Académie des Sciences.

Sée, P. 1919. *Les maladies du papier piqué: les champignons chromogènes qui les provoquent les modes de preservations*. Paris: O. Doin et Fils.

Sequeira, S.O., Cabrita, E.J. and Macedo, M.F. 2012. 'Antifungals on paper conservation: an overview', *International Biodeterioration and Biodegradation* 74: 67–86.

Sequeira, S.O., Cabrita, E.J. and Macedo, M.F. 2014. 'Fungal biodeterioration of paper: how are paper and book conservators dealing with it? An international survey', *Restaurator: International Journal for the Preservation of Library and Archival Material* 2(35): 181–99.

Sequeira, S.O., Phillips, A.J.L., Cabrita, E.J. and Macedo, M.F. 2017a. 'Ethanol as an antifungal treatment for paper: short-term and long-term effects', *Studies in Conservation* 62: 33–42.

Sequeira, S.O., Phillips, A.J.L., Cabrita, E.J. and Macedo, M.F. 2017b. 'Antifungal treatment of paper with calcium propionate and parabens: short-term and long-term effects', *International Biodeterioration and Biodegradation* 120: 203–15.

Sibilia, C. 1935. 'Le malattie crittogamiche dei libri', *Accademie e Biblioteche d'Italia* IX (3–4): 274-S0.

Singer, C. 1915. 'The dawn of microscopical discovery', *Journal of the Royal Microscopical Society* 35: 317–40, doi:10.1111/j.1365-2818.1915.tb04962.x.

Smith, A.W. 2012. 'Bleaching in paper conservation', *Restaurator: International Journal for the Preservation of Library and Archival Material* 33: 223–48.

Souza, P.N. da Costa, Grigoletto, T.L., de Moraes, L.A., Abreu, L.M., Guimarães, L.H., Santos, C., Galvão, L.R. and Cardoso, P.G. 2016. 'Production and chemical characterization of pigments in filamentous fungi', *Microbiology* 162(1): 12–22.

Sterflinger, K. 2010. 'Fungi: their role in deterioration of cultural heritage', *Fungal Biology Reviews* 24: 47–55.

Sterflinger, K. and Pinzari, F. 2012. 'The revenge of time: fungal deterioration of cultural heritage with particular reference to books, paper and parchment', *Environmental Microbiology* 14: 559–66.

Strlič, M., Kolar, J., Kočar, D., Drnovsek, T., Selih, V.S., Susić, R. and Pihlar, B. 2004. 'What is the pH of alkaline paper?', *e-Preservation Science* 1: 35–47.

Stupar, M., Lj. Grbić, M., Džamić, A., Unković, N., Ristić, M., Jelikić, A. and Vukojević, J. 2014. 'Antifungal activity of selected essential oils and biocide benzalkonium chloride against the fungi isolated from cultural heritage objects', *South African Journal of Botany* 93: 118–24.

Sussman, A.S. 1966. 'Longevity and survivability of fungi', in *The Fungi: An Advanced Treatise*, vol. II, G.C. Ainsworth and A.S. Sussman (eds), 477–86. London and New York: Academic Press.

Szczepanowska, H.M. and Lovett, Jr., C.M. 1992. 'A study of removal and prevention of fungal stains on paper', *Journal of the American Institute for Conservation* 31: 147–60.

Tanney, J.B, Nguyen, H.D.T., Pinzari, F. and Seifert, K.A. 2015. 'A century later: rediscovery, culturing and phylogenetic analysis of *Diploöspora rosea*, a rare onygenalean hyphomycete', *Antonie van Leeuwenhoek* 108: 1023–35.

Tavzes, C., Palcic, J., Fackler, K., Pohleven, F. and Koestler, R.J. 2013. 'Biomimetic system for removal of fungal melanin staining on paper', *International Biodeterioration and Biodegradation* 84: 307–13.

Teasdale, M.D., Fiddyment, S., Mattiangeli, V., Speller, C., Binois, A., Carver, M., Dand, C., Newfield, T.P., Webb, C.C., Bradley, D.G. and Collins, M.J. 2017. 'The York Gospels: a 1000-year biological palimpsest', *Royal Society Open Science* 4(10): 170988.

Troiano, F., Barbabietola, N., Colaizzi, P., Livi, P., Montanari, M. and Pinzari, F. 2012. 'La liofilizzazione quale intervento di recupero di volumi alluvionati ed attaccati da microfunghi', in *Prima, durante... invece del restauro CESMAR7*, C. Lodi and C. Sburlino (eds.) 209–14, Parma, 16–17 November 2012.

Troiano, F., Polo, A., Villa, F. and Cappitelli, F. 2014. 'Assessing the microbiological risk to stored sixteenth century parchment manuscripts: a holistic approach based on molecular and environmental studies', *Biofouling* 30: 299–311.

Vereecken, E. and Roels, S. 2012. 'Review of mould prediction models and their influence on mould risk evaluation', *Building Environment* 51: 296–310.

Verona, O. 1939. *Studio sulle cause microbiche che danneggiano la carta e i libri*. Rome: Ente Nazionale per la Cellulosa e per la Carta. Ed. culturali e di propaganda.

Zou, X., Uesaka, T. and Gurnagul, N. 1996. 'Prediction of paper permanence by accelerated aging. Part I: Kinetic analysis of the aging process', *Cellulose* 3: 243–67.

Zyska, B. 1997. 'Fungi isolated from library materials: a review of the literature', *International Biodeteriorioration and Biodegradation* 40: 43–51.

AUTHOR'S ADDRESS

Flavia Pinzari, Research Centre for Agriculture and Environment (CREA-AA), Via della Navicella, 2-4, 00184 Rome, Italy (flavia.pinzari@crea.gov.it)/ Department of Life Sciences, Natural History Museum of London, London, UK (f.pinzari@nhm.ac.uk)

BIODETERIORATION OF PHOTOGRAPHIC AND CINEMATOGRAPHIC MATERIALS: METHODS OF INVESTIGATION

Andrea Puškárová, Mária Bučková and Domenico Pangallo

PHOTOGRAPHIC AND FILM MATERIALS

Photographs and cinematographic films are part of our cultural heritage because they immediately capture the people, objects and moments that represent our history. Generally, these kinds of documentations are composed of organic and/or synthetic materials that are subjected to deterioration by physical agents (light, heat and moisture), chemical agents (atmospheric solution) and biological agents (bacteria, fungi and insects) (Maggi *et al.* 2000). Investigations of photographic and cinematographic materials are important for the purposes of the appropriate treatment and restoration.

Photographic materials have a stratigraphic structure and are basically composed of three components: a support, an image-forming material and a binder. The last two components comprise the photographic emulsion layer. Figure 1 illustrates the three structural layers found in 19th-century photographic print materials.

Paper, glass, cellulose nitrate, cellulose acetate and polyester have all been used at some time in history to manufacture supports (Lavédrine *et al.* 2003). The image-forming material selectively absorbs light that would otherwise be reflected from the paper support. The most important final image material found in the 19th century is metallic silver particles on black and white images and dyes on colour images. Metal toning – the treatment of the silver image with gold, platinum or combination of particles – was also used.

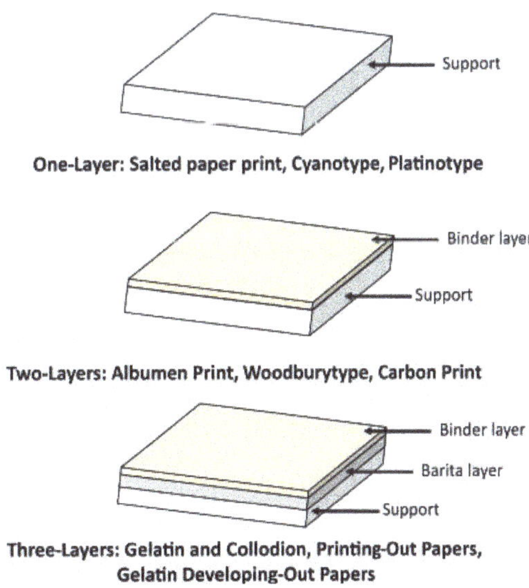

Figure 1 Schematic representation of the structural layers found in 19th-century photographic print materials according to Reilly and McCabe (1986).

Table 1 Types of photographic materials.

Photographic materials	Support	Binder
Daguerreotype	copper	absent
Calotype (negative)	paper	absent
Salted paper print (positive)	paper	absent
Albumen plate (negative)	glass	albumen
Collodion plate (negative)	glass	collodion
Gelatin plate (negative)	glass	gelatin
Ambrotype	glass	collodion
Tintype/ferrotype	iron	collodion
Albumen print	paper	albumen
Collodion print	paper	collodion
Gelatin print	paper	gelatin

The binder is the transparent layer in a photographic material in which the final image material is suspended and protected – albumen was popular from the mid-19th century to the early 20th century, while gelatin was the most commonly used binder during the 20th century together with some collodion. Albumen has a different protein structure from gelatin and does not form a reversible gel. Collodion, a nitrocellulose solution in a mixture of alcohol and ether containing ~12% nitrogen, is not sufficiently flexible or waterproof so plasticizers such as castor oil or glycerin are added. Gelatin is a water-soluble protein obtained by collagen hydrolysis. Replacement of collodion with gelatin was a significant technical innovation in the history of photography.

TYPES OF PHOTOGRAPHIC MATERIALS

Little is known about the role of microbes in the degradation of photographic and cinematographic materials. Knowledge of the composition of these materials will aid researchers when carrying out investigations into biodeterioration processes. The development of photography, as well as photographic negatives, is closely related to the camera obscura. In 1827, Joseph Nicephore Niepce developed the prototype for the modern photograph (heliograph) by allowing light to draw the image. Louis Daguerre went on to develop the daguerreotype process in 1839, the year largely considered as the birth of practical photography. The metal-based daguerreotype process was soon in competition with the paper-based calotype negative and salt print process invented by William Henry Fox Talbot. Monochrome photography was later improved by other methods including calotype, ambrotype, tintype, albumen print and gelatin print (Table 1).

In the daguerreotype process, a silver-coated copper plate was treated with iodine vapours and the exposed image was made by mercury vapour and stabilized with a salt solution. The daguerreotype's features are its highly polished silver support and the way it appears either as negative or positive depending on the angle of view and the direction from which light falls on it (Coe 1978; Reilly 1981). The daguerreotype required only minutes of exposure in the camera, producing clear and detailed results.

Calotype is a process for making a photographic negative. Paper negatives belong to a group of monolayer materials where instead of being dispersed in the binder layer (albumen, collodion or gelatin), the silver is incorporated directly between individual paper fibres (see Fig. 1). The term 'calotype' is used only in reference to a paper negative or a print made from a paper negative, and should not be used to denote salted paper prints in general (Wood 1971; Miller 2003). Salted paper print refers to a silver printing-out paper, which has been fixed in thiosulfate and consists of a silver image located in and among the paper fibres. The use of the salted paper printing process steadily grew throughout the 1840s, stimulated by techniques for making negatives (Reilly 1981).

Glass negatives are included in the group of two-layer materials (Fig. 1). Albumen glass negatives, despite producing a sharp image, were used relatively briefly. They were gradually replaced by the collodion wet plate process in which the negative consists of a collodion-based, light-sensitive emulsion coated on a sheet of glass. The disadvantage of wet collodion plates was that they had to be exposed and induced before the layer dried out. The collodion negative is distinguished by appearing as a positive image when viewed using reflected light on a dark background. Photographic technology experienced another period of rapid development during the 1870s due not only to improved photographic equipment, but also to the discovery of utilizing dry plates with silver bromide: gelatin silver emulsion, instead of collodion, was used to make glass plate negatives. Together with the advantage of dry exposure, the brittleness and high weight of the pad led to a more flexible and easier replacement (Whitman *et al.* 2007).

Derived from the collodion wet plate, ambrotypes were in use from 1855 to about 1865. Several opaque, non-reflective supports were employed with ambrotypes, including dark red or purple glass, and clear glass backed with black velvet or black varnish. The ambrotype normally appears as a positive image regardless of view angle (Maurice 1993).

Tintypes appeared in the late 1850s and were made well into the 20th century. Also known as ferrotypes (Davis 1984), tintypes have a support comprised of a thin sheet of lacquered iron to produce a positive image. Albumen photographs became the dominant form of photographic positives from 1855 to the turn of the 20th century (Reilly and McCabe 1986). In these photographs, a paper sheet was covered with albumen in order to bind the light-sensitive halide salts to the paper (Reilly 1981). The structure of the silver and gold image and the chemical characteristics of albumen itself are the primary reasons for the instability of this material. Because albumen prints have no baryta layer, individual paper fibres are visible through the albumen binder during microscopic examination (Reilly 1981; Reilly *et al.* 1982).

Glossy collodion print materials were used between the late 1880s and the 1920s. The range of image colours in these materials is typical of gold-toned printing-out papers. Under microscopic examination, the image has continuous tone and a smooth baryta coating obscuring the paper fibres of the support. Matte collodion prints were also used. The image colour of the semi-matte surface was usually neutral as a result of gold-and-platinum toning (Reilly 1981).

In gelatin prints, no paper fibres are visible as they are completely covered by the baryta layer. This print material is very difficult to distinguish from glossy collodion printing-out paper. Gelatin is important in the entire photographic process, from the formation of the silver halide emulsions through the manufacturing steps to the developing and printing operations. Because of these unique properties, gelatin is found in all photographic materials throughout cinematographic history. Efforts to replace gelatin with synthetic polymers have failed to produce a fully satisfactory material (Abrusci *et al.* 2005).

The development of colour was a milestone in the history of photography: autochrome plates allowed the production of a colour photograph. The principle was based on a fine raster of transparent grains of potato starch stained in red, green and blue. Negatives were exposed through a glass mat and the starch grains formed a colour filter. A colour photo was created by inserting the autochrome plate in the camera allowing the light that entered the lens to be filtered by the starch colour filter (Wilhelm and Brower 1993).

CINEMATOGRAPHIC MATERIAL

Film material used in cinematography is complex, consisting of a photographic emulsion layer overcoating a polymeric support (plasticized cellulose acetate) and a protective transparent cross-linked gelatin layer coating the emulsion (Gaspard *et al.* 2007). Each layer has different properties, the most important being the emulsion, the support, the protective layers and the antihalo (Ciliberto *et al.* 2013).

The first plastic material used in animation was celluloid, developed at the end of the 19th century as a support in photography. Celluloid is composed of cellulose nitrate plasticized with camphor (Allen *et al.* 1992; Shashoua 2006). Despite its good flexibility and dimensional stability, cellulose nitrate is highly flammable and subject to deterioration in the presence of moisture that can lead to the release of hazardous nitrogen-based gases (Giachet *et al.* 2014). These limitations led to the gradual replacement of cellulose nitrate by cellulose acetate, which was subsequently used throughout the 20th century. Cellulose acetate was originally considered to be a good archival material due to its lack of flammability but it has proved susceptible to deterioration through hydrolysis, migration, loss of plasticizer and photodegradation, which is associated with the generation of radicals due to the absorption of light (Reilly 1993; Shashoua 2006; Littlejohn *et al.* 2013).

In the 1970s, archivists began recognizing the decomposition process problem in cellulose-triacetate-based film. Chemical instability in the support caused significant loss, sometimes of entire films. This problem is easily detected when a film storage container is opened by the release of acetic acid vapours. In the film world this characteristic odour is known as the 'vinegar syndrome'. Acetic acid degradation is the primary cause of fading colour in chromogenic emulsions, which are employed universally in cinematography, and for the deformation and stiffening of the material (Abrusci *et al.* 2006). In addition, plasticizer migrates, leaving obvious viscous liquid or crystalline deposits on the film's surface. To address vinegar syndrome, cellulose triacetate materials have been replaced by polyethylene terephthalate (PET), which has superior chemical and dimensional stability (Reilly 1993; Bigourdan 2000). However, cellulose acetate is still in use, as many in animation production prefer the way cellulose acetate catches the black ink of the line drawings (Shashoua 2006; Littlejohn *et al.* 2013).

The photosensitive emulsion that coats the cellulose support consists mainly of gelatin. In the case of black and white films, the emulsion is comprised of light-sensitive silver halide salts accompanied by iron, lead, chrome and phosphorus lines: iron and lead are constituents found in film developers, chrome in the hardening agents and phosphorus is one of main components in the plasticizer used in the polymeric support (Gaspard *et al.* 2007). The photosensitive layer of colour films is somewhat more complicated as it is generally composed of green, red and blue sensitive emulsion layers sometimes separated by layers of clear gelatin.

Cinematographic film usually contains an antihalo layer that can be applied between the gelatin and support layers or on the back of the support. It absorbs reflection, thereby avoiding noise in the image (Ciliberto *et al.* 2013). Antihalo layers cover the cellulose triacetate support in order to avoid sharpness reduction in the image and light scattering, called halation, around bright objects in the images. For this purpose, in black and white negative films, a neutral-density grey dye is added to the layer of gelatin (Abrusci *et al.* 2006). The protective layer, localized at the top of the film, prevents the deposition of dust on the gelatin layer and functions to prevent scratches that can cause disturbances during projection (Ciliberto *et al.* 2013).

TECHNIQUES FOR ANALYSING PHOTOGRAPHIC AND FILM MATERIALS

The techniques used in investigating photographic and cinematographic materials are similar to those employed for investigating other historic materials: they aim to generate information on the classification and composition of historical materials in order to select suitable reparation methods for damaged items. Information on material identification and techniques used in photographs and cinematographic film is needed by both artists and conservators (Vila and Centeno 2013). A historical and critical approach when undertaking a chemical and physical investigation will allow a deeper knowledge to be gained of photographic and cinematographic materials and their preservation issues.

The characterization of photographic and cinematographic materials conservation status is usually assessed through visual analysis or microscopy. However, many materials cannot be completely characterized by a simple visual check and require more quantitative investigation (Cattaneo *et al.* 2008). The analytical techniques used for morphological and physicochemical identification of these materials are non-invasive. The surface morphology is studied by means of light microscopy and environmental scanning electron microscopy (ESEM). SEM coupled with energy dispersive X-ray (EDX) analysis provides information on the morphology and elemental composition of photographic and cinematographic materials (Doehne 2006). Other techniques used to determine the chemical composition of historical photographic and cinematographic materials include X-ray fluorescence (XRF) spectroscopy, inductively coupled plasma mass spectrometry (ICP-MS), Fourier transform infrared (FTIR) spectroscopy and Raman spectroscopy (Chabala *et al.* 1995; Hu *et al.* 1999; Creagh 2005; Gaspard *et al.* 2007). Sclocchi *et al.* (2017) used SEM-EDX, μ-Raman and XRF to distinguish the manufacture of two gelatin prints and the structural changes occurring in areas of foxing.

ICP-MS analysis provides low detection limits and high accuracy but requires sample preparation. FTIR and Raman spectroscopies allow the identification of molecular compounds present on the sample surface but are unable to analyse the composition of the layers underneath (Gaspard *et al.* 2007). Micro-invasive and non-invasive investigations are performed and evaluated in order to understand the physical and chemical degradation processes in photographic and cinematographic materials. Table 2 provides an overview of the most widely used techniques.

XRF analysis, a non-destructive technique requiring neither sampling nor direct contact with the photograph or film, provides qualitative and quantitative information on inorganic elements in materials (McCabe and Glinsman 1996; Stulik *et al.* 2002). FTIR spectroscopy represents a powerful technique in the diagnostics and analysis of cultural heritage to measure biodegradation (Carbó *et al.* 1997; Salvadó *et al.* 2005; Bersani *et al.* 2008). The method involves the absorption of infrared radiation by the sample, resulting in several molecular vibrations by the chemical bonds of molecular compounds (Mayo *et al.* 2004).

Raman spectroscopy is a non-invasive tool used for identifying pigments in image materials. For pigment-based photographs, identification of the organic components in the colloids is important for the artist in understanding the artistic process and for the conservator in preservation of the works. Organic compounds on daguerreotype have been identified by *in situ* Raman spectroscopy (Centeno *et al.* 2008, 2011), but the organic materials mixed with pigments in artworks are below the detection limit of the technique (Vila and Centeno 2013). Micro-Raman spectroscopy may be especially useful in cases where macroscopic evidence of specific photographic processes is lacking (e.g. the gelatin silver salt technique) (Marucci *et al.* 2014).

All these techniques can be used in combination in order to study photographic and film materials: Centeno *et al.* (2008) used Raman spectroscopy, complemented by *in situ* SEM-EDX, to non-destructively characterize image deterioration in eight representative daguerreotypes. This study demonstrated the suitability of non-invasive techniques in the study of daguerreotypes in order to preserve the integrity of a very delicate image structure. Marucci *et al.* (2014) investigated historical photographs using μ-Raman spectroscopy – the Raman spectra showed diagnostic peaks of silver halide as evidence that the silver salt gelatin technique had been employed.

Table 2 Techniques used for investigation of the photographic and cinematographic materials.

	ESEM-EDX	XRF	FTIR µ-FTIR ATR-FTIR	Raman µ-Raman	Reference
Photographic materials					
Daguerreotype	+	+	–	+	Centeno *et al.* 2008 Da Silva *et al.* 2010
Calotype (negative)	+	+	+	–	Eremin *et al.* 2002 Cattaneo *et al.* 2008
Salt paper print (positive)	+	+	+	–	Eremin *et al.* 2002 Cattaneo *et al.* 2008 Casoli and Fornaciari 2014 Čechák *et al.* 2015
Collodion plates (negative)	–		+	–	McCormick-Goodhart 1989
Ambrotype	+	–	+	+	Costa 2016
Tintype/ferrotype	+	–	+	+	Carretti *et al.* 2009 Costa 2016
Albumen print	+	+	+	–	Ricci *et al.* 2007 Puškárová *et al.* 2016 Čechák *et al.* 2015
Collodion print	+	+	+	–	Peres *et al.* 2013 Casoli and Fornaciari 2014 Čechák *et al.* 2015 Modica *et al.* 2017
Gelatin print	+	+	+	+	Bučková *et al.* 2014 Casoli and Fornaciari 2014 Čechák *et al.* 2015
Cinematograhic materials					
Cellulose nitrate	+	–	+	+	Edge *et al.* 1990, Walsh 1995, Ciliberto *et al.* 2013, Bučková *et al.* 2014
Cellulose diacetate	+	–	+	–	Ciliberto *et al.* 2013 Giachet *et al.* 2014
Cellulose triacetate	+	–	+	–	Edge *et al.* 1992 Ciliberto *et al.* 2013 Vivar *et al.* 2013
Polyester (polyethylene terephthalate)	+	–	–	–	Ciliberto *et al.* 2013

Centeno *et al.* (2014) characterized five historic platinum photographic papers using a combination of Raman, FTIR and XRF analyses. One paper, a Japine Platinotype for Sepia Tones, had a sensitized surface layer in the recto consistent with the paper substrate while the verso presented a slightly modified surface. Raman and FTIR measurements revealed that this surface layer was composed mainly of amorphous cellulose. The combined use of Raman and XRF provides information that can help to differentiate platinum photographic processes.

Costa (2016) evaluated chemical and physical characteristics on two ambrotypes and two tintypes, especially their degradation patterns. SEM-EDX identified the differences in image-forming particle morphology between the types. The presence of collodion was confirmed with µ-FTIR and the source of the protective varnishes was identified; µ-Raman detected the presence of metallic silver and silver chloride on the images.

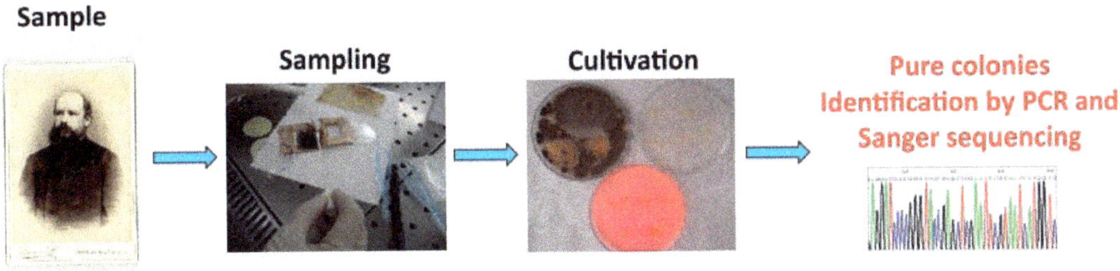

Figure 2 Example of analysis strategy. Sampling is often done with sterile swabs or needles, or by using membranes or adhesive tape. Organisms are then cultured in appropriate microbiological growth media. Subsequently, DNA is isolated from pure colonies and specific regions in the genome are amplified and sequenced by the Sanger approach.

BIODETERIORATION OF PHOTOGRAPHIC AND CINEMATOGRAPHIC MATERIALS

During the 19th and 20th centuries, photographs and films were created with different kinds of material and are considered multimaterial units. Archives, museums and galleries have numerous multilayer units in their collections. Materials that compose these items, such as paper, gelatin, albumin and cellulose, are vulnerable to biodeterioration caused by microorganisms.

Microorganisms on these materials appear as various spots and blemishes. In the case of fungi, blemishes can be small, concentric hyphae that are more or less embedded in the material. They may be white, black, brown, yellow or other colours (Czerwińska and Kowalik 1979). They can also show atypical growth forms, which can create difficulties in detecting and evaluating microbial contamination. Microorganisms disturb photographic images simply by their growth and colonization of the item. Damage is caused by the breaking down of surface material and production of enzymes, organic acids and pigmented metabolites. In the case of gelatin emulsions, secretion of enzymes and acids by microorganisms can cause the solubilization and disintegration of the image (Czerwińska and Kowalik 1979; Wilhelm and Brower 1993; Kodak 2002).

Another unaesthetic feature that affects photographs is foxing: the yellowish-rust colour staining of an image. Foxing caused by fungi is the result of a browning chemical reaction (Arai 2000; Zotti *et al.* 2011) between cello-oligosaccharides and aminobutyric and other amino acids, which are produced by the growth of tonophilic fungi on paper (Arai 1987). Florian and Manning (2000) studied the role of fungi in foxing using light and electron microscopy and found different fungal spots throughout the pages of a book, probably caused by members of the genus *Eurotium*. A combination of techniques, including culture-dependent and culture-independent approaches, led to the description of the microbial contamination dynamics of foxing stains present on the surface of two gelatin silver halide photographs (Sclocchi *et al.* 2017).

An analysis strategy is useful when dealing with contaminated photographic and cinematographic items. The first step is sampling using sterile swabs or needles, cellulose nitrate or acetate membranes and adhesive tape (Borrego *et al.* 2010; Sclocchi *et al.* 2013, 2017; Puškárová *et al.* 2016). Organisms on the sampling instrument are then cultured in appropriate microbiological growth media, as shown in Figure 2. When possible, direct cultivation of small pieces of photographic or cinematographic items is an alternative that isolates the microorganisms while reducing potential manipulation mistakes (Abrusci *et al.* 2005).

One sampling method, termed 'simulated inspection', was used to count the fungal spores released during inspection of cinematographic films (Bingley and Verran 2013). Sampling was performed inside a polymer isolation glove box of ~2.3 m^3: 50 cm of film was exposed inside the box and transferred from spool to spool. During this transfer, released spores were captured by an air sampler which had been placed inside the box.

Various and specific microbiological media can be used to increase detection and ensure the capture strain diversity. Fungi can generally be isolated using malt extract, sabouraud or potato dextrose agars. Dichloran rose bengal chloramphenicol and dichloran glycerol are suitable for the isolation of xerophilic species (Borrego *et al.* 2010; Vivar *et al.* 2013; Puškárová *et al.* 2016; Sclocchi *et al.* 2013, 2017). Media used to culture bacterial strains include tryptone soy, nutrient, Luria Bertani and R2A agars (Abrusci *et al.* 2005; Guiamet *et al.* 2011; Sclocchi *et al.* 2013; Puškárová *et al.* 2016).

In order to understand the bioactivity of microorganisms associated with photographs and film some researchers use isolation media, i.e. agar containing a specific photographic material substrate. Media supplemented with different kinds of substrates are used to demonstrate the biodegradative abilities of the isolated strains. Media containing gelatin are the most utilized because the majority of the studies investigated gelatin items (Abrusci *et al.* 2005; Borrego *et al.* 2010; Bingley and Verran 2013; Kwiatkowska *et al.* 2016; Puškárová *et al.* 2016). In fact, gelatin and starch agar were used to isolate proteolytic and amylolytic microorganisms (Borrego *et al.* 2010; Puškárová *et al.* 2016). Screening for other hydrolytic and oxidoreductive enzymes produced by isolated microorganisms was proposed by Puškárová *et al.* (2016). Analysis confirmed that the microbial community was responsible for the degradation of these items. Kwiatkowska *et al.* (2016) investigated the effect of the isolated fungi on the varnish applied to a glass plate negative; almost all the tested isolates grew well. Sclocchi *et al.* (2013) demonstrated an interesting characteristic of *Aspergillus* and *Eurotium* species colonizing a gelatin silver halide-based photographic layer. These microorganisms were able to absorb the silver present on a photograph – they reduced the silver ions and produced silver nanoparticles on their cell walls.

Molecular biology methods are frequently employed to correctly identify strains isolated from deteriorated materials. Typically, DNA is isolated from pure colonies and specific regions in the genome are amplified using eukaryotic or bacterial universal primers and sequenced using the Sanger approach (see Fig. 2). This strategy is faster and more reliable than biochemical tests or morphological analysis.

Table 3 lists microorganisms identified in various photographic and cinematographic objects, most of which contained gelatin as the binder. The investigations focused primarily on microfilamentous fungi as many researchers consider these to be the most deleterious organisms. The fungal isolates belong mainly to the genera *Penicillium, Aspergillus, Cladosporium*, *Cryptococcus* and *Trichoderma*. To our knowledge, few investigations have been carried out into identifying the microbiome associated with an albumin photograph. However, one study by Puškárová *et al.* (2016) found a large fungal diversity on these types of photograph. Bacterial strains were isolated in several studies and a comparison between gelatin- and albumin-based items showed some overlap regarding the genera recovered. The most prevalent bacteria belonged to the phyla Firmicutes (*Bacillus* and *Staphylococcus* spp.) and Actinobacteria (*Streptomyces* and *Kocuria* spp.). In some instances, members of the phylum Proteobacteria were also isolated.

Using molecular biology methods for the analysis of archival documents (Cappitelli *et al.* 2010; Sterflinger and Pinzari 2012), it was possible to also identify the non-cultivable portion of cultural heritage deteriogens. Culture-independent approaches are often considered complementary to cultivation analysis as they enhance microbial detection and provide a more complete picture of the microbial community present on a deteriorated object. These approaches were applied to gelatin-based negative and positive prints and to albumin-based photographs. Other materials related to photographic conservation were also analysed,

Table 3 Microorganisms identified from different photographic and cinematographic objects.

Material type	Identified microorganisms	Degradation abilities	Reference
Glass plate negatives	Fungi: *Cladosporium cladosporioides, Aspergillus versicolor, A. nidulans, A. repens, Penicillium expansum, P. canescens, P. cyclopium, P. janthinellum, Paecilomyces lilacinus, P. variotti, Rhodotorula* sp.	*A. versicolor, A. nidulans, A. repens, P. cyclopium, P. janthinellum* colonized the varnish surface.	Zyska *et al.* 1988
	Trichoderma longibrachiatum, Hypocrea lixii, Aspergillus versicolor, Penicillium sp.	*T. longibrachiatum* and *H. lixii* hydrolysed gelatin	Kwiatkowska *et al.* 2016
Black and white cinematographic films	Fungi: *Aspergillus* spp., *Penicillium* spp., *Alternaria alternata, Cladosporium cladosporioides, Mucor racemosus, Phoma glomerata, Trichoderma longibrachiatum, Cryptococcus albidus*	All fungal isolates except *C. albidus* hydrolysed gelatin	Abrusci *et al.* 2005
	Bacteria: *Staphylococcus* spp., *Bacillus* spp., *Sphingomonas paucimobilis, Kocuria kristinae, Pasteurella haemolytica*	*Bacillus* spp., *S. hominis* hydrolysed gelatin	
Gelatin photos on paper	Fungi: *Aspergillus* sp., *Penicillium* sp., *Candida* sp., *Cryptococcus* sp.. *Rhodotorula* sp.	Several fungal and bacterial isolates exhibited amylolytic and proteolysis	Borrego *et al.* 2010
Photos on glass plates	Bacteria: *Clostridium* sp., *Bacillus* sp., *Pseudomonas* sp.		
Ferrotype image			
Cellulose triacetate–based colour cinematographic films	Fungi: *Aspergillus* sp., *A. versicolor, A. penicillioides, A. sydowii, A. unguis, A. vitricola, Penicillium chrysogenum, Cladosporium* sp., *C. phaenocomae, Microascus* sp.	Unknown	Vivar *et al.* 2013
Film reels	Fungi: *Aspergillus versicolor, Penicillium brevicompactum, P. chrysogenum, P. citreonigrum, Trichoderma harzianum*	All isolates hydrolysed gelatin	Bingley and Verran 2013
Albumin photographs	Fungi: *Penicillium* spp., *Aspergillus versicolor, Phlebia* sp., *Bjerkandera adusta, Chaetomium elatum, Pleurotus pulmonarius, Mucor plumbeus, Cephalosporium* sp., *Trichothecium roseum, Acrostalagmus luteoalbus, Cladosporium macrocarpum*	Several microorganisms showed lipolytic, proteolytic and cellulolytic abilities	Puškárová *et al.* 2016
	Bacteria: *Bacillus* spp., *Kocuria* sp., *K. rhizophila, Streptomyces* spp., *Staphylococcus epidermidis*		

such as a cardboard frame and a glassine-paper envelope. Non-invasive sampling to obtain the surface microbiota, employing sterile cellulose acetate membranes, was performed on albumin prints (Puškárová *et al.* 2016) and on different materials forming the conservation unit: negative film, positive print, cardboard frame and a glassine-paper envelope (Bučková *et al.* 2014). In another study, biodeteriogens from foxing stains on black and white gelatin prints were recovered using adhesive tape strips (Sclocchi *et al.* 2017). From these sampling materials, it was possible to directly extract the microbial nucleic acids and identify organisms based on sequencing analysis and by comparison with restricted fragment length polymorphism (RFLP) of the clones (Bučková *et al.* 2014) or with denaturing gradient gel electrophoresis (DGGE) profiles produced by the single clones and the DGGE profile of the total microbial community (Puškárová *et al.* 2016; Sclocchi *et al.* 2017).

Nucleic acid analysis (NAA) offers a valuable way to investigate the microbiome. By performing both cultivation and culture-independent analyses results can be compared (Puškárová *et al.* 2016) as cultivation procedures are not always successful (Bučková *et al.* 2014) and may give very poor results (Sclocchi *et al.* 2017). Another important aspect of culture-independent investigation is that either DNA or RNA (cDNA) can be analysed to capture associated microbial species at the specific sampling time. DNA analysis provides information on the microbial species capable of colonizing the photograph multilayer material, while RNA analysis indicates the metabolically active species already colonizing the photograph surface (Sclocchi *et al.* 2017).

High-throughput sequencing (HTS) or next-generation sequencing (NGS) provides a powerful platform that allows for easy identification of large, diverse microbial communities. NGS platforms have been used to analyse the microbiota of cultural heritage items (Li *et al.* 2016; Adamiak *et al.* 2018), including paper documents (Kraková *et al.* 2018). Additionally, NGS can be used to evaluate the transcriptome of biodeteriogen communities, allowing insight into the living species in the population (Duran-Pinedo *et al.* 2014; Petrova *et al.* 2017). Using NGS to investigate microbial communities associated with photographic or film objects would seem to be a logical next step to understanding the communities present during biodeterioration.

CONCLUSIONS

The biodeterioration process of photographic and cinematographic materials is highly complex and many of these materials have not yet been investigated. Data are still needed to understand the biodeterioration process of other multilayered items. It is necessary to study the process of deterioration caused by microorganisms colonizing these objects to identify which strains are present in the communities and the extent of their activity. To successfully assess the cause of biodeterioration, a multidisciplinary approach and a step-by-step strategy provides the most reliable and robust interpretations of biodeterioration phenomena. This in turn leads to more suitable solutions to appropriately restore and conserve these fragile and multilayered items. This strategy includes non-destructive microscopic observation (stereomicroscope and SEM), spectrometric analyses of materials (FTIR, Raman and XRF) and microbial community analysis by using a combination of culture-dependent and culture-independent approaches complemented by bioactivity assays.

REFERENCES

Abrusci, C., Allen, N.S., Del Amo, A., Edge, M. and Martín-González, A. 2004. 'Biodegradation of motion picture film stocks', *Journal of Film Preservation* 67: 37.

Abrusci, C., Martín-González, A., Del Amo, A., Catalina, F., Collado, J. and Platas, G. 2005. 'Isolation and identification of bacteria and fungi from cinematographic films', *International Biodeterioration and Biodegradation* 56(1): 58–68.

Abrusci, C., Marquina, D., Del Amo, A., Corrales, T. and Catalina, F. 2006. 'A viscometric study of the biodegradation of photographic gelatin by fungi isolated from cinematographic films', *International Biodeterioration and Biodegradation* 58(3):142–9.

Adamiak, J., Otlewska, A., Tafer, H., Lopandic, K., Gutarowska, B., Sterflinger, K. and Piñar, G. 2018. 'First evaluation of the microbiome of built cultural heritage by using the Ion Torrent next generation sequencing platform', *International Biodeterioration and Biodegradation* 131: 11–18.

Ajò, D., Casellato, U., Fiorin, E. and Vigato, P.A. 2004. 'Ciro Ferri's frescoes: a study of painting materials and technique by SEM-EDS microscopy, X-ray diffraction, micro FT-IR and photoluminescence spectroscopy', *Journal of Cultural Heritage* 5(4): 333–48.

Allen, N.S., Edge, M., Jewitt, T.S. and Horie, C.V. 1992. 'Degradation and stabilization of cellulose triacetate base motion picture film', *Journal of Imaging Science and Technology* 36(1): 4–12.

Arai, H. 1987. 'Microbiological studies on the conservation of paper and related cultural properties. Part 5: Physiological and morphological characteristics of fungi isolated from foxing, formation mechanisms and countermeasures', *Science for Conservation* 26: 43–52.

Arai, H. 2000. 'Foxing caused by fungi: twenty-five years of study', *International Biodeterioration and Biodegradation* 46(3): 181–8.

Bersani, D., Campani, E., Casoli, A., Lottici, P.P. and Marino, I.G. 2008. 'Spectroscopic study of the degradation products in the holy water fonts in Santa Maria della Steccata church in Parma (Italy)', *Analytica Chimica Acta* 610(1): 74–9.

Bigourdan, J.L. 2000. 'Preservation of acetate base motion-picture film: from stability studies to film preservation in practice', in *The Vinegar Syndrome: A Handbook, Prevention, Remedies and the Use of New Technologies*. Bologna: GAMMA Group.

Bingley, G. and Verran, J. 2013. 'Counts of fungal spores released during inspection of mouldy cinematographic film and determination of the gelatinolytic activity of predominant isolates', *International Biodeterioration and Biodegradation* 84: 381–7.

Borrego, S., Guiamet, P., de Saravia, S.G., Batistini, P., Garcia, M., Lavin, P. and Perdomo, I. 2010. 'The quality of air at archives and the biodeterioration of photographs', *International Biodeterioration and Biodegradation* 64(2): 139–45.

Bruni, S., Cariati, F., Casadio, F. and Toniolo, L. 1999. 'Identification of pigments on a XV century illuminated parchment by Raman and FTIR microspectroscopies', *Spectrochimica Acta Part A: Molecular and Biomolecular Spectroscopy* 55(7): 1371–7.

Bučková, M., Puškárová, A., Sclocchi, M.C., Bicchieri, M., Colaizzi, P., Pinzari, F. and Pangallo, D. 2014. 'Co-occurrence of bacteria and fungi and spatial partitioning during photographic materials biodeterioration', *Polymer Degradation and Stability* 108: 1–11.

Cappitelli, F., Pasquariello, G., Tarsitani, G. and Sorlini, C. 2010. 'Scripta manent? Assessing microbial risk to paper heritage', *Trends in Microbiology* 18(12): 538–42.

Carbó, M.D., Martínez, V.P., Adelantado, J.G., Reig, F.B. and Moreno, M.M. 1997. 'Fourier transform infrared spectroscopy and the analytical study of sculptures and wall decoration', *Journal of Molecular Structure* 410: 559–63.

Carretti, E., Milano, M., Dei, L. and Baglioni, P. 2009. 'Noninvasive physicochemical characterization of two 19th century English ferrotypes', *Journal of Cultural Heritage* 10(4): 501–8.

Casoli, A. and Fornaciari, S. 2014. 'An analytical study on an early twentieth-century Italian photographs collection by means of microscopic and spectroscopic techniques', *Microchemical Journal* 116: 24–30.

Cattaneo, B., Chelazzi, D., Giorgi, R., Serena, T., Merlo, C. and Baglioni, P. 2008. 'Physico-chemical characterization and conservation issues of photographs dated between 1890 and 1910', *Journal of Cultural Heritage* 9(3): 277–84.

Čechák, T., Kopecká, I., Trojek, T., Štanzel, T. and Bártová, H. 2015. 'Application of X-ray fluorescence in an investigation of photographic heritage', *Radiation Physics and Chemistry* 116: 8–13.

Centeno, S.A., Meller, T., Kennedy, N. and Wypyski, M. 2008. 'The daguerreotype surface as a SERS substrate: characterization of image deterioration in plates from the 19th century studio of Southworth & Hawes', *Journal of Raman Spectroscopy* 39(7): 914–21.

Centeno, S.A., Schulte, F., Schrott, A. and Kennedy, N.W. 2011. 'Characterization of chlorine induced alterations in daguerreotype by SEM, XRF, and Raman spectroscopy', in *Art'11: 10th International Conference on Non-Destructive Investigations and Microanalysis for the Diagnostics and Conservation of Cultural and Environmental Heritage*, Italian Society for Non-destructive Testing Monitoring Diagnostics.

Centeno, S.A., Vila, A. and Barro, L. 2014. 'Characterization of unprocessed historic platinum photographic papers by Raman, FTIR and XRF', *Microchemical Journal* 114: 8–15.

Chabala, J.M., Levi-Setti, R. and Maternaghan, T.J. 1995. 'Photographic gelatin layer microstructures revealed by imaging secondary ion mass spectrometry', *Journal of Imaging Science and Technology* 39(3): 222–32.

Ciliberto, E., Gemmellaro, P., Iannuso, V., La Delfa, S., Urso, R.G. and Viscuso, E. 2013. 'Characterization and weathering of motion-picture films with support of cellulose nitrate, cellulose acetate and polyester', *Procedia Chemistry* 8: 175–84.

Coe, B. 1978. *Cameras: From Daguerreotypes to Instant Pictures*. Random House Value Publishing.

Costa, L. 2016. *Morphological and Chemical Characterization of Tintypes and Ambrotypes*, master's thesis, Universidade de Évora.

Creagh, D.C. 2005. 'The characterization of artefacts of cultural heritage significance using physical techniques', *Radiation Physics and Chemistry* 74(6): 426–42.

Czerwińska, E. and Kowalik, R. 1979. 'Microbiodeterioration of audiovisual collections', *Restaurator* 3(1–2): 63–80.

Da Silva, E., Robinson, M., Evans, C., Pejović-Milić, A. and Heyd, D.V. 2010. 'Monitoring the photographic process, degradation and restoration of 21st century daguerreotypes by wavelength-dispersive X-ray fluorescence spectrometry', *Journal of Analytical Atomic Spectrometry* 25(5): 654–61.

Davis, N. 1984. 'Tintypes: preliminary research and testing', *Art Conservation Training Programs Conference Publication*, 13–28.

Doehne, E. 2006. 'ESEM applications: from cultural heritage conservation to nano-behaviour', *Microchimica Acta* 155(1–2): 45–50.

Duran-Pinedo, A.E., Chen, T., Teles, R., Starr, J.R., Wang, X., Krishnan, K. and Frias-Lopez, J. 2014. 'Community-wide transcriptome of the oral microbiome in subjects with and without periodontitis', *ISME Journal* 8(8): 1659–72.

Eaton, G.T., Bard, C.C. and Lee, W.E. 1985. *Conservation of Photographs/F-40*. New York: Eastman Kodak Company.

Edge, M., Allen, N.S., Hayes, M., Riley, P.N.K., Horie, C.V. and Luc-Gardette, J. 1990. 'Mechanisms of deterioration in cellulose nitrate base archival cinematograph film', *European Polymer Journal* 26(6): 623–30.

Edge, M., Allen, N.S., Williams, D.A.R., Thompson, F. and Horie, C.V. 1992. 'Methods for predictive stability testing of archival polymers: a preliminary assessment of cellulose triacetate based motion picture film', *Polymer Degradation and Stability* 35(2): 147–55.

Eremin, K., Tate, J., Morrison-Low, A., Berry, J. and Stevenson, S. 2002. 'Non-destructive analysis of nineteenth century Scottish calotype negatives and salt prints', *MRS Online Proceedings Library Archive* 712.

Florian, M.L. and Manning, L. 2000. 'SEM analysis of irregular fungal fox spots in an 1854 book: population dynamics and species identification', *International Biodeterioration and Biodegradation* 46(3): 205–20.

Gaspard, S., Oujja, M., Rebollar, E., Abrusci, C., Catalina, F. and Castillejo, M. 2007. 'Characterization of cinematographic films by laser induced breakdown spectroscopy', *Spectrochimica Acta Part B: Atomic Spectroscopy* 62(12): 1612–17.

Giachet, M.T., Schilling, M., McCormick, K., Mazurek, J., Richardson, E., Khanjian, H. and Learner, T. 2014. 'Assessment of the composition and condition of animation cels made from cellulose acetate', *Polymer Degradation and Stability* 107: 223–30.

Guiamet, P., Borrego, S., Lavin, P., Perdomo, I. and de Saravia, S.G. 2011. 'Biofouling and biodeterioration in materials stored at the Historical Archive of the Museum of La Plata, Argentine and at the National Archive of the Republic of Cuba', *Colloids and Surfaces B: Biointerfaces* 85(2): 229–34.

Hu, Y., Vanhaecke, F., Moens, L., Dams, R. and Geuens, I. 1999. 'Determination of ruthenium in photographic materials using solid sampling electrothermal vaporization inductively coupled plasma mass spectrometry', *Journal of Analytical Atomic Spectrometry* 14(4): 589–92.

Kodak 2002. *Prevention and Removal of Fungus on Prints and Films*. Publication No. AE-22. Rochester, New York.

Kraková, L., Šoltys, K., Otlewska, A., Pietrzak, K., Purkrtová, S., Savická, D., Puškárová, A., Bučková, M., Szemes, T., Budiš, J. and Demnerová, K. 2018. 'Comparison of methods for identification of microbial communities in book collections: culture-dependent (sequencing and MALDI-TOF MS) and culture-independent (Illumina MiSeq)', *International Biodeterioration and Biodegradation* 131: 51–9.

Kwiatkowska, M., Ważny, R., Turnau, K. and Wójcik, A. 2016. 'Fungi as deterioration agents of historic glass plate negatives of Brandys family collection', *International Biodeterioration and Biodegradation* 115: 133–40.

Lavédrine, B., Gandolfo, J.P. and Monod, S. 2003. *A Guide to the Preventive Conservation of Photograph Collections*. Los Angeles: Getty Publications.

Li, Q., Zhang, B., He, Z. and Yang, X. 2016. 'Distribution and diversity of bacteria and fungi colonization in stone monuments analyzed by high-throughput sequencing', *PloS One* 11(9): e0163287.

Littlejohn, D., Pethrick, R.A., Quye, A. and Ballany, J.M. 2013. 'Investigation of the degradation of cellulose acetate museum artefacts', *Polymer Degradation and Stability* 98(1): 416–24.

Maggi, O., Persiani, A.M., Gallo, F., Valenti, P., Pasquariello, G., Sclocchi, M.C. and Scorrano, M. 2000. 'Airborne fungal spores in dust present in archives: proposal for a detection method, new for archival materials', *Aerobiologia* 16(3): 429–34.

Marucci, G., Monno, A. and van der Werf, I.D. 2014. 'Non invasive micro-Raman spectroscopy for investigation of historical silver salt gelatin photographs', *Microchemical Journal* 117: 220–24.

Maurice, P. 1993. 'Ambrotypes: positively capturing the past', *Material Culture Review/Revue de la culture matérielle* 38(1): 50–64.

Mayo, D.W., Miller, F.A. and Hannah, R.W. 2004. *Course Notes on the Interpretation of Infrared and Raman Spectra*. Hoboken NJ: John Wiley and Sons.

Mazzeo, R., Prati, S., Quaranta, M., Joseph, E., Kendix, E. and Galeotti, M. 2008. 'Attenuated total reflection micro FTIR characterisation of pigment–binder interaction in reconstructed paint films', *Analytical and Bioanalytical Chemistry* 392(1–2): 65–76.

McCabe, C. and Glinsman, L.D. 1996. 'Understanding Alfred Stieglitz' platinum and Palladium prints: examination by X-ray fluorescence spectrometry', in *Conference on Research Techniques in Photographic Conservation held at the School of Conservation, Royal Danish Academy of Fine Arts, Copenhagen, 14–19 May 1995*, 31–40.

McCormick-Goodhart, M. 1989. 'Research on collodion glass plate negatives: coating thickness and FTIR identification of varnishes', *Topics in Photographic Preservation* 3: 135–50.

Messier, P. 2011. *Photographic Papers in the 20th Century: Methodologies for Research, Authentication and Dating*. Westhaven CT: Yale University, Institute for the Preservation of Cultural Heritage.

Miller, H. 2003. 'The calotype', *History of Photography* 27(1): 7–12.

Modica, A., Alberghina, M.F., Brai, M., Bruno, M., Di Bella, M., Fontana, D. and Tranchina, L. 2017. 'XRF analysis to identify historical photographic processes: the case of some Interguglielmi Jr.'s images from the Palermo Municipal Archive', *Radiation Physics and Chemistry* 135: 76–80.

Nevin, A., Melia, J.L., Osticioli, I., Gautier, G. and Colombini, M.P. 2008. 'The identification of copper oxalates in a 16th century Cypriot exterior wall painting using micro FTIR, micro Raman spectroscopy and gas chromatography-mass spectrometry', *Journal of Cultural Heritage* 9(2): 154–61.

Peres, M., Costa, F.M., Gomes, A., Jardim, M.E., Ferreira, T., Dias, L., Mirão, J. and Carvalho, M.L. 2013. 'Analytical studies of 19th century photographs by non-destructive techniques', in *Science and Technology for the Conservation of Cultural Heritage*, M.A. Rogerio-Candelera, E. Cano and M. Lazzari (eds), 205–8. London: Taylor & Francis.

Petrova, O.E., Garcia-Alcalde, F., Zampaloni, C. and Sauer, K. 2017. 'Comparative evaluation of rRNA depletion procedures for the improved analysis of bacterial biofilm and mixed pathogen culture transcriptomes', *Scientific Reports* 7: 41114.

Puškárová, A., Bučková, M., Habalová, B., Kraková, L., Maková, A. and Pangallo, D. 2016. 'Microbial communities affecting albumen photography heritage: a methodological survey', *Scientific Reports* 6: 20810.

Reilly, J. 1981. *The Albumen and Salted Paper Book: The History and Practice of Photographic Printing, 1840–1895*. New York: Light Impressions.

Reilly, J.M. 1993. *IPI Storage Guide for Acetate Film*. Rochester: Image Permanence Institute.

Reilly, J.M. and McCabe, C. 1986. *Care and Identification of 19th Century Photographic Prints*. New York: Eastman Kodak Company.

Reilly, J.M., Severson, D.G. and McCabe, C. 1982. 'Image deterioration in albumen photographic prints', *Studies in Conservation* 27(1): 61–5.

Ricci, C., Bloxham, S. and Kazarian, S.G. 2007. 'ATR-FTIR imaging of albumen photographic prints', *Journal of Cultural Heritage* 8(4): 387–95.

Salvadó, N., Butí, S., Tobin, M.J., Pantos, E., Prag, A.J.N. and Pradell, T. 2005. 'Advantages of the use of SR-FT-IR microspectroscopy: applications to cultural heritage', *Analytical Chemistry* 77(11): 3444–51.

Salzer, R. and Siesler, H.W. (eds) 2009. *Infrared and Raman Spectroscopic Imaging*. Weinheim: Wiley-VCH Verlag.

Sclocchi, M.C., Damiano, E., Matè, D., Colaizzi, P. and Pinzari, F. 2013. 'Fungal biosorption of silver particles on 20th-century photographic documents,' *International Biodeterioration and Biodegradation* 84: 367–71.

Sclocchi, M.C., Kraková, L., Pinzari, F., Colaizzi, P., Bicchieri, M., Šaková, N. and Pangallo, D. 2017. 'Microbial life and death in a foxing stain: a suggested mechanism of photographic prints defacement,' *Microbial Ecology* 73(4): 815–26.

Shashoua, Y. 2006. 'Inhibiting the inevitable: current approaches to slowing the deterioration of plastics,' *Macromolecular Symposia* 238(1): 67–77.

Sterflinger, K. and Pinzari, F. 2012. 'The revenge of time: fungal deterioration of cultural heritage with particular reference to books, paper and parchment,' *Environmental Microbiology* 14(3): 559–66.

Stulik, D., Khanjian, H., de Tagle, A. and Botelho, A.M. 2002. 'Investigation of Jean-Louis-Marie-Eugene Durieu's toning and varnishing experiments: a non-destructive approach,' in *Preprints of the ICOM-CC 13th Triennial Meeting*, 658–63. London: James and James.

Vila, A. and Centeno, S.A. 2013. 'FTIR, Raman and XRF identification of the image materials in turn of the 20th century pigment-based photographs,' *Microchemical Journal* 106: 255–62.

Vivar, I., Borrego, S., Ellis, G., Moreno, D.A. and García, A.M. 2013. 'Fungal biodeterioration of color cinematographic films of the cultural heritage of Cuba,' *International Biodeterioration and Biodegradation* 84: 372–80.

Walsh, B. 1995. 'Identification of cellulose nitrate and acetate negatives by FTIR spectroscopy,' *Top Photography Preserve* 6: 80–97.

Whitman, K., Chen, J.J. and Osterman, M. 2007. *The History and Conservation of Glass Supported Photographs*. Rochester NY: George Eastman House.

Wilhelm, H. and Brower, C. 1993. *The Permanence and Care of Color Photographs*. Grinnel IA: Preservation Publishing.

Williams, S. and Brookes, E. 1986. 'The art of dress,' paper presented at the *Stockport History Society Conference on Textiles in the Nineteenth Century Stockport, 14 April 1967.*

Wood, R.D. 1971. 'JB Reade, FRS, and the early history of photography. Part II: Gallic acid and Talbot's calotype patent,' *Annals of Science* 27(1): 47–83.

Zotti, M., Ferroni, A. and Calvini, P. 2011. 'Mycological and FTIR analysis of biotic foxing on paper substrates,' *International Biodeterioration and Biodegradation* 65(4): 569–78.

Zyska, B.J., Cieplik, Z.T., Wojcik, A.R. and Kozlowska, R. 1988. 'Microbial deterioration of historic glass plate negatives,' in *Biodeterioration 7*, D.R. Houghton, R.N. Smith and H.O.W. Eggins (eds), 428–35. Springer, Dordrecht.

AUTHORS' ADDRESSES

- Andrea Puškárová, Institute of Molecular Biology, Slovak Academy of Sciences, Dubravská cesta 21, 84551, Bratislava, Slovakia (andrea.puskarova@savba.sk)
- Mária Bučková, Institute of Molecular Biology, Slovak Academy of Sciences, Dubravská cesta 21, 84551, Bratislava, Slovakia (maria.buckova@savba.sk)
- Domenico Pangallo, Institute of Molecular Biology, Slovak Academy of Sciences, Dúbravská cesta 21, 84551 Bratislava, Slovakia (domenico.pangallo@savba.sk)

BIODETERIORATION OF EASEL PAINTINGS: AN OVERVIEW

Ana Teresa Caldeira, Cátia Salvador, Tânia Rosado and António Candeias

INTRODUCTION

Since easel paintings emerged in the Middle Ages they have been an important art expression, endowing today's relevant cultural heritage assets with historic and cultural value. In terms of material composition, easel paints can be as simple as a mixture of pigments and binder, with the pigment providing the colour and the binder joining the particles of pigment together and ensuring adhesion to the support. Paints may also contain a variety of other materials to enhance optical or textural characteristics. The type of binder used results in different paint types (tempera, encaustic, oil, acrylic, vinylic) and enclosing different support types, making easel paintings materially diverse. They can also include organic compounds, which can be easily degraded by microbial agents. Figure 1 shows a schematic representation and a stratigraphic cross-section of easel paintings illustrating the different layers of a painting.

These artworks may contain a diversity of organic materials, namely protein compounds, polysaccharides and oils that are used as mediums and additives in the different paint layers as well as support materials such as wood, fabric, paper and parchment. These components are easily degraded by several physical, chemical and biological factors, leading to

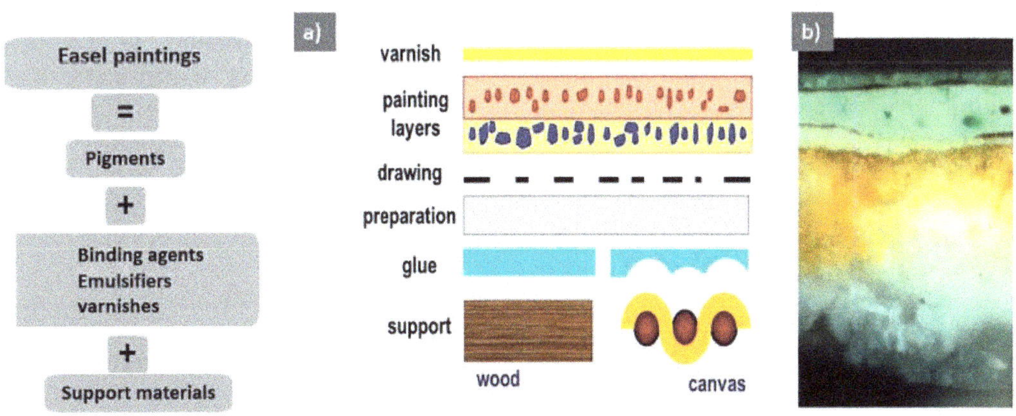

Figure 1 (a) Schematic representation and (b) stratigraphic cut of easel paintings, showing the different layers.

structural and aesthetic damage (Arslanoglu *et al.* 2010; Ahmed and Kolisis 2012).

Another important focus in easel painting materials is the presence of several organic constituents used as binders, adhesives and varnishes. These are commonly proteins (extracted from hens' eggs, bovine milk, animal skins and bones or fish bladders), siccative oils, plant resins, gums, mucilages, waxes (Arslanoglu *et al.* 2010) and, more recently, synthetic polymers such as vinylic and acrylic resins. The addition of these natural products results in an adherent film with variable elasticity and durability. When mixed with pigments, they form the basis for paints (binding media), in adhesives they allow solid joints in supports, and in varnishes they supply protective coatings (Stuart 2007; Arslanoglu *et al.* 2010; Ahmed and Kolosis 2012).

In recent years the integration of chemistry and physics into the study of patrimonial assets has allowed the application of analytical methodologies to heritage context. The material characterization of easel paintings can be assessed by *in situ* analysis using non-invasive techniques, such as energy dispersive X-ray fluorescence (EDXRF) spectrometry (elemental composition), spectrophotocolorimetry (colour parameters) and infrared spectrometry (molecular information). Analysis of microfragments removed from different areas, including biodegraded areas, can also be performed using various microanalysis techniques such as micro Raman (μ-Raman) spectrometry, micro Fourier transform infrared (μ-FTIR) spectrometry, liquid chromatography with mass spectrometry (LC-MS), gas chromatography with mass spectrometry (GC-MS), pyroloysis gas chromatography with mass spectrometry (Py-GC–MS) and scanning electron microscopy coupled to energy dispersive X-ray (SEM-EDX) spectrometry. These analyses allow a full characterization from a material point of view and are used to confirm the presence of microbial contamination.

Unfortunately, protein identification has been a particularly difficult issue to solve due to the degradation of materials over time, the presence of complex matrices, environmental contamination and any prior addition of materials during previous conservation interventions (Palmieri *et al.* 2011). Some chromatographic techniques have been used for proteinaceous materials identification in paintings but typically the return signals are difficult to interpret due to the presence of other organic and inorganic compounds that may mask the signals, making the correct analysis of these compounds impossible (Lluveras *et al.* 2010; Sciutto 2012; Albertini *et al.* 2011; Kazarian and Chan 2013). Immunological methods such as enzyme-linked immunosorbent assay (ELISA) offer advantages over other analytical approaches used in conservation science. ELISA provides selectivity regarding the biological source, the ability to resolve complex mixtures of proteins, high specificity of antibody-antigen binding and the capacity to detect protein in low quantities. This methodology is fast and affordable since it does not require expensive instrumentation, needs only minimal sample handling and is applicable to routine analysis (Tokarski *et al.* 2006; Cartechini *et al.* 2010; Schultz and Petersen 2011).

THE ROLE OF MICROORGANISMS IN THE BIODEGRADATION OF EASEL PAINTINGS

Microorganisms can play an important role in the biodegradation of easel paintings, together with aging, the chemical structure of the substrate and environmental conditions such as humidity, temperature, pH and light (Arslanoglu *et al.* 2010; Ahmed and Kolisis 2012). During long-term storage of objects of historical value, the materials can suffer chemical and structural changes due to exposure to different deteriorative conditions. For example, ultraviolet light causes oxidation, which results in the breaking of intermolecular bonds and facilitates penetration of microbial enzymes (Zotti *et al.* 2008; Kavkler *et al.* 2015) and humidity, which accelerates microbial attack and the consequential degradation processes. In the case of microbial decay of artwork, most of the problems result from the presence of water (Koestler 2000) which allows the paint to support microbial growth (Fig. 2), resulting in various kinds of damage.

As previously mentioned, easel paintings contain a diversity of organic materials, providing various ecological niches favourable to the growth of numerous microbial species. Taken together with aging and environmental conditions, several types of damage can occur, such as discoloration of pigments, stains, biofilm formation and/or degradation of the support and adhesive polymers. The microflora attacking paintings mainly include fungi because the many organic components of these artworks can provide a carbon source. In addition, fungi show a great tolerance for environmental conditions and can use condensation moisture to thrive. In contrast, the moisture content of easel paintings is rarely so high as to favour bacterial growth (Dhawan and Agrawal 1986). In the biodeterioration of painted materials, insects are also often reported to be a problem but they only occasionally damage the paint layer, preferentially attacking the wood of the frame or support and the vegetable glue (Tiano 2002).

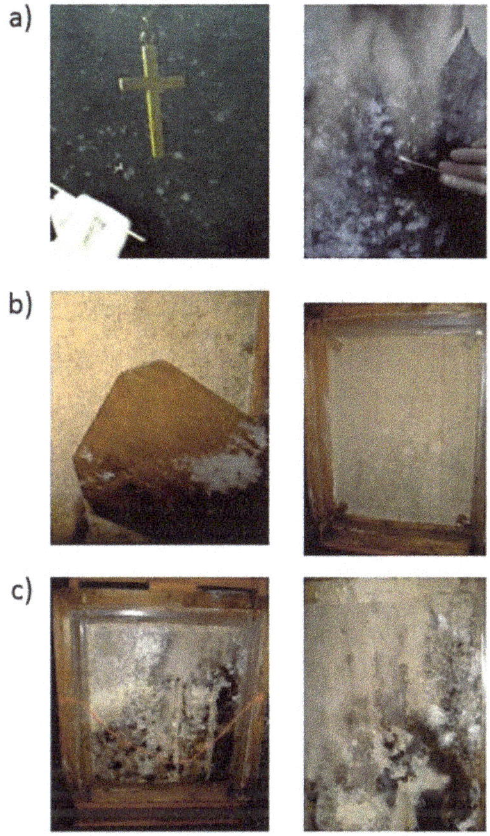

Figure 2 Extreme biodegradation evidencing the presence of fungal biofilms in (a) a painting surface, (b) a wooden frame and (c) a canvas (painting reverse).

Easel paintings contain diverse components that are easily degraded by microorganisms, causing both structural and aesthetic damage. The interaction between physical, chemical and biological factors leads to degradation of the support and adhesive polymers, resulting in cracking and detachment of the paint layers (Capodicasa *et al.* 2010; Elias *et al.* 2011; Ahmed and Kolisis 2012; Salvador *et al.* 2017). Figure 2 shows some examples of damage promoted by the presence of microorganisms inside the painting layers and associated support materials. Generally, the first element subjected to microbial deterioration is the support. The attack usually starts from the reverse side where the microorganisms penetrate inside, causing cracks and detachment of the paint layers. The rapid degradation process is in part accelerated by the humidity associated with storage conditions.

Biological attack on the paint layer is less frequent than on the support and depends on the nature of pigments and binders. Those compounds most susceptible are casein and egg distemper, emulsion distemper and linseed oil. In contrast, the presence of heavy metals in some pigments, such as lead, zinc or chromium, can increase the resistance of the paint layer to microbe invasion (Tiano 2002).

Easel paintings are very important artworks and their preservation is essential. An extensive study on these materials in order to enable their correct identification and characterization is a vital step in understanding the technique used by the artist, studio or guild, providing relevant information for a focused strategy for an eventual conservation and restoration intervention (Albertini *et al.* 2011; Elias *et al.* 2011).

The material study of easel paintings and patrimonial assets using *in situ* analysis techniques or laboratory analysis is well established

Table 1 Comparison of the different methods used to evaluate microbial communities associated with paintings (adapted from Rosado *et al.* 2015 and Salvador *et al.* 2017).

Methods	Advantages	Disadvantages
Culture-dependent methods	Provides high cell density, useful for *in vitro* assays	Sequencing is needed for identification Some microorganisms are not cultivable Incomplete screening
DGGE	Uses metagenomic DNA Useful to compare microbial contamination between different samples	Sequencing of bands is needed for identification Incomplete screening
Next-generation sequencing	Uses metagenomic DNA Full characterization of the microbial population Large number of samples can be analysed quickly	Expensive
Raman microspectrometry	No sample preparation required Non-destructive method High sensitivity Rapid screening for detection of microbial metabolites	Sample heating by laser radiation can destroy the sample Incomplete screening
FTIR-ATR	No sample preparation is required High reproducibility Easy to use, rapid and versatile technique Solids, pastes, gels, liquids and powders can be analysed	Samples could be destroyed by the pressure Incomplete screening
SEM and SEM-EDX	Image with high magnification and resolution Elemental composition of the materials allowing their localization in the samples Detects proliferation of the microorganisms in the samples	Coating samples with gold or carbon can destroy them
Enzymatic dehydrogenase	Detects cell viability Cell viability could correlate with contamination levels Determines biochemical marker function	Destructive Time-consuming

(see above). Despite small differences observed, it is usually possible to obtain information on constituent materials and degradation products resulting from physical and chemical processes. However, much remains to be discovered about biodegradation processes – not only in the study of biomechanisms, but also the extent of their effects – to allow new preventive and treatment methodologies to be developed (Caldeira *et al.* 2015). Although the involvement of microorganisms in the degradation process is well known, the specific role of the different species that comprise the microbial communities is not yet well understood. This is due primarily to the fact that current methodologies tend to favour the identification of organisms that are easily cultivable. Identification of the microbial diversity present on easel paintings is needed to develop and apply appropriate conservation and mitigation actions to prevent further contaminations (Stuart 2007).

Fungi are particularly harmful because their hyphae may proliferate extensively and spores are generally present and available for germination. In addition, various carboxylic acids or other metabolites are produced during fungal metabolism and these chemicals may play a significant role in attacks on easel paintings (Tran *et al.* 2012; Rosado *et al.* 2013a, b). Fungi can induce the discoloration and deterioration of paint surfaces, leading to the appearance of spots, which alter the colour. Furthermore,

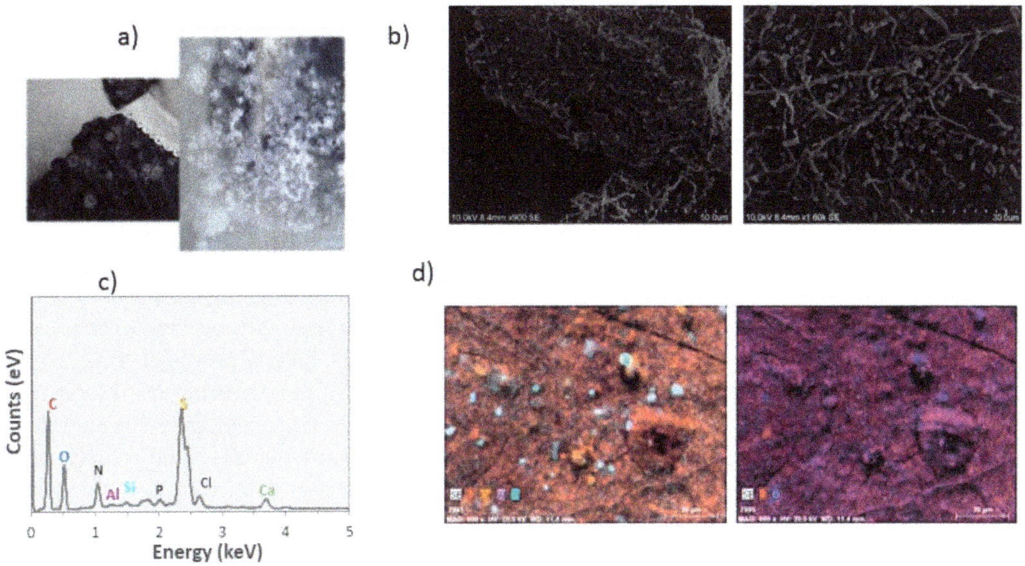

Figure 3 Biodeterioration evaluated by (a) the visual presence of microbial contamination in a painting surface contamination assessed by (b) SEM and (c) SEM-EDX point analysis and (d) 2D mapping.

hyphal penetration can result in the fragmentation of the colonized paint. Bacterial growth, on the other hand, is often associated with the formation of biofilms, which also promotes discoloration (Milanesi *et al.* 2006; Rosado *et al.* 2014; Caldeira *et al.* 2015).

Traditional identification of the microbial diversity is based on the cultivation of microorganisms in specific nutrient media but this approach only detects a very low percentage of the microbial species actually present (González and Saiz-Jiménez 2005). In order to understand the mechanisms that promote the degradation of easel paintings, it is important to identify the microbial populations that colonize them. Culture-independent methods enable detection of uncultivable microorganisms, giving a more complete view of the microbial communities present in a certain sample as compared to traditional cultivation techniques (Schabereiter-Gurtner *et al.* 2001; Justé *et al.* 2008). The use of culture-independent molecular techniques, such as those based on specific nucleic acid sequence detection, allows the differentiation of microorganisms within complex microbial communities (Portillo and Gonzalez 2009).

The detection of microorganisms is based mainly on sequencing the small subunit ribosomal DNA (rDNA) genes: 16S rDNA for prokaryotes and 18S rDNA or ITS for eukaryotes. Molecular fingerprinting techniques such as denaturing gradient gel electrophoresis (DGGE) are used to determine and identify the genetic diversity of natural microbial communities present in paintings (González and Saiz-Jiménez 2004, 2005; Malik *et al.* 2008; Pepe *et al.* 2011). DGGE separates amplified rDNA fragments of the same length but with different base pair sequences (Rölleke *et al.* 1996; Malik *et al.* 2008). This methodology has been applied to the study of microbial diversity in several environments (Jones *et al.* 2009; Roh *et al.* 2009; Nam *et al.* 2011) and to access microbial diversity in a heritage context (Rosado *et al.* 2014). In recent years, next-generation sequencing technology has allowed high-throughput sequencing, revolutionizing the study of microbial diversity. Both approaches allow a more thorough understanding of the microbial communities associated with artworks and their dynamics, transforming the knowledge of microbial diversity in this field (Caldeira *et al.* 2015).

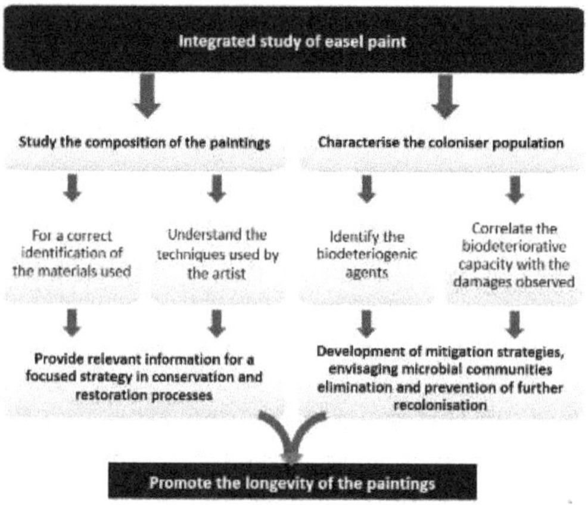

Figure 4 A schematic strategy for the integrated study and conservation of easel paintings.

The characterization of the microbial population is an important tool and is crucial to understanding the role of the microorganisms in the decay of the artworks, but the materials and the environmental context cannot be isolated from biogenic processes. An integrated study of these interactions can effectively allow a broader understanding of the degradation processes of easel paintings. Table 1 summarizes several approaches used to give an understanding of the biogenic deterioration of easel paintings, combining biochemical and genetic markers, complemented with detection of metabolic byproducts induced by microbial presence.

Micro Raman spectrometry and Fourier transformed infrared with attenuated total reflectance (FTIR-ATR) analysis can yield valuable data on the presence of microbes through the detection and identification of metabolic products. Enzymatic dehydrogenase assays give crucial information on the presence of living cells in the artwork as these enzymes reflect the total oxidative activity and are only found in active cells. Therefore, this information can be used as an indicator of the presence of metabolically active cells and as a measure of the metabolic rate (Rosado *et al.* 2013b). SEM-EDX analyses of paint microfragments allow confirmation of microbial contamination and distribution in samples, and understanding of the presence of organic materials identified by the carbon, sulfur, oxygen and nitrogen detected in two-dimensional (2D) elemental composition mapping. Figure 3 shows examples of the presence of microbial contamination assessed by SEM-EDX. These analyses provide evidence of biological species and spatial distribution of the inorganic constituents (aluminium, silicon, calcium) and organic material (carbon, nitrogen, oxygen, sulfur) that are strong indicators of microbiological contamination.

The different methods referred to above are very useful and have a powerful impact due to their capacity for covering the biotic and abiotic aspects being an added-value approach to characterize the materials and to diagnose/identify the problem that induce the decay processes on easel paintings. The application of complementary methodologies to characterize easel paint decay constitutes a starting point for successful conservation intervention plans: Figure 4 illustrates a schematic strategy for an integrated approach. Understanding this multivariable process provides the relevant information needed for a focused strategy to promote the longevity of the paintings.

CONCLUSIONS

In summary, the microbial community plays an essential role in the degradation of easel paintings. Modern chemical and genetic techniques permit both early detection of biodeterioration and detailed investigation of the degradation processes. Remediation and preservation technologies are still in the early stages of development.

REFERENCES

Ahmed, H.E. and Kolisis, F.N. 2012. 'A study on using of protease for removal of animal glue adhesive in textile conservation,' *Journal of Applied Polymer Science* 124: 3565–76.

Albertini, E., Raggi, L., Vagnini, M., Sassolini, A., Achilli, A., Marconi, G., Cartechini, L., Veronesi, F., Falcinelli, M., Brunetti, B. and Miliani, C. 2011. 'Tracing the biological origin of animal glues used in paintings through mitochondrial DNA analysis', *Analytical and Bioanalytical Chemistry* 399: 2987–95.

Arslanoglu, J., Schultz, J., Loike, J. and Peterson, K. 2010. 'Immunology and art: using antibody-based techniques to identify proteins and gums in artworks', *Journal of Biosciences* 35(1): 3–10.

Caldeira, A.T., Rosado, T., Silva, M., Gonzalez, M. and Candeias, A. 2015. 'Micrорganismos e património: novas abordagens', *Magazine da SPM* 4: 10.03.

Capodicasa, S., Fedi, S., Porcelli, A.M. and Zannoni, D. 2010. 'The microbial community dwelling on a biodeteriorated 16th century painting', *International Biodeterioration and Biodegradation* 64(8): 727–33.

Cartechini, L., Vagnini, M., Palmieri, M., Pitzurra, L., Mello, T., Mazurek, J. and Chiari, G. 2010. 'Immunodetection of proteins in ancient paint media', *Accounts of Chemical Research* 43: 867–76.

Dhawan, S. and Agrawal, O.P. 1986. 'Fungal flora of miniature paintings and lithographs', *International Biodeterioration and Biodegradation* 22(2): 95–9.

Elias, M., Mas, N. and Cotte, P. 2011. 'Review of several optical non-destructive analyses of an easel painting: complementarity and crosschecking of the results', *Journal of Cultural Heritage* 12(4): 335–45.

González, J.M. and Saiz-Jiménez, C. 2004. 'Microbial diversity in biodeteriorated monuments as studied by denaturing gradient gel electrophoresis', *Journal of Separation Science* 27(3): 174–80.

González, J.M. and Saiz-Jiménez, C. 2005. 'Application of molecular nucleic acid-based techniques for the study of microbial communities in monuments and artworks', *International Microbiology* 8: 189–94.

Jones, R.T., Robeson, M.S., Lauber, C.L., Hamady, M., Knight, R. and Fierer, N. 2009. 'A comprehensive survey of soil acidobacterial diversity using pyrosequencing and clone library analyses', *ISME Journal* 3(4): 442–53.

Justé, A., Thomma, B. and Lievens, B. 2008. 'Recent advances in molecular techniques to study microbial communities in food-associated matrices and processes', *Food Microbiology* 25(6): 745–61.

Kavkler, K., Gunde-Cimerman, N., Zalar, P. and Demsar, A. 2015. 'Fungal contamination of textile objects preserved in Slovene museums and religious institutions', *International Biodeterioration and Biodegradation* 97: 51–9.

Kazarian, S.G. and Chan, K.L.A. 2013. 'ATR-FTIR spectroscopic imaging: recent advances and applications to biological systems', *Analyst* 138: 1940–51.

Koestler, R.J. 2000. 'When bad things happen to good art', *International Biodeterioration and Biodegradation* 46: 259–60.

Lluveras, A., Bonaduce, I., Andreotti, A. and Colombini, M.P. 2010. 'GC/MS analytical procedure for the characterization of glycerolipids, natural waxes, terpenoid resins, proteinaceous and polysaccharide materials in the same paint microsample avoiding interferences from inorganic media', *Analytical Chemistry* 82: 376–86.

Malik, S., Beer, M., Megharaj, M. and Naidu, R. 2008. 'The use of molecular techniques to characterize the microbial communities in contaminated soil and water', *Environment International* 34(2): 265–76.

Milanesi, C., Baldi, F., Borin, S., Vignani, R., Ciampolini, F., Faleri, C. and Cresti, M. 2006. 'Biodeterioration of a fresco by biofilm forming bacteria', *International Biodeterioration and Biodegradation* 57: 168–73.

Nam, Y.D., Jung, M.J., Roh, S.W., Kim, M.S. and Bae, J.W. 2011. 'Comparative analysis of Korean human gut microbiota by barcoded pyrosequencing', *PLoS One* 6(7): e22109.

Palmieri, M., Vagnini, M., Pitzurra, L., Rocchi, P., Brunetti, B., Sgamellotti, A. and Cartechini, L. 2011. 'Development of an analytical protocol for a fast, sensitive and specific protein recognition in paintings by enzyme-linked immunosorbent assay (ELISA)', *Analytical and Bioanalytical Chemistry* 399: 3011–23.

Pepe, O., Palomba, S., Sannino, L., Blaiotta, G., Ventorino, V., Moschetti, G. and Villani, F. 2011. 'Characterization in the archaeological excavation site of heterotrophic bacteria and fungi of deteriorated wall painting of Herculaneum in Italy', *Journal of Environmental Biology* 32: 241–50

Portillo, M.C. and Gonzalez, J.M. 2009. 'Comparing bacterial community fingerprints from white colonizations in Altamira Cave (Spain)', *World Journal of Microbiology and Biotechnology* 25(8): 1347–52.

Roh, S.W., Kim, K.H., Nam, Y.D., Chang, H.W., Park, E.J. and Bae, J.W. 2009. 'Investigation of archaeal and bacterial diversity in fermented seafood using barcoded pyrosequencing', *ISME Journal* 4(1): 1–16, 442–53.

Rölleke, S., Muyzer, G., Wawer, C., Wanner, G. and Lubitz, W. 1996. 'Identification of bacteria in a biodegraded wall painting by denaturing gradient gel electrophoresis of PCR-amplified gene fragments coding for 16S rRNA', *Applied and Environmental Microbiology* 62(6): 2059–65.

Rosado, T., Gil, M., Mirão, J., Candeias, A. and Caldeira, A.T. 2013a. 'Oxalate biofilm formation in mural paintings due to microorganisms: a comprehensive study', *International Biodeterioration and Biodegradation* 85: 1–7.

Rosado, T., Martins, M.R., Pires, M., Mirão, J., Candeias, A. and Caldeira, A.T. 2013b. 'Enzymatic monitorization of mural paintings biodegradation and biodeterioration', *International Journal of Conservation Science* 4: 603–12.

Rosado, T., Mirão, J., Candeias, A. and Caldeira, A.T. 2014. 'Microbial communities analysis assessed by pyrosequencing: a new approach applied to conservation state studies of mural paintings', *Analytical and Bioanalytical Chemistry* 406: 887–95.

Rosado, T., Mirão, J., Candeias, A. and Caldeira, A.T. 2015. 'Characterizing microbial diversity and damage in mural paintings', *Microscopy and Microanalysis* 21: 78–83.

Salvador, C., Bordalo, R., Silva, M., Rosado, T., Candeias, A. and Caldeira, A.T. 2017. 'On the conservation of easel paintings: evaluation of microbial contamination and artists materials', *Applied Physics A* 123: 80, doi:10.1007/s00339-016-0704-5.

Schabereiter-Gurtner, C., Piñar, G., Lubitz, W. and Rölleke, S. 2001. 'Analysis of fungal communities on historical church window glass by denaturing gradient gel electrophoresis and phylogenetic 18S rDNA sequence analysis', *Journal of Microbiological Methods* 47: 345–54.

Schultz, J. and Petersen, K. 2011. 'Antibody-based techniques to distinguish proteins and identify sturgeon glue in works of art', in *Proceedings of Symposium 2011 – Adhesives and Consolidants for Conservation*, 1–13. Ottawa: Canadian Conservation Institute.

Sciutto, G., Dolci, L.S., Guardigli, M., Zangheri, M., Prati, S., Mazzco, R. and Roda, A. 2012. 'Single and multiplexed immunoassays for the chemiluminescent imaging detection of animal glues in historical paint cross sections', *Analytical and Bioanalytical Chemistry* 405(2–3): 933–40.

Stuart, B.H. 2007. *Analytical Techniques in Materials Conservation*. London: Wiley.

Tiano, P. 2002. 'Biodegradation of cultural heritage: decay, mechanisms and control methods', seminar article, New University of Lisbon, Department of Conservation and Restoration, 7–12 January 2002.

Tokarski, C., Martin, E., Rolando, C. and Cren-Olive, C. 2006. 'Identification of proteins in Renaissance paintings by proteomics', *Analytical Chemistry* 78(5): 1494–502.

Tran, T.H., Govin, A., Guyonnet, R., Grosseau, P., Lors, C., Garcia-Diaz, E., Damidot, D., Deves, O. and Ruot, B. 2012. 'Influence of the intrinsic characteristics of mortars on biofouling by *Klebsormidium flaccidum*', *International Biodeterioration and Biodegradation* 70: 31–9.

Zotti, M., Ferroni, A. and Calvini, P. 2008. 'Microfungal biodeterioration of historic paper: preliminary FTIR and microbiological analyses', *International Biodeterioration and Biodegradation* 62: 186–94.

AUTHORS' ADDRESSES

- Ana Teresa Caldeira, Chemistry Department, School of Sciences and Technology, Rua Romão Ramalho 59, 7000-761 Évora, Portugal/ HERCULES Laboratory, University of Évora, Largo Marquês de Marialva, 8, Palácio do Vimioso, 7000-089 Évora, Portugal (atc@uevora.pt)
- Cátia Salvador, HERCULES Laboratory, University of Évora, Largo Marquês de Marialva, 8, Palácio do Vimioso, 7000-089 Évora, Portugal
- Tânia Rosado, HERCULES Laboratory, University of Évora, Largo Marquês de Marialva, 8, Palácio do Vimioso, 7000-089 Évora, Portugal
- António Candeias, Chemistry Department, School of Sciences and Technology, HERCULES Laboratory, University of Évora, Largo Marquês de Marialva, 8, Palácio do Vimioso, 7000-089 Évora, Portugal (candeias@uevora.pt)

MODERN MATERIALS AND CONTEMPORARY ART

Francesca Cappitelli and Federica Villa

INTRODUCTION

All the materials employed for hundreds of years prior to the 20th century are still in common use today, both in the art world and its conservation, and such materials show the same decay patterns as in previous centuries. For example, the presence of abundant biofilm consisting mainly of cyanobacteria, algae, fungi and heterotrophic bacteria can be seen on the deteriorated brick surfaces of the 20th-century cathedral of La Plata in Argentina and two studied Mayan sites (Herrera and Videla 2003). Similarly, a contemporary oil painting from Nigeria showing stains on the canvas has been studied to characterize and evaluate the deterioration potential of its microbial community. A cellulolytic assay revealed that all fungal strains and seven bacterial strains isolated from the painting itself could attack cellulose (Okpalanozie *et al.* 2016).

The 1956 canvas painting *The Battle of Kosovo* by Petar Lubarda, which has suffered from biodeterioration, was studied by Pavić *et al.* (2015). All the bacterial strains isolated from this painting were tested to observe their ability to grow in the presence of the pigments found in the artwork. Indeed, all the isolates grew with ivory, black, red and yellow ochre as the sole source of phosphate and iron, resulting in faded pigments. Most of the isolates induced the solubilization of zinc white and cobalt deep green pigments. In contrast, the most toxic pigment was cadmium red which inhibited the growth of half the isolates.

In addition to conventional materials, 20th-century artists had a plethora of new materials from which to choose, but modern artists turned to unconventional materials because these provided a departure from traditional artmaking, permitted the exploration of new techniques and technologies, and an engagement with new materials (Barnett 2015). An example of this can be seen in the curent use of synthetic inorganic and organic pigments such as titanium dioxide, cadmium red and lithol rubine (see De Keijzer 2002 for a complete list). Chocolate and cheese have also become a medium, and artists now craft elaborate sculptures from edible materials (Barnett 2015). There is significant literature dedicated to the spoilage of food by micro-organisms (e.g. Tucker 2016), so these pathways of biodeterioration are well understood. Artworks are created with the understanding that the constituent materials will change and decay, and have to be replaced over time (Ryan 2011).

Art is also made with bacteria. Pigmented bacterial spores or cells can be used as paint, applied through brushstrokes or excreted pigments as the raw artistic material is used (Charkoudian *et al.* 2010). Recently, Fernandez-Rodriguez *et al.* (2017) designed a genetically encoded system that gives *Escherichia coli* the ability to distinguish between red, green and blue light, to which it responds by changing gene expression. This system has been employed to produce 'colour photographs' on bacterial culture plates by controlling pigment production.

Synthetic polymeric materials, the development of which began in the mid-19th century in an effort to find alternatives to materials that were either very expensive or in short supply, have been used exponentially for decades. Synthetic polymers, composite materials and nanoparticles have become increasingly important in the conservation of cultural heritage. A composite material (or simply 'a composite') is a multiphase material that exhibits a significant proportion of the properties of the constituent materials. The wood cell wall is a naturally occurring fibre-reinforced composite in which the cellulose 'fibres' are embedded in an amorphous matrix of hemicelluloses and lignin (Obataya 2009). Modern composite materials derive from the combination of metals, ceramics and polymers to produce superior materials in terms of properties such as toughness, stiffness and high temperature resistance. For example, natural fibres can be used in combination with polymers to obtain improved composites which, before the final curing, are in a viscous plastic state that can be shaped as a plaster on the surface to be treated (Chidichimo *et al.* 2015). An epoxy-clay nanocomposite (epoxy-montmorillonite clay) tested by Spathis *et al.* (2013) had improved anticorrosive properties for the protection of metal surfaces, compared to a pristine glassy epoxy polymer that was still permeable to corroding agents such as water and oxygen. Nanoparticles are defined as a material with at least one dimension less than 100 nm (Handy *et al.* 2008). Nanoparticles alone, or with polymers, are applied to cultural heritage surfaces where they play a crucial protective role, especially from biodeterioration (Sierra-Fernandez *et al.* 2017).

MODERN OBJECTS

The first synthetic materials used in artworks were naturally occurring polymers, for example cellulose, that had undergone chemical modification. Soon after, the use of synthetic polymeric materials brought about radical changes in the art world as they greatly influenced artists and those involved in conservation. Today, many synthetic polymers in the form of adhesives, consolidants and protective coatings are used not only in the conservation of cultural heritage items, but also in their creation. Indeed, synthetic polymers are a part of everyday life, and plastic objects have been considered collectable pieces since the 1980s (Grattan 1993; Hummelen and Sille 2006). Many museums and art galleries feature plastic objects in their 20th- and 21st-century collections. Heritage institutions are continually updating and growing their collections, acquiring plastic objects of historical and technological significance. Plastic artefacts include furniture, equipment, jewellery, items used in information technology and photography, toys and contemporary art (Shashoua 2006). In addition to materials used to make cultural heritage objects, synthetic polymers can be found in the form of adhesives, consolidants and protective coatings used to preserve artworks from further deterioration (Miliani *et al.* 2002; Favaro *et al.* 2006).

Plastics

Private collectors and museum curators have observed that plastic objects degrade with time, and some start to deteriorate within just a few years. Some synthetic polymers tend to deteriorate before others, requiring the use of constant controls. Shashoua (2008) noted that approximately 1% of objects containing plastics are a high conservation priority and 12% require essential conservation. The deterioration process of synthetic polymeric materials also differs: chemical (oxidation), physical (constant bending) and/or biological. Biotic and abiotic factors affect organic matter synergistically, and for synthetic polymers, exposure to these factors is highly undesirable as it impacts both material integrity and function. Biofilm development occurs on the material surface, which acts as substrata for microbial adhesion. Biofilm damage can result in: a biological coating that masks surface properties; increased leaching of additives (plasticizers, stabilizers, antioxidants, antistatics, lubricants, etc.) and of monomeric components of the substrate that act as nutrients; metabolite production by organisms comprising the biofilm, such as acids, that can damage the object; enzymatic action leading to depolymerization, physical penetration and disruption of the item; water accumulation within the item; and pigment secretion that leads to discoloration (Flemming 1998). Furthermore, microbial growth resulting in biofilm formation is ubiquitous, occurring in surface dirt and debris and in the cracks of synthetic polymers.

Museums today have many collections of objects made of, or containing, polyvinyl chloride (PVC) such as the eponymous Barbie doll, toys, clothes and electrical insulation (Shashoua 2001). Plasticized PVC, which can account for up to a third of the total material, is very unstable due to the migration of the plasticizer. Moriyama *et al.* (1993) used cryo scanning electron microscopy (cryo-SEM) to examine PVC deterioration patterns caused by fungi isolated from scratches, cracks and holes in bathrooms. Plasticizers commonly found in PVC formulations can act as nutrients for fungi, particularly black fungi (Hamilton 1983; Roberts and Davidson 1986; Webb *et al.* 2000). In fact, *Aspergillus niger* deterioration of PVC surfaces was reported by El-Aghoury *et al.* (2006) to be much faster than plasticizer loss.

Polyurethane polymer (PUR) is a common constituent used in furniture, coatings, construction materials, adhesives, fibres and paints as well as in contemporary art (Howard 2002; Rowe and Howard 2002; Hummelen and Sille 2006). Researchers examining 20th-century museum textiles have reported PUR deterioration by microorganisms (Kerr and Batcheller 1993; Tarnowskia *et al.* 2007), involving polymer backbone breakdown that leads to the loss of structural integrity (Dylingowski and Hamel 2004). PUR synthesis using polyester as the polyol is called a polyester PUR, while that using polyether is referred to as polyether PUR. Polyether-based polyurethanes are less susceptible to microbial attack than polyester ones. It has been suggested that PUR biodeterioration, leading to polyol and di- or polyisocyanate products, takes place through two types of polyurethane enzymatic activity occurring at the cell membrane and in the extracellular space (Evans and Levisohn 1968; Rowe and Howard 2002; Shah *et al.* 2008). In addition, the synthetic polymer polyamide-6 (nylon-6) is subject to degradation as reported by Friedrich *et al.* (2007). In this study, 58 fungi were tested and of these the Basidiomycete *Bjerkandera adusta* caused the most damage, presumably through enzymatic degradation by manganese peroxidase.

Synthetic adhesives

Conservation techniques currently use synthetic adhesives, consolidants and protective compounds more frequently than the traditional, natural products. The biodeterioration of polyvinyl acetate resins was reported as early as the 1950s by the Istituto Centrale del Restauro in Rome (Giacobini 1957). In 1968, the Ostia Antica superintendents in Rome began using acrylic compounds in conservation treatments to replace the easily degraded natural compounds – Paraloid-detached frescoes showed no biodeterioration problems for the first three years after application (Bassi and Giacobini 1973). Although some bacteria, yeasts, algae and lichens able to grow on synthetic products were detected by Favali *et al.* (1978), and much later by Cappitelli *et al.* (2004), the main causative agent of the synthetic polymeric deterioration, especially in early experiments, was found to be filamentous fungi (Domaslowski and Strzelczyk 1986; Salvadori and Nugari 1988; Leznicka *et al.* 1991; Sorlini *et al.* 1991). The filamentous fungi were identified traditionally on the basis of macroscopic colony features and reproductive structure morphology. However, morphological traits can be ambiguous due to intermediate forms and phenotypic overlapping of the different taxa, precluding reliable identification of fungi at low taxonomic levels (Ahmed 2016). Additionally, fungal biomass estimation relies on time-consuming and (occasionally) inaccurate cultivation methods. Microscopic examination of fungal structures is more accurate, but this requires considerable microbiological knowledge and experience, therefore there is an urgent need for rapid, precise and sensitive identification tools for biodeterioration studies. A fluorometric assay was developed to detect minute quantities of fungal biomass on cultural heritage materials (Konkol *et al.* 2010). This technology allows the detection of vegetative fungi and their reproductive structures, and can facilitate the early detection of fungal contamination in museums and private collections, allowing curators and collectors to take preventive measures against biodeterioration.

Novel, fast and efficient sequencing methods that allow the identification and quantification of microbial community members are also available, providing greater insight into the fungal communities that colonize artwork. Today, there are two main methodologies for the study of fungal communities: targeted metagenomics based on polymerase chain reaction (PCR) amplification and sequencing of one or more molecular markers (Ahmed 2016), and shotgun metagenomic analyses which give an overview of the microbial genome sequencing (Cuadros-Orellana *et al.* 2013). Metagenomic data are complex and analysis is challenging, but they allow the identification and quantification of microbial taxa and genes, thereby revealing community structure and function. Recently, these techniques were applied to studies of biodeterioration in ancient art (Gutarowska *et al.* 2015; Li *et al.* 2016; Xu *et al.* 2017), but there is still little in today's scientific literature with regard to metagenomic investigations of microbial communities associated with modern art collections.

It is notable that shotgun metagenomic analyses involve the unrestricted sequencing of the genome of all the microorganisms present in a sample, thus allowing investigation of the bacterial, archaeal and eukaryotic communities. There are case studies related to biodeterioration agents other than filamentous fungi. Bacteria also play a role: *Pseudomonas fluorescens*, for example, can cause deterioration of polyurethane-based products such as Impranil (Saad *et al.* 2004). Impranil was explored as an impregnating agent for the consolidation of PUR foam-based objects (van Oosten and Lorne 2003). In another study, biofilms dominated by the bacteria *P. aeruginosa*, *Ochrobactrum anthropi*, *Alcaligenes denitrificans*, *Xanthomonas maltophila* and *Vibrio harveyi* formed readily on the surfaces of synthetic materials for use in space missions (Gu 2007).

Synthetic polymers were used to treat the Buddhist Klippe statues in Hangzhou, China, which date back to the Yuan Dynasty and are considered to be one of the best representations of bas-relief on precipices. Li *et al.* (2017) correlated the production of lipases and esterases to the degradation of synthetic materials containing ester bonds. The authors used an innovative method to screen the lipolytic and ester-hydrolytic activity of the microorganisms isolated from these ancient artworks. The result was that more than half of the bacterial and fungal isolates showed the ability to hydrolyse esters, and an even greater percentage exhibited lipolytic activity.

Outdoor materials

Outdoor metal sculptures and objects are subject to corrosion therefore organic coatings, such as the acrylic-based coating Incralac, are widely used for protection. Scanning electron microscopy (SEM) and electrochemical impedance spectroscopy (EIS) revealed that a yeast strain isolated from a bronze statue treated with Incralac accelerated the deterioration of the coating itself (McNamara *et al.* 2004). In another study, algae and fungi recolonized a specimen treated with a silicone resin, Ahydrosil Z, more quickly than on untreated specimens (Leznicka 1992). At the Roman archaeological site at Luni, Italy, stones impregnated with an epoxy resin and an acrylic-siliconic resin treatment showed rapid recolonization by the alga *Stichococcus bacillaris* (Evans and Levisohn 1968).

In 2007, the Ausonia and Hungaria Hotel on the Venice Lido underwent conservation treatment as some of the art nouveau polychrome tiles on its facade exhibited stains putatively attributed to microbial growth. During this intervention, biocides and Paraloid B72, a synthetic polymer-based product commonly used in conservation treatments, were applied. Soon after the treatment, a massive growth of a cryptoendolithic cyanobacterial biofilm was detected on the tiles (Giacomucci *et al.* 2011) (Fig. 1). Interestingly, the biofilm was located in much deeper areas in the presence of Paraloid B72 than in samples without the synthetic resin, making its removal more difficult. Another example of such colonization can be seen in the forum of the Roman city of Baelo Claudia in Spain where lichens colonized areas treated with a synthetic polyester resin used to consolidate stucco walls and column capitals (Ariño and Saiz-Jimenez 1996).

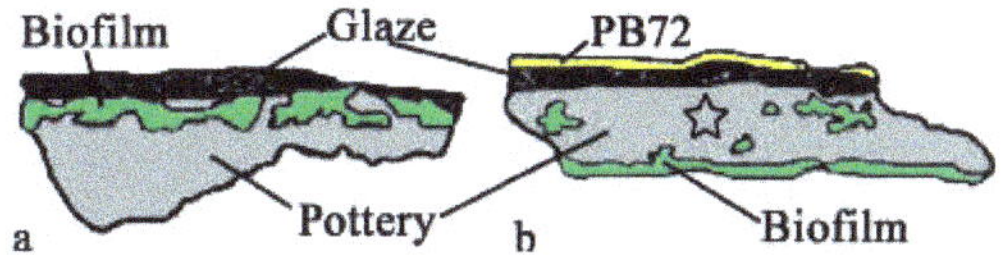

Figure 1 A schematic representation of samples that are characterized by (a) the absence of the acrylic resin and (b) treatment with the synthetic polymer, Paraloid B72 (PB72). The star indicates a richer microbial community, including the presence of black fungi such as *Phoma*, and bacterial species such as *Methylibium* sp., which are absent in (a) (image from Giacomucci *et al.* 2011.)

Gaylarde *et al.* (2011) published an extensive review of the biodeterioration occurring beneath exterior architectural paint films. In an earlier study, Shirakawa *et al.* (2002) reported the ecological succession of fungi, over a 10-month period, on two Brazilian buildings painted with a white acrylic paint. The walls were treated with hypochlorite prior to being painted. Just a week later, they showed well-developed colonies of yeast and *Cladosporium* species. After the third week, the yeast population decreased to undetectable levels, but eight months later they had again increased to high levels.

MODERN MATERIALS IN CONSERVATION

Over the years, tests have been carried out on many freshly dried synthetic polymer-based products used in conservation treatments to ascertain their resistance to microbial attack (Nugari and Priori 1985; Koestler and Santoro 1988; Salvadori and Nugari 1988). Laboratory tests for the degree to which acrylics, polyvinyl acetates and alkyds are susceptible to fungal colonization revealed acrylics to be the most and alkyds the least resistant (Cappitelli *et al.* 2005). Surprisingly, Kigawa *et al.* (2005) found that fungi isolated from inside or around the 7th- and 8th-century Takamatsuzuka and Kitora tumuli in Japan could grow on Paraloid B72 samples. It was noted that the behaviour of products in response to fungal colonization varied depending on how they were synthesized. In Paraloid B72, obtained by the copolymerization reaction of ethyl methacrylate and methyl acrylate, the presence of an acrylic hydrogen atom in alpha to the carbonyl group can start photochemical degradation reactions that are able to promote fungal growth (Sabatini *et al.* 2018). Additionally, acrylics with long lateral chains, namely poly(stearyl methacrylate) and poly(lauryl methacrylate), were more susceptible to biodeterioration than the short side chain homologous polymers (Cappitelli *et al.* 2004, 2007b). This could be a consequence of the minor enzymatic steric impediment of compounds with long lateral chains. Finally, the presence of other compounds in the formulation (particularly in emulsion products) and of added biocides could change the behaviour of synthetic polymer-based products.

Polyvinyl acetate emulsions were widely used in the 20th century as binding media and protective coatings. A significant reduction in the pH of the polyvinyl acetate emulsions Mowilith DM5 and Mowilith 20 was observed when they were inoculated with *Aspergillus versicolor*, *A. niger*, *Cladosporium herbarum*, *C. sphaerospermum*, *Engyodontium album*, *Penicillium aurantiogriseum*, *P. chrysogenum*, *Trichoderma longibrachiatum*, *Debaryomyces hansenii*, *Rhodotorula mucilaginosa*, *Bacillus subtilis*, *B. licheniformis* and *Rhodococcus fascians* (Heyn *et al.* 1995). Fungal and bacterial attack on Mowilith DM5 (a copolymer of vinyl acetate and *n*-butyl acrylate), Mowilith DMC2 (a copolymer of vinyl acetate and dibutyl maleate), and Conrayt PVA emulsions has been assessed by pyrolysis gas chromatography with mass spectrometry (Py-GC-MS), based on online derivatization with hexamethyldisilazane (Doménech-Carbó *et al.* 2009). Py-GC-MS analyses performed with the fungi *A. niger*, *P. chrysogenum*, *T. pseudokoningii*,

C. cladosporioides, *Chaetomium globosum*, *Rhizopus oryzae*, *Aureobasidium pullulans* and the bacteria *Streptomyces cellulofans*, *B. amyloliquefaciens*, *Arthrobacter oxydans* and *Burkholderia cepacia* revealed a decrease in the content of the phthalate external plasticizer used in these PVA emulsions.

In an effort to protect the Trevi Fountain in Rome from the growth of phototrophic organisms, use was made of the water repellents Imlar CPC 1175T (acrylic polymer), Rhodorsil RC80 (ethylsilicate polysiloxane), Wacker 290L and Wacker 280 (oligomeric polysiloxanes) (Nugari and Pietrini 1997). Treated stone specimens were partially immersed for a year in the fountain basin under natural weathering conditions. Biological analyses were then carried out on both the treated samples and the untreated controls. The results were very similar: considerable growth of photoautotrophic microflora was observed on the immersed surface of the samples but there was no algal growth on the non-immersed areas. The comparison of the phototrophic microflora grown on the test samples with that on the monument before the conservation treatment revealed them to be identical, demonstrating that the tested protective polymers did not inhibit algal growth.

Three commercial anti-graffiti products, Fluorolink P56 (fluoropolymer), Weather Seal Blok-Guard and Graffiti Control (silicone elastomer), and Protectosil Antigraffiti (silane), were contaminated by common soil microbes, and biodeterioration was measured by EIS and Fourier transform infrared spectroscopy in attenuated total reflection mode (FTIR-ATR) (Tarnowskia *et al.* 2007). The EIS results showed that there was no attack on these products by the soil microflora. Blok-Guard, upon investigation using FTIR-ATR, did appear to be affected by biodeterioration. Polyimides, supplied in the form of resins, fibres, films and coatings, were also tested for biosusceptibility using EIS (Gu *et al.* 1999). *A. versicolor*, *C. cladosporioides* and a *Chaetomium* sp. were isolated from a studied polyimide specimen. These fungi have also been isolated from electronic components and magnetic discs (McCain and Mirocha 1994).

Investigation into the biodeterioration of freshly dried synthetic macromolecules might not be enough to predict the actual behaviour of the applied polymer. Apart from the chemical composition of the polymer itself, at least two other factors must be taken into account: the mode of application (Cappitelli *et al.* 2006a) and natural aging (Cappitelli *et al.* 2007a). For example, the processed wood of the white poplar (*Populus alba*) and Norway spruce (*Picea abies*), consolidated by a combination of Paraloid B72 and Regalrez 1126, was resistant to white and brown rot fungi in the short term, but this treatment was much less effective in longer term simulated aging experiments (Clausi *et al.* 2011). Furthermore, when cotton and linen consolidated with synthetic polymers were subjected to fungal deterioration treatment it was found that those consolidated by grafting ethyl acrylate and methyl methacrylate monomers onto cellulose chains were much less susceptible to biodeterioration than the same textiles coated with Paraloid B72 and ethyl acrylate/methyl methacrylate preformed copolymer, a synthetic resin (Cappitelli *et al.* 2006b). This result was very important as it established, for the first time, that a synthetic resin was susceptible to biodeterioration to different degrees depending upon the way the polymer was applied to the substratum.

Little is known about the susceptibility of naturally aged acrylic resins to biological degradation, the only exception being the facade of the Tempio Malatestiano in Italy that was treated with acrylic resins. This treatment led to black fungal growth in the cracks and fissures of the structure (Pinna and Salvadori 1999). Heavy colonization by black fungi was also found on the polyisobutylmethacrylate used in the 1970s during conservation of Milan Cathedral. These black fungi were identified by denaturing gradient gel electrophoresis (DGGE) coupled with partial 18S rRNA gene sequencing, and both the fungal presence and spatial distribution were evaluated using immunofluorescence staining with melanin-binding antibodies. In contrast, freshly dried polyisobutylmethacrylate subjected to biodeterioration treatment showed only traces of fungal growth (Cappitelli *et al.*

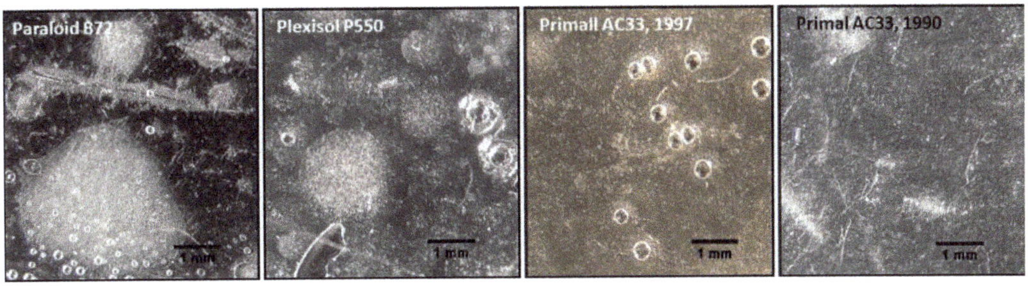

Figure 2 Fungal growth on the naturally aged synthetic binders Paraloid B72, Plexisol P550 40TB, Primal AC33 (produced in 1997), and Primal AC33 (produced in the early 1990s). The most susceptible was Paraloid B72, while the least susceptible was Primal AC33 (binders supplied by Dr Paolo Cremonesi).

2007a). It is possible that the chemical and physical degradation of the polymer worked synergistically with the biological deterioration (Cappitelli *et al.* 2007a).

It is not only synthetic polymeric materials exposed to outdoor conditions that undergo chemical-physical transformations which favour microbial growth – unexposed polymeric materials can become susceptible to biodeterioration. In 2008, the capacity of fungi to grow on the naturally aged synthetic binders Primal AC33 (produced in 1997, from Barcelona, Spain), Primal AC33 (produced in the early 1990s, from Bergamo, Italy), Plexisol P550 40TB (produced in 1997), and Paraloid B72 (produced in 1966) was evaluated (Villa *et al.* 2009). The most susceptible was Paraloid B72, heavily biodeteriorated by fungi, while the least susceptible was Primal AC33 from Bergamo that showed only traces of growth (Fig. 2).

PLASTIC INFORMATION CARRIERS

The use of plastic to create artworks and documentary materials has increased since the 1960s. Plastic is used in many ways to record information: audiotapes, computer diskettes, compact discs (CDs) and digital video discs (DVDs) are now commonly stored in archival and library collections. In addition, photographic materials are often made of plastic (Cappitelli and Sorlini 2005). Plastic audiovisual materials, including CDs, are subject to biodeterioration (García-Guinea *et al.* 2001). Initially, the fungal colonization of plastics in audio materials is manifested as deterioration in signal, while further colonization can lead to degradation of the plastic itself (McCain and Mirocha 1994; Brothers 1997). A popular optical support for storing digital data, the CD is a 12-cm-diameter polycarbonate disc (usually poly(bisphenol A carbonate), where information is stored as pits coated with a layer of a reflective metal, usually aluminium, and a top lacquer layer. A *Geotrichum*-like fungus was isolated from a biodeteriorated CD by García-Guinea *et al.* (2001). Analysis revealed important alterations in both the metal and plastic layers. Recently, the fungus was shown to degrade both the CD pits and the polycarbonate layer, and produce aryl-alcohol oxidase and Mn^{2+}-oxidizing peroxidase, oxidoreductases typical of ligninolytic Basidiomycetes (Romero *et al.* 2007).

Photographic supports were traditionally made of cellulose or its esters, cellulose acetate and cellulose nitrate (nitrocellulose), but later they were composed primarily of poly(ethylene terephtalate) (Cappitelli and Sorlini 2005). The degree of substitution of cellulose acetate is the average number of acetyl groups per monomer, and ranges from zero, in the case of cellulose, to three for cellulose triacetate. Cellulose acetate biodeterioration involves the deacetylation of side chains (leading to a decrease in degree of substitution) and hydrolysis of the polymer backbone. Moriyoshi *et al.* (2003) purified a cellulose acetate esterase and an enzyme hydrolysing the β-1,4 linkages in cellulose acetate

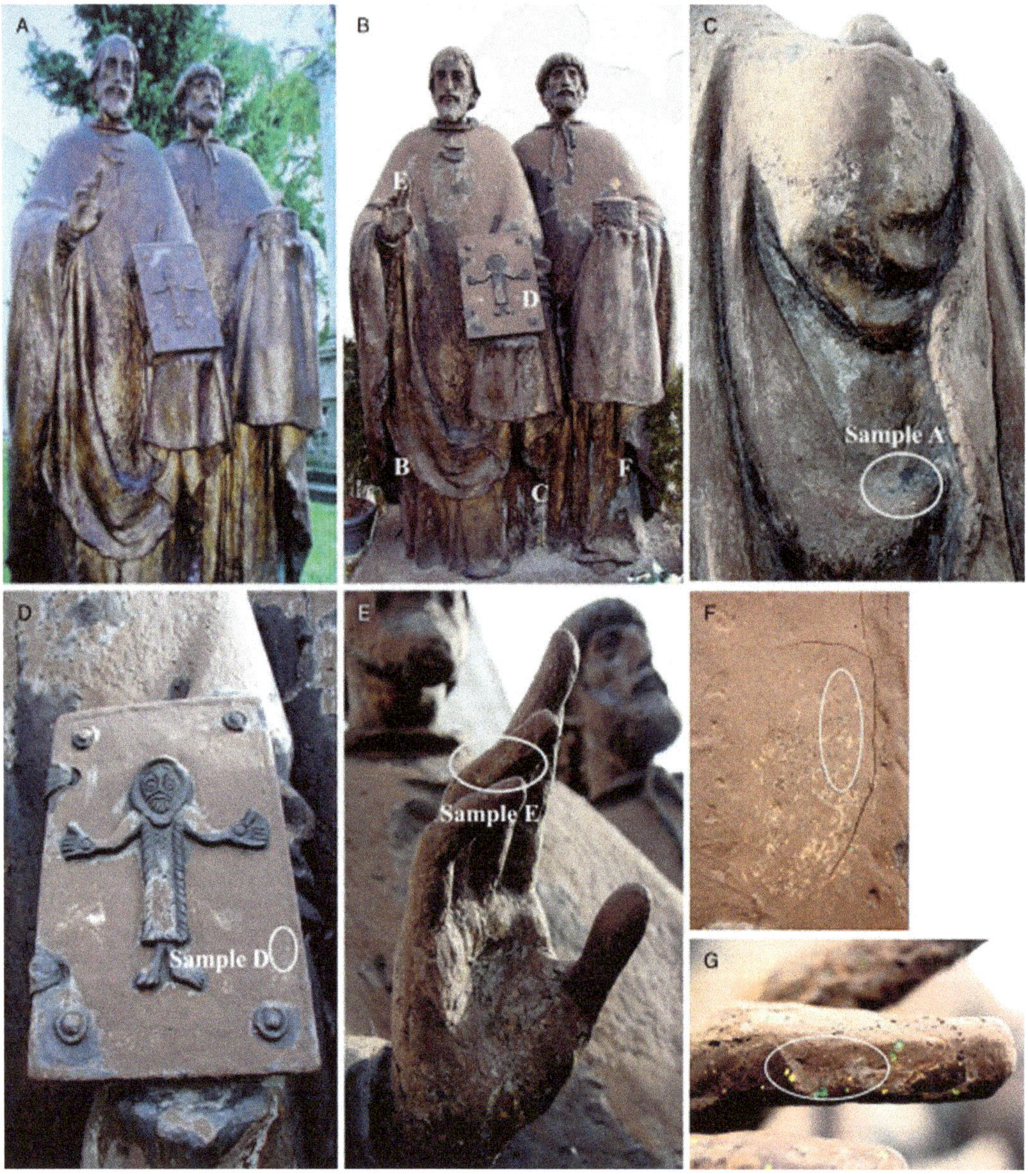

Figure 3 The epoxy statue of Cyril and Methodius: (A) the statue when installed in 1989 and (B) as it is now with sampling areas (B–F) marked; the change of colour with respect to the statue in 1989 is evident. (C) Sample A located on the back of the sculpture; (D) a close-up view of sample D on the book near a beige stain; (E) a sampling area located on the finger of Cyril (sample E); (F) detail of sampling from the book in panel D in which the different holes and green spots on the surface can be seen; (G) the algal-cyanobacterial-lichen colonization on sampling area E can be clearly seen: several holes and yellow-green-orange spots characterize this area (images from Pangallo *et al.* 2015).

molecules from the soil bacterium *Neisseria sicca*. The limiting step is the deacetylation: the lower the degree of substitution, the more biodegradable the cellulose acetate (Buchanan *et al.* 1993).

Nitrocellulose, an early synthetic polymer first produced in the 1840s (Stuart 2007), led to the development of photography and the film industry, and in 1869 was plasticized with camphor and patented as celluloid. Auer *et al.*

(2005) demonstrated that, in medium containing nitrocellulose as the sole nitrogen source, the fungus *Chaetomium elatum* degraded by as much as 43% in 28 days. In the presence of amino acids and starch as co-substrate, *Coprinus cinereus* caused a 37% decrease. Fungal attack was also reported on a 1960s nitrocellulose carrying case for a film projector (Silva 2006). Biodeterioration of the plastic found in information carriers is most concerning as it leads to loss of information and of the container itself. Sulfate-reducing bacteria can also transform nitrogen-containing compounds: in a study by Petrova *et al.* (2002), nitrocellulose was degraded in the presence of a *Desulfovibrio* species. Zhang *et al.* (2004) investigated the degradation of polyethylene terephthalate (PET) fibres by microbes and lipase, and discovered resultant cracks on the fibre surfaces.

The Space History Division of the National Air and Space Museum at the Smithsonian Institution in the USA manages the most comprehensive collection of artefacts documenting the world's space exploration such as spacesuits, including those from the Apollo lunar missions, and other astronaut-related equipment. Fungi belonging to the genera *Paecilomyces* and *Cladosporium* were isolated from two synthetic polymers in the Apollo suits. Fungal growth occurred rapidly, and these isolates were found to be capable of attacking and degrading those same space suit synthetic polymers (Breuker *et al.* 2003; Gu 2007).

The 'Futuro' house, conceived and built as a portable ski cabin in the mid-1960s by the Finnish architect Matti Suuronne, was constructed of fibreglass-reinforced polyester, poly(esterpolyurethane) and poly(methylmethacrylate). Studies on some examples have revealed significant degradation by the massive growth of microorganisms, which were identified by fluorescent *in situ* hybridization as Cyanobacteria and Archaea (Cappitelli *et al.* 2006a). To the best of our knowledge, these are the first biomolecular technique studies on the biodeterioration of modern materials in a contemporary art object.

For several decades, epoxy resins were frequently used in the former Czechoslovakia to create original sculptures and statue replicas. Pangallo *et al.* (2015) examined the microflora responsible for the deterioration of the epoxy statues of the saints Cyril and Methodius by Ludmila Cvengrošova that were installed outdoors in 1989 (Fig. 3). The biodiversity found was high and included rock-inhabiting taxa, algal photobionts (*Trebouxia* spp., *Chloroidium ellipsoideum* and *Chlorella angustoellipsoidea*), Cyanobacteria (*Leptolyngbya* sp., *Phormidium* sp., *Cylindrospermum stagnale*, *Hassallia byssoidea* and *Geitlerinema* sp.), black yeasts (related to the species *Friedmanniomyces endolithicus*, *Pseudotaeniolina globosa*, *Phaeococcomyces catenatus* and *Catenulostroma germanicum*) and several plant-associated fungi.

NEW APPROACHES TO BIODETERIORATION CONTROL

Biocides

Today, biocides are included in synthetic polymer-based product formulations to protect the product from degradation by microorganisms. Alternatively, they can be 'painted' onto the surface to provide external antimicrobial protection. The cost of the biocide can be indicative of the cost of biodeterioration, as its use will impact on the final product's cost, becoming an essential or value-added component. Conservators of today's modern materials are of the opinion that most synthetic polymers are resistant to biodeterioration because of their chemical nature and inclusion of effective, long-lasting biocides in the formulations (Shashoua 2008).

Little is known about the commercial introduction of antimicrobial treated plastic. Dylingowski and Hamel (2004) reported that a 1960s review by Turner concerning antimicrobials in synthetic materials included the following fungicides available at that time and were being used in a range from 0.25 to 5%: N-(trichloromethylthio)-4-4-cyclohexene-1,2-dicarboximide (Captan), copper naphthenate, copper-8-quinolinolate, organomercurial, pentachlorophenol, phenyl mercuric formate and tetramethylthiuram disulfide. Today, such concentrations would be considered high. The authors claimed that the first antimicrobial agent tailored to plastics, and employed mainly for flexible PVC applications, was 10,10'-oxybisphenoxarsine (OBPA) supplied by the Scientific Oil and Chemical Company Inc. since 1965. Flexible PVC is probably the plastic treated most frequently with biocides.

Shirakawa *et al.* (2002) compared the fungal communities on surfaces coated with acrylic paints with and without biocides, including carbamate, N-octyl-2H-isothiazolin-3-one and N-(3,4-dichlorophenyl)-N,N-dimethyl urea. They claimed that biocides had no effect on fungal biodiversity but did impact fungal abundance. Conservation of photographic materials involves polymeric materials, the most interesting being polymeric emulsions (Gaikalova and D'yakonov 2005). In fact, these same researchers proposed the introduction of biocides in order to improve the latex formulation used in the conservation of archival documents.

Experiments conducted by Tiralová and Reinprecht (2004) showed that acrylate impregnation of the wood did not protect beech and spruce from biodeterioration by *Coniophora puteana* and *Gloeophyllum trabeum*, although it did increase the mechanical properties of the wood. However, there was less susceptibility when the wood was also treated with the biocide 2-tiocyanomethylthiobenztiazol, administered in combination or before the polymer impregnation. Work by Henriques *et al.* (2014) investigated the protection of maritime pinewood from fungal biodeterioration by treating it with a water-based boron biocide followed by impregnation with acrylic and epoxy-based consolidants. The authors observed an increase in the mechanical strength of the wood with both consolidants and no significant influence with the biocide product alone.

Until the 1990s, there was an increasing trend towards the use of biocides in materials such as plastics, water-based paints and coatings to prevent microbial deterioration. However, the biocide had to be both physically and chemically compatible with the polymer formulation (Howard 2002). For instance, discoloration of an item can result from biocides chemically incompatible with additives (Dylingowski and Hamel 2004). Although biocides are included in formulations, their effect is relatively short term for the conservation of cultural heritage items. Finally, and most importantly, environmental regulations in European countries and elsewhere have severely restricted the use of biocides – as they are eventually released into the environment and, because generally they are not specifically targeted against biodeteriorating microorganisms, they present health and safety issues both for humans and the environment. Environmental concerns have led to the progressive withdrawal of antifouling paints containing organotin derivatives, which are both stable and highly toxic (Vallee-Rehel *et al.* 1998). It is worth noting that the use of biocides might not always delay the polymer deterioration process: Abdel-Kareem (2005) claimed that of the fungicides tested on linen textiles impregnated with synthetic polymers, only dichlorophene could prevent biodeterioration. At the same time, however, dichlorophene accelerated deterioration caused by heat and light. Organic biocides can act as nutrients, so they should be used with caution in conservation practices (Krumbein and Gross 1992; Leznicka 1992; Warscheid 2000). Furthermore, cells that have been killed might provide nutrients for subsequent colonization therefore leaving dead cells on the surface is not advisable (Flemming 1998).

Surface treatments

Recently, several environmentally friendly alternatives have been proposed (Flemming *et al.* 1996) for the preservation of synthetic macromolecules in cultural heritage (Cappitelli *et al.* 2006a). One biocide-free strategy mimics the lotus effect, a self-cleansing property of the surface of lotus leaves. By transferring this fascinating effect to artificial surfaces, it is possible to obtain surfaces that can be cleaned by simple rainfall (Neinhuis and Barthlott 1997; Cheng and Rodak 2005).

Natural antifouling compounds and cell-to-cell communication blockers represent another attractive strategy (Cappitelli and Sorlini 2008). Such antifouling compounds have been isolated mainly from marine organisms and plants (Terlezzi *et al.* 2000). For example, zosteric acid or p-(sulphooxy) cinnamic acid, a secondary metabolite from the seagrass *Zostera marina*, has emerged as a powerful anti-biofilm compound effective against both bacteria and fungi, reducing microbial adhesion by over 90% at sub-lethal concentrations and having a strong impact on biofilm thickness and morphology (Villa *et al.* 2010, 2011). The potential of this molecule lies in its very low toxicity towards organisms and the environment: it has shown no measurable LD50 for larval fish, an acute toxicity profile similar to table sugar, a half-life of a few days in seawater and cytocompatibility with soft and hard tissues (Polo *et al.* 2014).

Salicylic acid, a secondary metabolite widely distributed throughout the plant kingdom and well known as the active component of aspirin, has been observed to significantly reduce biofilm formation in both human and other environments. Indeed, in *E. coli*, salicylic acid reduces bacterial adhesion, affects biofilm development, and promotes the detachment of an established biofilm by interacting with proteins that play a role in quorum sensing, reactive oxygen species accumulation, motility, extracellular polymeric matrix components, transport and metabolism (Lemos *et al.* 2014; Cattò *et al.* 2017). In accordance with the idea of using natural compounds, a study by Mansour and Salem (2015) assessed the growth inhibition of the fungus *Trichoderma harzianum* by different plant extracts mixed with Paraloid B72 applied on wood using environmental electron scanning microscopy with energy dispersive X-ray spectrometry (ESEM-EDX).

USE OF NANOMATERIALS AS ANTIMICROBIALS

Nanocomposites have been explored recently in the field of contemporary art as an innovative strategy for producing new materials with distinctive physical, chemical and mechanical properties. Their unique features have brought to light new techniques and materials with which artists today are experimenting. However, nanocomposites cannot escape being colonized by microorganisms. Suslin *et al.* (2016) claim that epoxy/layered silicate nanocomposites have low stability against microorganisms and therefore biodeteriorate easily. This has led to the modification of epoxy-organo-montmorillonite nanocomposites by various quaternary ammonium ions to enhance their resistance to fungi such as *A. niger*, *P. chrysogenum* and *Trichoderma viridescens*, which are among the most aggressive polymer degrading microorganisms.

In order to protect nanocomposites from biodeterioration, nanostructured metal oxides with antimicrobial activities have been incorporated into modern materials and applied in conservation treatments. Copper nanoparticles embedded in polymer matrices have been reported by different authors to act as a preservative treatment, thereby inhibiting biological growth on stone heritage (Quaresima *et al.* 1997; Ditaranto *et al.* 2011; Pinna *et al.* 2012). For instance, Ditaranto *et al.* (2011) developed core-shell copper nanoparticles mixed with the silicon-based product Estel1100 to obtain a composite material for application on stone substrates. It was demonstrated that the resultant coating exerted a long-lasting biological activity due to the continuous and controlled release of copper ions. *In situ* applications have also been performed recently: copper nanoparticles mixed with four types of commercially available consolidant/water-repellent products were applied to different stone and plaster surfaces located in the archaeological area of Fiesole in Italy (Pinna *et al.* 2012). The innovative mixture of these particles with a water repellent yielded good results in terms of preventing biological colonization (Pinna *et al.* 2012).

Essa and Khallaf (2016) incorporated copper nanoparticles into polymer matrices, producing nanocopper composites with notable antimicrobial capability. The functionalized consolidation polymers were applied on limestone and sandstone blocks, and tested for antimicrobial activity against *E. coli*, *Streptomyces parvulus* and *B. subtilis*. The functionalized consolidation polymers could be used not only to inhibit the colonization of stone monuments, but also to improve the physical and mechanical properties of the treated substrata.

Despite the promising results obtained with copper nanoparticles, the dark colour of the nanostructured metal limits its application to light coloured stone, unless very low levels are used. In this context, zinc oxide nanoparticles meet the requirements for application on artwork, being biocompatible, chemically stable, bioactive and white which will not affect the colour of the substrate (Janaki *et al.* 2015; Gambino *et al.* 2017). Taking into account the many favourable characteristics of zinc oxide nanoparticles, van der Werf *et al.* (2015) and Ditaranto *et al.* (2015) embedded them in tetraethoxysilane and siloxane-based materials, demonstrating that higher nanoparticle loading can be achieved without affecting the colour of the stone substrates. The zinc oxide-based nanocomposites were applied to calcareous stone, and the biological activity of the nanocomposites was evaluated against *A. niger*, demonstrating high biostatic activity of the tested nanocomposites, even at rather low ZnO loadings.

For stone conservation purposes, Ruffolo *et al.* (2010) dispersed ZnO and $ZnTiO_3$ in different polymeric matrices (acrylic and fluorinated) in order to obtain a new coating technology with hydrophobic, consolidant and biocidal properties for the restoration of stone material. The $ZnTiO_3$ mixture applied on marble samples exhibited better antifungal activity than ZnO. In addition, protective coatings based on $Ca[Zn(OH)_3]_2{\cdot}2H_2O$ were prepared by the sol-gel method and evaluated on three different limestone materials (Gómez-Ortíz *et al.* 2013, 2014) (Fig. 4). The nanocomposite inhibited the growth of *P. oxalicum* and *A. niger* under alternating periods of dark and light, maintaining its antifungal properties after 21 days of inoculation. Gambino *et al.* (2017) tested the effectiveness of ZnO nanoparticles (ZnO-NPs) on the subaerial biofilms of four fungi isolated from the tomb of Tausert and Setnakht in the Valley of the Kings, Egypt. The researchers observed that although ZnO-NPs could be a valid option to prevent fungal growth on mural paintings, the fungal biofilm responded with an early production of coloured compounds and spores, two events potentially detrimental to cultural heritage conservation efforts.

Silver and titanium nanocomposites are the most used modern materials in conservation science (Munafò *et al.* 2015). Aflori *et al.* (2013) designed a silsesquioxane-based hybrid material with methacrylate units containing titanium and silver nanoparticles. These authors concluded that the silsesquioxane-based hybrid nanocomposites containing titanium nanoparticles showed a higher antibacterial/antifungal activity against *E. coli* and *Candida albicans*. Polo *et al.* (2011) investigated the photocatalytic and antimicrobial properties of a titanium dioxide nanopowder (TiO_2) thin film made by sol-gel deposition. The TiO_2 thin film almost completely eradicated *P. aeruginosa* planktonic cells in 24 hours (initial concentration 10^8 cells mL^{-1}), compared to a 1000-fold reduction caused by UV-A light alone. In contrast, photocatalytic treatment with neither TiO_2 film nor TiO_2 nanopowder had no effect on *P. aeruginosa* biofilm. Ruffolo *et al.* (2013) studied the efficacy of TiO_2 and silver-spiked TiO_2 synthesized by the sol-gel method as antifouling agents in stone conservation. Both types of nanoparticles were dispersed in a polymer and applied to marble surfaces, resulting in a decrease of antifungal activity on the treated stone surfaces. Goffredo and Munafò (2015) found that, in the short term, the photocatalysis of TiO_2 applied on stone was very efficient, and was able to degrade most of the soiling on the treated surfaces. On the other hand, after an accelerated weathering process, nanoparticles seem to lose photoactivity. La Russa *et al.* (2014) studied the antifungal activity of alkaline earth metal ions (silver, iron, strontium and a mixture) conjugated or spiked in TiO_2. Findings showed that silver, iron and strontium increased the TiO_2 activity against biodeterioration, revealing their successful application for stone conservation of outdoor, indoor and marine archaeological items as a potential biocide.

The ability of silver nanoparticles grafted to Serena stone surfaces to inhibit bacterial viability was evaluated against *B. subtilis* (Bellissima *et al.* 2014). The results indicated that the new solution, generated through the condensation

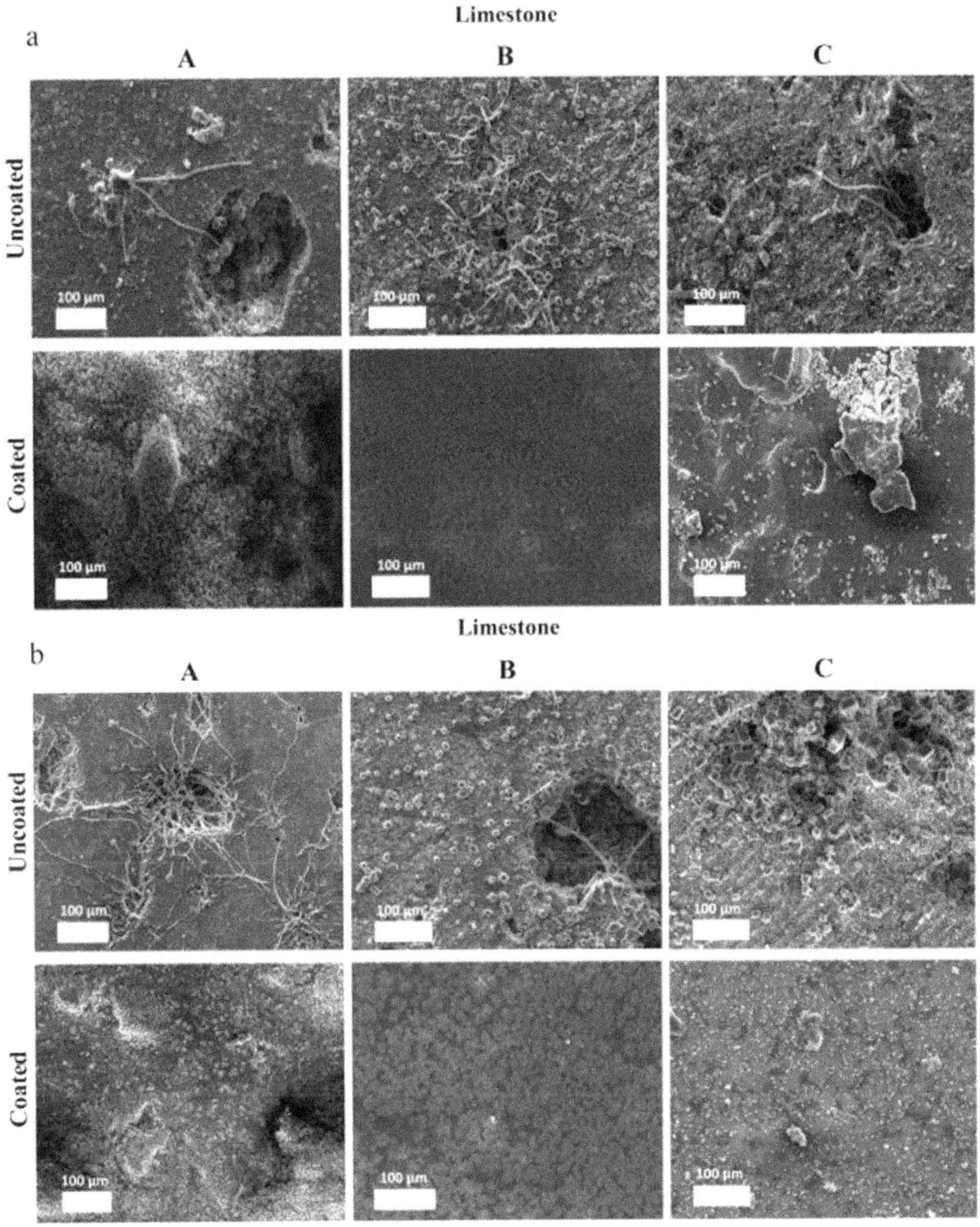

Figure 4 Limestone substrates A, B and C inoculated with (a) *Aspergillus niger* and (b) *Penicillium oxalicum* treated with and without calcium zinc hydroxide dihydrate coatings. Samples were incubated for 21 days under periods of light and dark (images from Gómez-Ortíz *et al.* 2014).

of a silane precursor (tetraethylorthosilicate, TEOS) on the surface of silver nanoparticles, inhibited microbial colonization by reducing cell viability by up to 80%. More recently, Carrillo-González *et al.* (2016) produced silver nanoparticles using a more environmentally acceptable synthesis method. The potential of silver nanoparticles as a preventive and/or corrective treatment to protect stucco materials from biodeterioration, as well as a microbial inhibitor on three stone materials (stucco, basalt and calcite), was analysed. Based on *in vitro* results, the researchers concluded that silver nanoparticles exhibit broad-spectrum activity against bacteria and fungi, and are dose-dependent

BIOCLEANING

Although biocleaning is discussed by Bosch-Roig and Ranelli (in this volume), we would like to highlight here the biological procedures for removing synthetic polymeric substances. Currently, the traditional method of removing a degraded synthetic polymer is mainly accomplished through the use of solvent mixtures (Lomax and Fisher 1990; Granowski *et al.* 2007). This is a chemical procedure that has several drawbacks, including difficulty in removing cross-linked polymers and health and safety risks. Alternatively, a lipase was shown to successfully remove the Paraloid B72 used to treat 15th- and 19th-century paintings (Bellucci *et al.* 1999). A methodology to select bacteria with such capability was also published recently (Troiano *et al.* 2014).

Paints used for modern graffiti quite often contain synthetic polymers as binders. Bacteria have been proposed and assayed for the removal of binders in graffiti (Sanmartín *et al.* 2014, 2015). In a study by Giacomucci *et al.* (2012) the ability of *Desulfovibrio desulfuricans* ATCC 13541, probably using the ammonification pathway, to degrade a nitrocellulose binder was successfully tested. Germinario *et al.* (2017) investigated an innovative method based on lipase use to remove graffiti containing acrylic resins from stone. The system proposed was oil, which is used for polymer removal, in water, which is necessary for lipase activity. Cleaning was evaluated using visible reflectance spectra, colour measurement and the percentage of white in the images of the treated areas using IMAGE J open source software. The enzymes used for the cleaning tests were a mixture of commercially available lipases, and lipases from *Candida rugosa* and *C. antarctica*. Although in general the treatments were successful, there was no single formulation effective for all the coloured marker pens used in the graffiti. Therefore, before commencing treatment, various solutions need to be tested.

REFERENCES

Abdel-Kareem, O.M.A. 2005. 'The long-term effect of selected conservation materials used in the treatment of museum artifacts on some properties of textiles', *Polymer Degradation and Stability* 87(1): 121–30.

Aflori, M., Simionescu, B., Bordiani I. and Olaru, M. 2013. 'Silesquioxane-based hybrid nanocomposite with methacrylate units containing titania and/or silver nanoparticles as antibacterial/antifungal coatings for monumental stones', *Materials Science and Engineering B* 178(19): 1339–46.

Ahmed, A. 2016. 'Analysis of metagenomics next generation sequence data for fungal ITS barcoding: do you need advance bioinformatics experience?', *Frontiers in Microbiology* 7: 1061.

Ariño, X. and Saiz-Jimenez, C. 1996. 'Lichen deterioration of consolidants used in the conservation of stone monuments', *Lichenologist* 28(4): 391–4.

ASTM G21-96. 2002. Standard Practice for Determining Resistance of Synthetic Polymeric Materials to Fungi, ASTM International, West Conshohocken, PA, doi: 10.1520/G0021-96R02.

Auer, N., Hedger, J. and Evans, C. 2005. 'Degradation of nitrocellulose by fungi', *Biodegradation* 16(3): 229–36.

Barnett, C. 2015. 'Purposeful impermanence: biodegradable art and its challenge to conservation', *Interventions* 4(1). Available at: https://interventionsjournal.wordpress.com/2015/01/21/purposeful-impermanence-biodegradable-art-and-its-challenge-to-conservation/ (accessed 6 June 2018).

Bassi, M. and Giacobini, C. 1973. 'Scanning electron microscopy: a new technique in the study of the microbiology of works of art', *International Biodeterioration Bulletin* 9(3): 57–68.

Bellissima, F., Bonini, M., Giorgi, R., Baglioni, P., Barresi, P., Mastromei, G. and Perito, B. 2014. 'Antibacterial activity of silver nanoparticles grafted on stone surface', *Environmental Science and Pollution Research* 21(23): 13278–86.

Bellucci, R., Cremonesi, P. and Pignagnoli, G. 1999. 'A preliminary note on the use of enzymes in conservation: the removal of aged acrylic resin coatings with lipase', *Studies in Conservation* 44(4): 278–81.

Breuker, M., McNamara, C., Young, L., Perry, T., Young, A. and Mitchell, R. 2003. 'Fungal growth on synthetic cloth from Apollo spacesuits', *Annals of Microbiology* 53: 47–54.

Brothers, P. 1997. 'Moldy audiotape', *Abbey Newsletter* 21: 106.

Buchanan, C.M., Gardner, R.M. and Komarek, R.J. 1993. 'Aerobic biodegradation of cellulose acetate', *Journal of Applied Polymers Science* 47(10): 1709–19.

Cappitelli, F. and Sorlini, C. 2005. 'From papyrus to compact disc: the microbial deterioration of documentary heritage', *Critical Reviews in Microbiology* 31(1): 1–10.

Cappitelli, F. and Sorlini, C. 2008. 'Microorganisms attack synthetic polymers in items representing our cultural heritage', *Applied and Environmental Microbiology* 74(3): 564–9.

Cappitelli, F., Zanardini, E. and Sorlini, C. 2004. 'The biodeterioration of synthetic resins used in conservation', *Macromolecular Bioscience* 4(4): 399–406.

Cappitelli, F., Vicini, S., Piaggio, P., Abbruscato, P., Princi, E., Casadevall, A., Nosanchuk, J.D. and Zanardini, E. 2005. 'Investigation of fungal deterioration of synthetic paint binders using vibrational spectroscopic techniques', *Macromolecular Bioscience* 5(1): 49–57.

Cappitelli, F., Sorlini, C., Pedemonte, E., Princi, E. and Vicini, S. 2006a. 'Effectiveness of graft synthetic polymers in preventing biodeterioration of cellulose-based materials', *Macromolecular Symposia* 238: 84–91.

Cappitelli, F., Principi, P. and Sorlini, C. 2006b. 'Biodeterioration of modern materials in contemporary collections: can biotechnology help?', *Trends in Biotechnology* 24(8): 350–54.

Cappitelli, F., Principi, P., Pedrazzani, R., Toniolo, L. and Sorlini, C. 2007a. 'Bacterial and fungal deterioration of the Milan Cathedral marble treated with protective synthetic resins', *Science of the Total Environment* 385(1–3): 172–81.

Cappitelli, F., Nosanchuk, J.D., Casadevall, A., Toniolo, L., Brusetti, L., Florio, S., Principi, P., Borin, S. and Sorlini, C. 2007b. 'Synthetic consolidants attacked by melanin-producing fungi: case study of the biodeterioration of Milan Cathedral (Italy) marble treated with acrylics', *Applied and Environmental Microbiology* 73(23): 271–7.

Carrillo-González, R., Martínez-Gómez, M.A., González-Chávez, M.D. and Mendoza Hernández, J.C. 2016. 'Inhibition of microorganisms involved in deterioration of an archaeological site by silver nanoparticles produced by a green synthesis method', *Science of the Total Environment* 16: 30320–25.

Cattò, C., Grazioso, G., Dell'Orto, S., Gelain, A., Villa, S., Marzano, V., Vitali, A., Villa, F., Cappitelli, F. and Forlani, F. 2017. 'The response of *Escherichia coli* biofilm to salicylic acid', *Biofouling* 33(3): 235–51.

Charkoudian, L.K., Fitzgerald, J.T., Khosla, C. and Champlin, A. 2010. 'In living color: bacterial pigments as an untapped resource in the classroom and beyond', *PLoS Biology* 8(10): e1000510.

Cheng, Y.T. and Rodak, D.E. 2005. 'Is the lotus leaf superhydrophobic?', *Applied Physics Letters* 86(4): 144101.

Chidichimo, G., Beneduci, A., Gattuso, C., Senatore, A., Dalena, F. and Roviello, V. 2015. 'New fiber composite materials for cultural heritage conservation', in *Heritage and Technology, XIII International Forum, Le vie deiMercanti, Aversa, Capri (Italy) June 2015*.

Clausi, M., Crisci, G.M., La Russa, M.F., Malagodi, M., Palermo, A. and Ruffolo, S.A. 2011. 'Protective action against fungal growth of two consolidating products applied to wood', *Journal of Cultural Heritage* 12(1): 28–33.

Cuadros-Orellana, S., Leite, L.R., Smith, A., Medeiros, J.D., Badotti, F., Fonseca, P.L.C., Vaz, A.B.M., Oliveira, G. and Góes-Neto, A. 2013. 'Assessment of fungal diversity in the environment using metagenomics: a decade in review', *Fungal Genomics and Biology* 3: 110.

De Keijzer, M. 2002. 'The history of modern synthetic inorganic and organic artists' pigments', in *Contributions to Conservation: Research in Conservation at the Netherlands Institute for Cultural Heritage (ICN Instituut Collectie Nederland)*, 42–54. London: James and James.

Ditaranto, N., Loperfido, S., van der Werf, I., Mangone, A., Cioffi, N. and Sabbatini, L. 2011. 'Synthesis and analytical characterisation of copper-based nanocoatings for bioactive stone artworks treatment', *Analytical and Bioanalytical Chemistry* 399(1): 473–81.

Ditaranto, N., van der Werf, I.D., Picca, R.A., Sportelli, M.C., Giannossa, L.C., Bonerba, E., Tantillo, G.

and Sabbatini, L. 2015. 'Characterization and behaviour of ZnO-based nanocomposites designed for the control of biodeterioration of patrimonial stoneworks', *New Journal of Chemistry* 39: 6836–43.

Domaslowski, W. and Strzelczyk, A. 1986. 'Evaluation of applicability of epoxy resins to conservation of stone historic monuments', in *Case Studies in the Conservation of Stone and Wall Paintings*, N.S. Brommelle and P. Smith (eds), 126–32. London: IIC.

Doménech-Carbó, M.T., Bitossi, G., de la Cruz-Cañizares, J., Bolívar-Galiano, F., del Mar López-Miras, M., Romero-Noguera, J. and Martín-Sánchez, I. 2009. 'Microbial deterioration of Mowilith DMC 2, Mowilith DM5 and Conrayt poly(vinyl acetate) emulsions used as binding media of paintings by pyrolysis-silylation-gas chromatography-mass spectrometry', *Journal of Analytical and Applied Pyrolysis* 85(1–2): 480–86.

Dylingowski, P.J. and Hamel, R.G. 2004. 'Microbial degradation of plastics', in *Directory of Microbicides for the Protection of Materials: A Handbook*, 325–42. Dordrecht: Springer.

El-Aghoury, A., Vasudeva, R.K., Banu, D., Elektorowicz, M. and Feldman, D. 2006. 'Contribution to the study of fungal attack on some plasticized vinyl formulations', *Journal of Polymers and the Environment* 14: 135–47.

Essa, A.M. and Khallaf, M.K. 2016. 'Antimicrobial potential of consolidation polymers loaded with biological copper nanoparticles', *BMC Microbiology* 16(1): 144.

Evans, D.M. and Levisohn, I. 1968. 'Biodeterioration of polyester-based polyurethane', *International Biodeterioration Bulletin* 4: 89–92.

Favali, M.A., Barbieri, N. and Bassi, M. 1978. 'A green alga growing on a plastic film used to protect archaeological remains', *International Biodeterioration Bulletin* 14: 89–93.

Favaro, M., Mendichi, R., Ossola, F., Russo, U., Simon, S., Tomasin, P. and Vigato, P.A. 2006. 'Evaluation of polymers for conservation treatments of outdoor exposed stone monuments. Part I: Photo-oxidative weathering', *Polymer Degradation and Stability* 91(12): 3083–96.

Fernandez-Rodriguez, J., Moser, F., Song, M. and Voigt, C.A. 2017. 'Engineering RGBRGB color vision into *Escherichia coli*', *Nature Chemical Biology* 13(7): 706–8.

Flemming, H.C. 1998. 'Relevance of biofilms for the biodeterioration of surfaces of polymeric materials', *Polymer Degradation and Stability* 59(1–3): 309–15.

Flemming, H.C., Griebe, T. and Schaule, G. 1996. 'Antifouling strategies in technical systems: a short review', *Water Science and Technology* 34(5–6): 517–24.

Friedrich, J., Zalar, P., Mohorčič, M., Klun, U. and Krză, A. 2007. 'Ability of fungi to degrade synthetic polymer nylon-6', *Chemosphere* 67(10): 2089–95.

Gaikalova, E.S. and D'yakonov, A.N. 2005. 'Influence of the components of the conserving polymeric composition on preservation of a photographic image', *Russian Journal of Applied Chemistry* 78(7): 1199–201.

Gambino, M., Ahmed, M.A.A., Villa, F. and Cappitelli, F. 2017. 'Zinc oxide nanoparticles hinder fungal biofilm development in an ancient Egyptian tomb', *International Biodeterioration and Biodegradation* 122: 92–9.

Garcia-Guinea, J., Cardenes, V., Martinez, A.T. and Martinez, M.J. 2001. 'Fungal bioturbation paths in a compact disk', *Naturwissenschaften* 88(8): 351–4.

Gaylarde, C.C., Morton, L.H.G., Loh, K. and Shirakawa, M.A. 2011. 'Biodeterioration of external architectural paint films: a review', *International Biodeterioration and Biodegradation* 65(8): 1189–98.

Germinario, G., van der Werf, I.D., Palazzo, G., Regidor, R.J.L., Montes-Estelles, R.M. and Sabbatini, L. 2017. 'Bioremoval of marker pen inks by exploiting lipase hydrolysis', *Progress in Organic Coatings* 110: 162–71.

Giacobini, C. 1957. 'Sostanze plastiche sintetiche quali fonte di carbonio per alcuni microrganismi', *Bollettino dell'Istituto Centrale del Restauro* 31–32: 108–13.

Giacomucci, L., Bertoncello, R., Salvadori, O., Martini, I., Favaro, M., Villa, F., Sorlini, C. and Cappitelli, F. 2011. 'Microbial deterioration of artistic tiles from the facade of the Grande Albergo Ausonia and Hungaria (Venice, Italy)', *Microbial Ecology* 62(2): 287–98.

Giacomucci, L., Toja, F., Sanmartín, P., Toniolo, L., Prieto, B., Villa, F. and Cappitelli, F. 2012. 'Degradation of nitrocellulose-based paint by *Desulfovibrio desulfuricans* ATCC 13541', *Biodegradation* 23(5): 705–16.

Goffredo, G.B. and Munafò, P. 2015. 'Preservation of historical stone surfaces by TiO_2 nanocoatings', *Coatings* 5(2): 222–31.

Gómez-Ortíz, N., De la Rosa-García, S., González-Gómez, W., Soria-Castro, M., Quintana, P., Oskam, G. and Ortega Morales, B. 2013. 'Antifungal coatings based on $Ca(OH)_2$ mixed with $ZnO\text{-}TiO_2$ nanomaterials for protection of limestone monuments', *ACS Applied Materials and Interfaces* 5(5): 1556–65.

Gómez-Ortiz, N.M., González-Gómez, W.S., De la Rosa-García, S.C., Oskam, G., Quintana, P., Soria-Castro, M., Gómez-Cornelio S. and Ortega Morales, B.O. 2014. 'Antifungal activity of $Ca[Zn(OH)_3]_2{\cdot}2H_2O$ coatings preservation of limestone monuments: an *in vitro* study', *International Biodeterioration and Biodegradation* 91: 1–8.

Granowski, C., Wise, A., Drover, C. and Stachurski, Z. 2007. 'A preliminary investigation in acrylic glazing breakdown', *Topics in Photographic Preservation* 12(17): 99–113.

Grattan, D. (ed.) 1993. *Saving the Twentieth Century: The Conservation of Modern Materials.* Ottawa: Canadian Conservation Institute.

Gu, J.D. 2007. 'Microbial colonization of polymeric materials for space applications and mechanisms of biodeterioration: a review', *International Biodeterioration and Biodegradation* 59: 170–79.

Gu, J.D., Mitton, D.B., Ford, T.E. and Mitchell, R. 1999. 'Microbial degradation of polymeric coatings measured by electrochemical impedance spectroscopy', *Biodegradation* 9: 39–45.

Gutarowska, B., Celikkol-Aydin, S., Bonifay, V., Otlewska, A., Aydin, E., Oldham, A.L., Brauer, J.I., Duncan, K.E., Adamiak, J., Sunner, J.A. and Beech, I.B. 2015. 'Metabolomic and high-throughput sequencing analysis: modern approach for the assessment of biodeterioration of materials from historic buildings', *Frontiers in Microbiology* 6: 979.

Hamilton, N.F. 1983. 'Biodeterioration of flexible polyvinyl chloride films by fungal organisms', in *Biodeterioration 5*, T.A. Oxley and S. Barry (eds), 663–78. Chichester: John Wiley and Sons.

Handy, R.D., von der Kammer, F., Lead, J.R., Hassellov, M., Owen, R. and Crane, M. 2008. 'The ecotoxicology and chemistry of manufactured nanoparticles', *Ecotoxicology* 17: 287–314.

Henriques, D., de Brito, J., Duarte, S. and Nunes L. 2014. 'Consolidating preservative-treated wood: combined mechanical performance of boron and polymeric products in wood degraded by *Coniophora puteana*', *Journal of Cultural Heritage* 15(1): 10–17.

Herrera, L.K. and Videla H.A. 2003. 'Biodeterioration and weathering of three different sites of the Latin American cultural heritage', in *Conference on Microbial Impact on Building Materials, 8–9 September 2003, Lisbon, Portugal*, M. Ribas Silva (ed.), 67–76. Bagneaux: RILEM publication SARL.

Heyn, C., Petersen, K. and Krumbein, W.E. 1995. 'Investigations on the microbial degradation of synthetic polymers used in the conservation and restoration of art objects', in *Biodeterioration and Biodegradation*, A. Bousher, M. Chandra and R. Edyvean (eds), 73–9. Rugby: IChemE.

Howard, G.T. 2002. 'Biodegradation of polyurethane: a review', *International Biodeterioration and Biodegradation* 49: 245–52.

Hummelen, I. and Sille, D. (eds) 2006. *Modern Art: Who Cares?* London: Archetype Publications.

Janaki, A.C., Sailatha, E. and Gunasekaran, S. 2015. 'Synthesis, characteristics and antimicrobial activity of ZnO nanoparticles', *Spectrochimica Acta Part A: Molecular and Biomolecular Spectroscopy* 144: 17–22.

Kerr, N. and Batcheller, J. 1993. 'Degradation of polyurethanes in 20th century museum textiles', in *Saving the Twentieth Century: The Conservation of Modern Materials*, D. Grattan (ed.), 189–206. Ottawa: Canadian Conservation Institute.

Kigawa, R., Hayakawa, N., Yamamoto, N., Kawanobe, W., Sano, C. and Aoki S. 2005. 'Evaluation of mould resistance of various synthetic resins used in conservation of historic sites', *Hozon Kagaku* 44: 149–56.

Koestler, R.J. and Santoro, E.D. 1988. *Assessment of the Susceptibility to Biodeterioration of Selected Polymers and Resins: Final Report.* Los Angeles: Getty Conservation Institute.

Konkol, N., McNamara, C.J. and Mitchell, R. 2010. 'Fluorometric detection and estimation of fungal biomass on cultural heritage materials', *Journal of Microbiological Methods* 80(2): 178–82.

Krumbein, W.E. and Gross, M. 1992. 'Interactions of biocides with cultures of biodeteriorating microbiota in agar diffusion and rock tube tests', in *Proceedings of the Seventh International Congress on Deterioration and Conservation of Stone*, J.D. Rodrigues, F. Henriques and F.T. Jeremias (eds), 501–9. Lisbon: Laboratorio Nacional de Engenharia Civil.

La Russa, M.F., Macchia, A., Ruffolo, S.A., De Leo, F., Barberio, M., Barone, P., Crisci, G.M. and Urzì, C. 2014. 'Testing the antibacterial activity of doped TiO_2 for preventing biodeterioration of

cultural heritage building materials', *International Biodeterioration and Biodegradation* 96: 87–96.
Lemos, M., Borges, A., Teodósio, J., Araújo, P., Mergulhão, F., Melo, L. and Simões, M. 2014. 'The effects of ferulic and salicylic acids on *Bacillus cereus* and *Pseudomonas fluorescens* single- and dual-species biofilms', *International Biodeterioration and Biodegradation* 86: 42–51.
Leznicka, S. 1992. 'Antimicrobial protection of stone monuments with phydroxybenzoic acid esters and silicone resin', in *Proceedings of the Seventh International Congress on Deterioration and Conservation of Stone*, J.D. Rodrigues, F. Henriques and F.T. Jeremias (eds), 481–90. Lisbon: Laboratorio Nacional de Engenharia Civil.
Leznicka, S., Kuroczkin, J., Krumbein, W.E., Strelczyk, A.B. and Petersen, K. 1991. 'Studies on the growth of selected fungal strains on limestones impregnated with silicone resins (Steinfestiger H and Elastosil E41)', *International Biodeterioration* 28: 91–111.
Li, Q., Zhang, B., He, Z. and Yang, X. 2016. 'Distribution and diversity of bacteria and fungi colonization in stone monuments analyzed by high-throughput sequencing', *PLoS One* 11(9): e0163287.
Li, Q., Zhang, B., Wang, L. and Ge, Q. 2017. 'Distribution and diversity of bacteria and fungi colonizing ancient Buddhist statues analyzed by high-throughput sequencing', *International Biodeterioration and Biodegradation* 117: 245–54.
Lomax, S.Q. and Fisher, S.L. 1990. 'An investigation of the removability of naturally aged synthetic picture varnishes', *Journal of the American Institute for Conservation* 29(2): 181–91.
Mansour, M.M.A. and Salem, M.Z.M. 2015. 'Evaluation of wood treated with some natural extracts and Paraloid B-72 against the fungus *Trichoderma harzianum*: wood elemental composition, *in-vitro* and application evidence', *International Biodeterioration and Biodegradation* 100: 62–9.
McCain, J.W. and Mirocha, C.J. 1994. 'Screening computer diskettes and other magnetic media for susceptibility to fungal colonization', *International Biodeterioration and Biodegradation* 33: 255–68.
McNamara, C.J., Breuker, M., Helms, M., Perry, T.D. and Mitchell, R. 2004. 'Biodeterioration of Incralac used for the protection of bronze monuments', *Journal of Cultural Heritage* 5: 361–4.
Miliani, C., Ombelli, M., Morresi, A. and Romani, A. 2002. 'Spectroscopic study of acrylic resins in solid matrices', *Surface and Coatings Technology* 151–2: 276–80.
Moriyama, Y., Kimura, N., Inoue, R. and Kawaguchi, A. 1993. 'Examination of fungal deterioration on plasticized polyvinyl chloride by cryo-scanning electron microscopy', *International Biodeterioration and Biodegradation* 31(3): 231–9.
Moriyoshi, K., Ohmoto, T., Ohe, T. and Sakai, K. 2003. 'Role of endo-1,4-β-glucanases from *Neisseria sicca* SB in synergistic degradation of cellulose acetate', *Bioscience, Biotechnology, and Biochemistry* 67(2): 250–57.
Munafò, P., Goffredo, G.B. and Quagliarini. E. 2015. 'TiO-based nanocoatings for preserving architectural stone surfaces: an overview', *Construction and Building*, doi:10.1016/j.conbuildmat.2015.02.083.
Neinhuis, C. and Barthlott, W. 1997. 'Characterization and distribution of water-repellent, self-cleaning plant surfaces', *Annals of Botany* 79(6): 667–77.
Nugari, M.P. and Pietrini, A.M. 1997. 'Trevi Fountain: an evaluation of inhibition effect of water-repellents on Cyanobacteria and algae', *International Biodeterioration and Biodegradation* 40(24): 247–53.
Nugari, M.P. and Priori, G.F. 1985. 'Resistance of acrylic polymers (Paraloid B72, Primal AC33) to microorganisms. Part 1', in *The Fifth International Congress on Deterioration and Conservation*, vol. 2, 685. Lausanne: Press Polytechniques Romandes.
Obataya, E. 2009. 'Effects of aging and heating on the mechanical properties of wood', in *Wood Science for Conservation of Cultural Heritage*, L. Uzielli (ed.). Florence: Firenze University Press.
Okpalanozie, O.E., Adebusoye, S.A., Troiano, F., Polo, A., Cappitelli, F. and Ilori, M.O. 2016. 'Evaluating the microbiological risk to a contemporary Nigerian painting: molecular and biodegradative studies', *International Biodeterioration and Biodegradation* 114: 184–92.
Pangallo, D., Bučková, M., Kraková, L., Puškárová, A., Šaková, N., Grivalský, T., Chovanová, K. and Zemánková, M. 2015. 'Biodeterioration of epoxy resin: a microbial survey through culture-independent and culture-dependent approaches', *Environmental Microbiology* 17(2): 462–79.
Pavić, A., Ilić-Tomić, T., Pačevski, A., Nedeljković, T., Vasiljević, B. and Morić, I. 2015. 'Diversity and biodeteriorative potential of bacterial isolates from deteriorated modern combined-technique canvas painting', *International Biodeterioration and Biodegradation* 97: 40–50.

Petrova, O.E., Tarasova, N.B. and Davydova, M.N. 2002. 'Biotechnological potential of sulfate-reducing bacteria for transformation of nitrocellulose', *Anaerobe* 8: 315–17.

Pinna, D. and Salvadori, O. 1999. 'Biological growth on Italian monuments restored with organic or carbonatic compounds', in *Of Microbes and Art: The Role of Microbial Communities in the Degradation and Protection of Cultural Heritage*, O. Ciferri, P. Tiano and G. Mastromei (eds), 149–54. New York: Kluwer Academic/ Plenum Publishers.

Pinna, D., Salvadori, B. and Galeotti, M. 2012. 'Monitoring the performance of innovative and traditional biocides mixed with consolidants and water-repellents for the prevention of biological growth on stone', *Science of the Total Environment* 423: 132–41.

Polo, A., Diamanti, M.V., Bjarnsholt, T., Høiby, N., Villa, F., Pedeferri, M.P. and Cappitelli, F. 2011. 'Effects of photoactivated titanium dioxide nanopowders and coating on planktonic and biofilm growth of *Pseudomonas aeruginosa*', *Photochemistry and Photobiology* 87: 1387–94.

Polo, A., Foladori, P., Ponti, B., Bettinetti, R., Gambino, M., Villa, F. and Cappitelli, F. 2014. 'Evaluation of zosteric acid for mitigating biofilm formation of *Pseudomonas putida* isolated from a membrane bioreactor system', *International Journal of Molecular Science* 15(6): 9497–518.

Quaresima, R., Baccante, A., Volpe, R. and Corain, B. 1997. 'Realization and possibility of polymeric matalo-organic matrixes with biocides activity', in *New Concepts, Technologies and Materials for the Conservation and Management of Historic Cities, Sites and Complexes*, vol. 3. A. Moropoulou, F. Zezza, E. Kollias and I. Papachristodoulou (eds), 323–35. Athens.

Roberts, W.T. and Davidson, P.M. 1986. 'Growth characteristics of selected fungi on polyvinyl chloride film', *Applied and Environmental Microbiology* 51: 673–6.

Romero, E., Speranza, M., García-Guinea, J., Martínez, Á.T. and Martínez, M.J. 2007. 'Ananamorph of the white-rot fungus *Bjerkandera adusta* capable of colonizing and degrading compact disc components', *FEMS Microbiology Letters* 275: 122–9.

Rowe, L. and Howard, G.T. 2002. 'Growth of *Bacillus subtilis* on polyurethane and the purification and characterization of a polyurethanase-lipase enzyme', *International Biodeterioration and Biodegradation* 50: 33–40.

Ruffolo, S.A., La Russa, M.F., Malagodi, M.O., Oliviero Rossi, C., Palermo, A.M. and Crisci, G.M. 2010. 'ZnO and $ZnTiO_3$ nanopowders for antimicrobial stone coating', *Applied Physics A: Materials Science and Processing* 100: 829–34.

Ruffolo, S.A., Macchia, A., La Russa, M.F., Mazza, L., Urzì, C., De Leo, F., Barberio, M. and Crisci, G.M. 2013. 'Marine antifouling for underwater archaeological sites: TiO_2 and Ag-Doped TiO_2', *International Journal of Photoenergy* 2013: article ID 251647.

Ryan, G. 2011. 'Variable materials, variable roles: the shifting skills required in contemporary art conservation', in *Objects Specialty Group Postprints* 18: 105–12.

Saad, D.S., Kinsey, G.C., Kim, S. and Gaylarde, C.C. 2004. 'Extraction of genomic DNA from filamentous fungi in biofilms on water-based paint coatings', *International Biodeterioration and Biodegradation* 54: 99–103.

Sabatini, V., Cattò, C., Cappelletti, G., Cappitelli, F., Antenucci, S., Farina, H., Ortenzi, M.A., Camazzo, S. and Di Silvestro, G. 2018. 'Protective features, durability and biodegration study of acrylic and methacrylic fluorinated polymer coatings for marble protection', *Progress in Organic Coatings* 114: 47–57.

Salvadori, O. and Nugari, M.P. 1988. 'The effect of microbial growth on synthetic polymers used on works of art', in *Biodeterioration 7*, D.R. Houghton, R.N. Smith and H.O.W. Smith (eds), 424–7. Amsterdam: Elsevier.

Sanmartín, P., Cappitelli, F. and Mitchell, R. 2014. 'Current methods of graffiti removal: a review', *Construction and Building Materials* 71: 363–74.

Sanmartín, P., DeAraujo, A., Vasanthakumar, A. and Mitchell, R. 2015. 'Feasibility study involving the search for natural strains of microorganisms capable of degrading graffiti from heritage materials', *International Biodeterioration and Biodegradation* 103: 186–90.

Shah, A.A., Hasan, F., Hameed, A. and Ahmed, S. 2008. 'Biological degradation of plastics: a comprehensive review', *Biotechnology Advances* 26: 246–65.

Shashoua, Y. 2001. *Inhibiting the Deterioration of Plasticized Poly(vinyl chloride): A Museum Perspective*, PhD dissertation, Technical University of Denmark and the National Museum of Denmark, Copenhagen.

Shashoua, Y. 2006. 'Inhibiting the inevitable: current approaches to slowing the deterioration of plastics', *Macromolecular Symposia* 238: 67–77.

Shashoua, Y. 2008. *Conservation of Plastics: Materials Science, Degradation and Preservation*. Oxford: Butterworth-Heinemann.

Shirakawa, M.A., Gaylarde, C.C., Gaylarde, P.M., John, V. and Gambale, W. 2002. 'Fungal colonization and succession on newly painted buildings and the effect of biocide', *FEMS Microbiology Ecology* 39: 165–73.

Sierra-Fernandez, A., De La Rosa-García, S.C., Gomez-Villalba, L.S., Gómez-Cornelio, S., Rabanal, M.E., Fort, R. and Quintana, P. 2017. 'Synthesis, photocatalytic, and antifungal properties of MgO, ZnO and Zn/Mg oxide nanoparticles for the protection of calcareous stone heritage', *ACS Applied Materials and Interfaces* 9(29): 24873–86.

Silva, M. 2006. 'Investigating cellulose nitrate degradation caused by fungal attack', in *The Future of the 20th Century: Collecting, Interpreting and Conserving Modern Materials*, C. Rogerson and P. Garside (eds), 72–6. London: Archetype Publications.

Sorlini, C., Falappi, D. and Ranalli, G. 1991. 'Biodeterioration preliminary tests on samples of Serena stones treated with resin', *Annals of Microbiology* 41: 71–9.

Spathis, P., Merachtsaki, D., Triantafyllidis, K., Giannakoudakis, P. and Xidas, P. 2013. 'Electrochemical impedance spectroscopy measurements for the corrosion behaviour evaluation of epoxy: (Organo) clays nanocomposite coatings', in *Conservation Science for the Cultural Heritage: Applications of Instrumental Analysis*, E.A. Varella (ed.), 273–85. Berlin and Heidelberg: Springer Verlag.

Stuart, B.H. 2007. *Analytical Techniques in Materials Conservation*. Chichester: John Wiley and Sons.

Suslin, M., Nedilko, O. and Mishurov, D. 2016. 'The influence of alkylammonium modified clays on the fungal resistance and biodeterioration of epoxy-clay nanocomposites', *International Biodeterioration and Biodegradation* 110: 136–40.

Tarnowskia, A., Zhang, X., McNamara, C., Martin, S.T. and Mitchell, R. 2007. 'Biodeterioration and performance of anti-graffiti coatings on sandstone and marble', *Journal of ACCR* 32: 3–16.

Terlezzi, A., Conte, E., Zupo, V. and Mazzella, L. 2000. 'Biological succession on silicone fouling-release surfaces: long term exposure tests in the harbour of Ischia, Italy', *Biofouling* 15(4): 327–42.

Tiralová, Z. and Reinprecht, L. 2004. 'Fungal decay of acrylate treated wood', *International Research Group on Wood Preservation*, document no. IRG/WP 04e30357.

Troiano, F., Vicini, S., Gioventù, E., Lorenzi, P.F., Improta, C.M. and Cappitelli, F. 2014. 'A methodology to select bacteria able to remove synthetic polymers', *Polymer Degradation and Stability* 107: 321–7.

Tucker, G.S. 2016. *Food Preservation and Biodeterioration*, 2nd edn. Oxford: Wiley-Blackwell.

Vallee-Rehel, K., Mariette, B., Hoarau, P.A., Guerin, P., Langlois, V. and Langlois, J.Y. 1998. 'A new approach in the development and testing of antifouling paints without organotin derivatives', *Journal of Coatings Technology* 70(880): 55–63.

van der Werf, I.D., Ditaranto, N., Picca, R.A., Sportelli, M.C. and Sabbatini, L. 2015. 'Development of a novel conservation treatment of stone monuments with bioactive nanocomposites', *Heritage Science* 3: 29.

van Oosten, T. and Lorne, A. 2003. 'The influence of Impranil and Plextol B as impregnating agents on the ageing of PUR foams', in *Conference Kunststoff-Kunst 18, 19 September 2003, Bern.*

Villa, F., Sorlini, C. and Cappitelli, F. 2009. 'Biodeterioramento di consolidanti e adesivi naturali e sintetici', in *CESMAR7, L'Attenzione alle Superfici Pittoriche. Materiali e Metodi per il Consolidamento e Metodi Scientifici per Valutarne l'efficacia 2*, D. Kunzelman (ed.), 33–7. Padua: Il Prato.

Villa, F., Albanese, D., Giussani, B., Stewart, P.S., Daffonchio, D. and Cappitelli, F. 2010. 'Hindering biofilm formation with zosteric acid', *Biofouling* 26(6): 739–52.

Villa, F., Pitts, B., Stewart, P.S., Giussani, B., Roncoroni, S., Albanese, D., Giordano, C., Tunesi, M. and Cappitelli, F. 2011. 'Efficacy of zosteric acid sodium salt on the yeast biofilm model *Candida albicans*', *Microbial Ecology* 62(3): 584–98.

Warscheid, T. 2000. 'Integrated concepts for the protection of cultural artifacts against biodeterioration', in *Of Microbes and Art: The Role of Microbial Communities in the Degradation and Protection of Cultural Heritage*, O. Ciferri, P. Tiano and G. Mastromei (eds), 185–202. New York: Kluwer Academic/Plenum Publishers.

Webb, J.S., Nixon, M., Eastwood, I.M., Greenhalgh, M., Robson, G.D. and Handley, P.S. 2000. 'Fungal colonization and biodeterioration of plasticized polyvinyl chloride', *Applied Environmental Microbiology* 66(8): 3194–200.

Xu, J., Wei, Y., Jia, H., Xiao, L. and Gong, D. 2017. 'A new perspective on studying burial environment before archaeological excavation: analyzing bacterial community distribution by high-throughput sequencing', *Scientific Reports* 7: 41691.

Zhang, J., Wang, X., Gong, J. and Gu, Z. 2004. 'A study on the biodegradability of polyethylene terephthalate fiber and diethylene glycol terephthalate', *Journal of Applied Polymer Science* 93(3): 1089–96.

AUTHORS' ADDRESSES

- Francesca Cappitelli, Department of Food, Environmental and Nutritional Sciences, University of Milan, via Mangiagalli 25, 20133 Milan, Italy (francesca.cappitelli@unimi.it)
- Federica Villa, Department of Food, Environmental and Nutritional Sciences, niversity of Milan, via Mangiagalli 25, 20133 Milan, Italy (federica.villa@unimi.it)

THE USE OF DYES AS A METHOD TO CONTROL TEXTILE BIODETERIORATION

Barbara Błyskal

INTRODUCTION

Throughout history, in both the Old World and the New, fibres, textiles and dyes have played a tremendously important role as coveted articles of trade, symbols of wealth and high status, and expressions of human identity, ingenuity and art. Textiles probably figured more significantly than pottery in the economic and religious life of prehistoric cultures, influencing social relationships within and between groups. Textile evidence, however, will always remain incomplete and disproportionately recovered. The history of human occupation of a given region may extend back 12,000 years, but most textile specimens date to the last 2000 years. Their absence in older assemblages does not indicate that they were not made and used, but rather reflects the lower probability of historic textiles being preserved. Fragments and some complete fabric examples have survived in considerable quantity, predominantly in regions of extremely arid climate such as the American Southwest, Peru, the Nile Valley and Inner Mongolia, because of the practice of using caves and protected overhangs as habitation, shrines or burial places. Other sites, such as the Indus Valley, have yielded few textiles, yet there can be little doubt that weaving, along with other arts, must have been well developed among the sophisticated inhabitants (Kent 1983; Teague 1998).

Textile objects of historical value, now preserved in private and public collections, can provide valuable information on the technology of their manufacture and subsequent use. They also have a great potential to contribute to archaeologists' interpretation of prehistoric cultures. Therefore, as rare and valuable records of human achievement, ancient textiles deserve careful attention and preservation efforts. Identifying the component materials of historic textiles and determining their degradation properties is important, since these factors influence their long-term preservation (Tímár-Balázsy and Eastop 1998).

FIBRE CLASSIFICATION

The earliest people relied on the natural resources of a region for all their needs. Specialization in the production of particular fibres, yarns or fabrics depended on the accessibility of materials: nomadic herders specialized in wool, while farmers focused on the production of flax or cotton fabrics. Differences in fabrics were also functional: fur dominated in colder climates, and light cotton fabrics in hot regions. Changes in the prevalence of different fibres in textiles could indicate climate shifts, the introduction of new fibre sources, the dissolution of an existing trade network, or alterations in the cultural contexts in which fibres were used. Fibres of natural origin can be divided into two major categories: those derived from plants and those from animals. Approximately 550–700 different plants have been used in the production of textiles worldwide. Plants provide cellulose fibres, such as bast fibres that derive from their woody stems. These include flax (linen), jute, hemp, ramie (the strongest vegetable fibre, eight times stronger than cotton), nettle and kenaf among others. Cellulose fibres are also obtained from leaves – including sisal and henequen (agave), yucca, New Zealand flax (muka) and raffia – as well as seed such as kapok, cotton and milkweed. Animals provide a variety of useful protein fibres, such as wool (sheep, llama, alpaca and vicuna), hair (dog, camel), cashmere, angora, mohair and silk.

Although a wide variety of raw materials was employed in the manufacture of ancient textiles, the most historically important textile fibres were cotton, wool and silk. From archaeological finds, it is known that cotton was cultivated around 3000 BC in the Indus Valley, 2500 BC in Peru and 500 BC in Egypt. Domestication of wild sheep took place around 9000 BC in the Middle East, but the wool of wild sheep and goat could have been woven before their domestication, more than 11,000 years ago. The earliest archaeological discovery of silk dates to the second millennium BC. Natural fibres were processed in various ways before being made into yarns. Spinning, a technique believed to be 20,000 years old, is the production of a potentially endless yarn from a bundle of fibres of various lengths by twisting them together so that their ends overlap. A spun yarn is stronger than the sum of the individual fibres from which it is made up. Flax, either wild or cultivated, provided the first spinnable fibres and weavable thread, probably 10,000 years ago. The fibre component is of fundamental importance in textile preservation. Understanding the chemical and physical properties of fibres is a prerequisite to formulate and undertake appropriate conservation treatments, whether preventive or remedial (Böhmer 2002; Teague 1998; Timár-Balázsy and Eastop 1998).

CHEMICAL COMPOSITION OF FIBRE POLYMERS

Chemically, fibres are polymers consisting of small repeating units (monomers). Some fibres are homopolymers, comprising mainly of one kind of unit, one example being cellulose, which contains β-D-glucose units alone. The predominant polymers of some other fibres are heteropolymers, which are formed from two or more types of unit. Examples include protein fibres which can be built up from 20 different amino acids. The primary structure of polymers is determined by the type and sequence of their constituent units. As fibre polymers are of organic origin, their main constituent atoms are carbon and hydrogen. Heteroatoms, such as oxygen, nitrogen, sulfur and halogens, can also be present in the backbone of the polymer chain or as end and/or side groups (functional groups). The role of functional groups is of crucial importance in polymer chains as they determine polarity (and hence bond strength), solubility in non-water solvents, water affinity, pH, colour and chemical reactivity. The cohesion of the fibre material is provided by the covalent bonds between their constituent atoms and by secondary bonds. Secondary bonds can be formed between suitable atoms or groups of atoms on the same chain or between different chains, stabilizing the secondary and tertiary structure of polymer chains. Under ordinary conditions, primary bonds, once destroyed, are not usually re-formed. Secondary bonds, being weaker and temporary, can be pulled apart and re-formed many times.

The polymers of natural fibres contain crystalline (well-ordered) and amorphous (non-ordered) regions. The ratio of crystalline to amorphous regions is basically determined by the presence and size of the side groups. If side groups are absent, very close fitting of the chains is possible, resulting in a highly crystalline structure. However, if the polymer chain contains many large side groups, close fitting is not possible and the polymer will have a highly amorphous structure. The mobility of polymer chains in the crystalline regions is limited under ambient conditions because these regions are rigid and stiff, thus preventing the penetration of water, chemicals and biological agents. Such fibres are stronger, stiffer, difficult to dye and more resistant to chemicals and aging. On the other hand, in the amorphous regions greater segment movement is possible, providing flexibility and elasticity. These regions are more accessible to water and other agents and therefore can be dyed more easily. Swelling upon water absorption and the chemical reactions of deterioration, including enzymatic activity by microorganisms, is often limited to the amorphous regions of fibres, although with time these processes can extend to the crystalline regions if the secondary bonds between the chains rupture (Warner 1995; Feughelman and James 1998; Tímár-Balázsy and Eastop 1998).

CELLULOSE-BASED FIBRES

Cellulosic fibres contain different proportions of cellulose, non-cellulosic polysaccharides (hemicellulose and pectin), proteins, waxes, organic acids (e.g. oxalic acid), lignin (not present in cotton), colouring matter (pigments and dyes), minerals and other trace materials. The proportion of cellulose, hemicellulose, pectin and lignin varies among different plant-based materials. For example, the highest percentage of cellulose is found in cotton at 83%, and of hemicellulose, pectin and lignin in kapok at 32%, 7% and 15%, respectively. Cellulose, hemicellulose and pectin are more prone to deterioration than lignin, and usually degrade first. High hemicellulose content promotes penetration of water and aqueous solutions due to its hydrophilicity. Pectins function as a source of food for microbes, and thereby contribute and sustain microbiological attack and colonization. High lignin content retards microbiological deterioration because penetration of water is hindered (lignin has hydrophobic properties) so microbes are not readily introduced. When microbial degradation occurs in high lignin fibres (e.g. jute), only lignin remains.

Plant fibres lack the insulating properties of many animal fibres, but produce excellent fabrics when retention of body heat is not essential. Fibres found on seed surfaces are finer and shorter, while those obtained from plant stems and leaves are coarser and longer. Approximately 70% of cotton fibre is crystalline, while the crystallinity of linen can be as high as 90%. This explains why linen fragments may survive in an archaeological context whereas cotton fragments deteriorate completely. Cellulose plant fibres exhibit good strength and moisture absorption with moderate to low elasticity. At 65% relative humidity (RH) and 22 °C, cotton, linen and hemp, and jute fibres have a moisture regain of 8.5%, 12% and 13.8%, respectively. Severe loss of moisture can lead to the escape of bound water from the fibre, causing it to shrink and create hydrogen bonds between the microfibrils, ultimately leading to the brittleness of the fabric. The most appropriate RH level for maintaining the softness and flexibility of cellulosic textiles is within the range of 40–65%. Below a RH of 40%, desiccation of the fibre may occur, and above 65%, swelling and subsequent colonization by microorganisms may take place (Teague 1998; Tímár-Balázsy and Eastop 1998; Böhmer 2002).

PROTEIN-BASED FIBRES

Proteins are chemically most stable in their isoelectric region (the pH at which the basic and acidic characters of their amino acids are in equilibrium). Both the solubility and swelling of a protein are limited at its isoelectric region. For conservation purposes, the isoelectric region of protein materials is considered to be pH 3–7 for silk fibroin and pH 5–7 for wool keratin. Within these pH ranges the material is less vulnerable, the probability of damage is reduced and its extent is more likely to be limited.

Protein-based fibres normally have good strength, insulating capacity, moisture absorption and elasticity. Their size varies from very fine fibres useful for delicate cloth to coarse hairs and wools. Sheep's wool is the most important and highly valued of the natural fibres: it is flexible, an excellent insulator, is aesthetically pleasing and dyes easily in addition to other advantageous properties. The protein of sheep's wool, keratin, is composed of a considerable number of amino acids with large side groups, including the sulfur-containing amino acid cysteine. Due to the presence of many large side groups on the protein chains and disulfide cross-links between the chains, only a small percentage of the α-keratin is crystalline (*c.*30% or less, depending on the source of wool). Wool fibre binds between 16% and 18% water by weight at ambient RH. Washed wool fibre contains 80% α-keratin, 0.5–1% lanolin and some non-keratinous proteins. Three different morphological parts, each with a different cell structure, can be distinguished in wool fibre: the cuticle (scales), the cortex (built of cortical cells) and the amorphous matrix. Providing the cuticle is intact, wool fibre is protected against many types of deterioration due to the presence of the resistant disulfide bonds. These bonds resist acid hydrolysis, but can undergo rupture as a result of alkaline treatment. The solubility of wool only increases after the reduction of the disulfide bonds, making it vulnerable to attack by proteolytic enzymes.

Silk is made of twin filaments of the protein fibroin cemented together with a non-fibrous protein called sericin. In degummed silk, the sericin gum that gives the fibre a stiff quality is removed and the twin fibroin filaments are separated into two fibres possessing a unique sheen. The water-soluble sericin attracts microorganisms, but degummed silk, with its closely packed supramolecular structure, is more resistant to the enzymes of microorganisms. Fibroin is able to bind water at up to 30% of its dry weight at 65% RH. As silk bonds strongly with water, it retains its flexibility even at 40% RH. The silk fibre may desiccate when stored in an environment with an RH below 40% or at high temperatures. Dehydrated silk is rigid, brittle and has decreased softness. Silk fibres have a crystallinity of ~60% and these highly crystalline regions render them mechanically very strong and resistant to chemical agents. Under similar conditions, wool, which is highly amorphous, may degrade completely (Teague 1998; Tímár-Balázsy and Eastop 1998; Böhmer 2002).

COLOUR IN FIBRES

Colour in textiles may be either the product of the pigmentation of the original fibre or added to the raw fibre, the yarn or the finished fabric. It is often difficult to distinguish between natural fibre colours, dyes and discoloration, particularly in the case of archaeological textiles, due to the environment in which a textile has been maintained. The imperfect penetration of colour through the diameter of yarns is a common indication that dyes have been used. However, dyeing before the yarn is spun or a lengthy immersion bath at appropriate temperatures can produce a uniform depth of colour that approximates the appearance of natural fibre colour. Chemical and physical tests are often necessary to confirm whether a natural or artificial colour is present (Teague 1998).

NATURAL PIGMENTATION OF FIBRES

Fibre colour arising from natural pigmentation can be modified by the growing conditions and the processing type and intensity. It is also possible to manipulate natural colour through the genetic selection of an appropriate plant or animal source. Such practices have ancient roots in indigenous farming communities. Five thousand years ago, farmers in central South America began to deliberately select naturally pigmented cotton varieties for their colour. Laying out native cotton in the sun for several days after picking increased colour saturation and homogeneity in the pigmented varieties, and boiling light-sensitive colours rendered them more lightfast and stable. Plant fibres derived from stems and leaves normally possess yellow, green, grey or brown natural pigmentation that dries to brown or brownish-grey tones. Cotton provides a greater natural colour range than most of the bast and leaf fibres, but coloured varieties are not found in all areas. Natural fibres usually require extensive processing prior to use. The final colour varies with the amount of extraneous (non-fibrous) material removed from the original matrix, as well as with the intensity of treatments (e.g. bleaching) that remove pigment from individual fibres.

Animal fibres, such as sheep's wool, goat's hair and silk, have their own natural colouring with a palette that extends from orange to brown, grey and black. Pigments that are distributed in the cortical and medullary cells of the fibre produce colour in wool. Stock has been bred for its distinctive colouring, black sheep being an example. Often, however, their naturally black wools are not lightfast, and they fade under intense sunlight even before being shorn and later when used in textiles. The only race of sheep with non-fading black wool inhabits the High Atlas of Morocco. In contrast, silk is coloured largely in response to the diet of the silkworm, although species' characteristics also have an impact (Teague 1998; Böhmer 2002).

NATURAL DYES ON FIBRES

Colour has been used to ornament and enhance fabrics for almost as long as humanity has produced textiles: it denotes significance and designates gender, age, rank, power and emotion. Early textiles, historical documents and archaeological artefacts all provide evidence of the skills with which colours were used in antiquity. It is now accepted that in the distant past different groups of people independently discovered how to dye fibres with various plants. Both the materials used in colouring textiles and the ways in which they were applied varied regionally, revealing differences in both the availability of resources and long-standing cultural preferences. It is not clear how the relatively complicated process of dyeing was developed, but in all likelihood the initial discovery of dye substances may have resulted from other activities. For example, cooking a dye plant for food or medicine may have led to the serendipitous discovery of the colouring potential of the substance as a result of the accidental staining of clothing. Logically, the next step would be to explore nature to find materials that would give brilliant and permanent colours with which to decorate textiles. The craftsmanship of production, the use of natural dyes and the ability to apply the chemicals needed to achieve permanent coloration were developed by trial and error and become an art shrouded in mystery. Travels of discovery and trade, over land and sea, helped to disseminate knowledge from one civilization to another, and for textile dyeing a new range of dyestuffs became available.

Dyes from plants can be obtained from their leaves, flowers, fruits, berries, stigmas or their dried and ground roots or bark. Some well-known and significant historical examples of plant dyestuffs include indigo, weld, madder, sandalwood, logwood, henna, turmeric, walnut and oak. Apart from dyes derived from plants, animals also played a significant role in pre-industrial textile dyes around the world. Dyes of animal origin include the secretions of various molluscs, from which different hues of purple (e.g. Tyrian purple) can be obtained, and the dried bodies of insects that provide kermes, cochineal and lac dyes. Minerals, despite being the first colours used by humans to produce the famous cave paintings 30,000 years ago, are not suitable for dyeing organic fibres permanently with a few exceptions, such as chromate and lead (Liles 1990; Teague 1998; Tímár-Balázsy and Eastop 1998; Böhmer 2002; Schoeser 2003; Hofenk de Graaff 2004; Balfour-Paul 2006).

Dyes are pigmented compounds that can bond to fibres to provide colour. The special structures causing colour in organic materials are called chromophoric groups or chromophores. Longer conjugated chromophoric systems can absorb higher light wavelengths, resulting in different, usually deeper, colours. Auxochromes, the functional groups that modify the light absorbance of the organic compound, may shift the absorption of the compound to longer wavelengths of light or cause a reverse shift. Therefore, the true colour of an organic molecule is determined by its chromophoric and auxochromic groups. The presence of polar auxochromic groups makes a dyestuff water-soluble and suitable for fast dyeing. Dyestuffs are held in place mainly by secondary bonds (in certain cases covalent bonds or salt linkages) between the polar groups of the fibre and dye. Textiles are dyed with dyestuffs directly or with the aid of another ingredient. The molecules of the dyestuff can be bound either on the surface or within the fibres, often via a chemical bridge provided by a mordant (Tímár-Balázsy and Eastop 1998; Böhmer 2002; Hofenk de Graaff 2004).

Dyes vary greatly in their resistance to colour change: no dye is absolutely fast in the sense that it retains the original colour precisely as it was created, regardless of exposure to light, water, heat, chemical substances or time. The soft colours of old textiles are often the product of the gradual decomposition of colouring matter and do not represent the original intent of the maker. Eventually many colours shift to a dull grey brown, and yarns that are now brown may well have been a yellow, gold, red or green hue when dyed originally. Moreover, in the case of mordant dyes used in burial textiles, there is a possibility of metal ion uptake or exchange of the metal ions of a mordant with other metal ions from the ground, giving rise to colour changes. Therefore, caution is needed when interpreting the colours of textile finds to be their original appearance (Teague 1998; Tímár-Balázsy and Eastop 1998; Hofenk de Graaff 2004).

Different dye substances and processes are often necessary, depending upon the type of fibre to be coloured; cellulose and protein fibres react differently to dyes. Animal fibres are complicated proteins that are not chemically neutral but have many chemically active sites, such as carboxyl and amino groups available for bonding dyestuffs to the fibres. All types of wool, silk and goat's hair dye easily. Plant fibres consist of chemically neutral cellulose with few unbound chemically active sites. Cellulose contains no basic or acidic groups and therefore no adhesion points that allow molecules from the dyestuff to bind readily to the fibres. It is therefore more difficult to dye plant fibres; time-consuming processes are required, and the final colours are seldom as brilliant as those achieved from the dyeing of wool and silk (Teague 1998; Böhmer 2002).

An understanding of the manner in which colours have been obtained, processed and applied to all types of fibres throughout history increases the appreciation of textiles, as well as essential for the conservation of historic pieces. Dyes can be classified depending on their chemical structure and method of application. Classification according to chemical structure can be based on the characteristic chromophoric group of the dye (e.g. azo dyes), the chromophoric structure (e.g. anthraquinones, such as alizarin, purpurin, carminic acid and kermesic acid; flavonoids, such as luteolin, apigenin and quercetin; naphthoquinones, such as lawsone and juglone; and carotenoids, such as crocetin and bixin), the derivatives of the dye (e.g. indigoids, such as indigotin) or other characteristic components.

Dyes can be grouped commensurate with the method of application. Water-soluble dyes remain so even after dyeing, the wash and light fastness being relatively low. Examples include direct dyes on cellulosic and protein fibres, acid and basic dyes on protein fibres, and disperse dyes. Dyes that dissolve in water form non-water-soluble dyestuffs in the fibre after dyeing. These include sulfur and azo dyes on cellulosic fibres, vat dyes on cellulosic and protein fibres, and oxidation dyes. Dyes in this group may have good wash fastness properties. Vat dyes are relatively complex to prepare, but produce very long-lasting colour on textiles, exhibiting high light and relatively good wash fastness, although fastness against rubbing is low. As a testament to their longevity, vat dyes have even been found in early prehistoric contexts.

Mordant dyes form the third group. Mordants are water-soluble salts applied before, during or after dyeing and are used to fix the colour on the fibre. Mordants not only increase the dye affinity for the fibre and affect colour fastness, but also help to improve colour depth. Mordant dyestuffs are very suitable for dyeing protein fibres. Cellulose fibres such as linen and cotton have less favourable dye bonding characteristics and are more difficult to dye with mordant dyestuffs. For centuries the most widely used mordant was alum: it gives the best results with respect to dye stability as it does not radically affect the colour. Iron and copper (historically the most readily available), and tin salts are also satisfactory and recommended mordants. Each of these minerals interacts differently with the dye to produce variations in colour: alum and tin produce bright, vivid shades, while iron dulls colour and copper lends greenish overtones. Some mordants can weaken fibres and cause premature deterioration; iron and tin are especially notorious in this respect. Colours obtained using mordant dyes usually exhibit excellent wash fastness for two reasons: the strong chemical bond created between the dye and fibre by the metal ion and the formation of a large complex molecule. However, this bond can be destroyed by the prolonged action of hot concentrated bases or acids.

The last group, the reactive dyes, adhere to cellulosic and protein fibres by covalent bonds and have excellent wash fastness (Bird 1972; Vigo 1994; Teague 1998; Tímár-Balázsy and Eastop 1998; Böhmer 2002; Hofenk de Graaff 2004; Balfour-Paul 2006).

BIODETERIORATION PROCESSES

Objects of historical value are subject to a number of chemical and microbial agents. Therefore, they are characterized by their preservation state and require an individual approach to conservation, storage and display. One of the fundamental tasks in the preservation of valuable artefacts is to identify the factors that enhance the degradation processes. The most important of these are photochemical reactions and deterioration induced by high temperature, aging, acids and alkalis, oxidizing and reducing agents, various chemical treatments and deterioration caused by living organisms, especially microorganisms. Biodeterioration induced by the presence of microbes is a worldwide problem as they possess a broad range of mechanisms capable of promoting damage and are found in all, including extreme, environments.

Historic fabrics of natural origin, such as cotton and wool, are particularly susceptible to microbial degradation. Interdependent factors that affect the rate of textile biodeterioration include the chemical composition and quality of the material, substrate pH and the presence of auxiliary substances such as dyes; physical and chemical changes in fibres that occur during manufacturing processes, use and storage; nutrient and moisture content and the accessibility to microorganisms of areas where such content is high; production of microbial metabolites, such as enzymes, acids and pigments, capable of textile digestion or stabilizing the moisture content; and environmental factors, including temperature, RH, exposure to light and oxygen availability (Błyskal 2012).

Inappropriate storage is the most common cause of biological damage to textiles. Of all the microorganisms, fungi pose the greatest threat to antique objects. Fungal germination structures enter the material during production or use; they can also emanate from the air or other infected surfaces in a given environment. Together with dust particles, fungal spores in dust can be deposited on the material and, when they encounter favourable conditions, begin to grow. The growth of fungi often starts on an easily available nutrition source, which may be organic and inorganic compounds in dust, such as textile filament fragments, cuticle cell debris or starch grains. These substances, along with absorbed water, may create a favourable microenvironment on the material surface. Once initiated, growth can expand into the textile fibres from which microorganisms draw nutrients, causing the product to deteriorate. Microbial deterioration occurs both chemically and mechanically. The chemical action of fungi on a polymer can take the form of assimilation; that is, the use of substrate components as a nutrient source. Assimilation is a process mediated by specific enzymes which can split high molecular weight compounds into smaller ones that can be readily consumed by microorganisms. In addition, materials can be destroyed by microbiological corrosion caused by active chemical compounds released by microbes, such as acids. Substrate penetration by fungal hyphae also results in mechanical degradation due to exertion on the fibres as a result of the growth and colonization of live cells within the textile.

The most favourable conditions for fungal growth leading to subsequent fabric

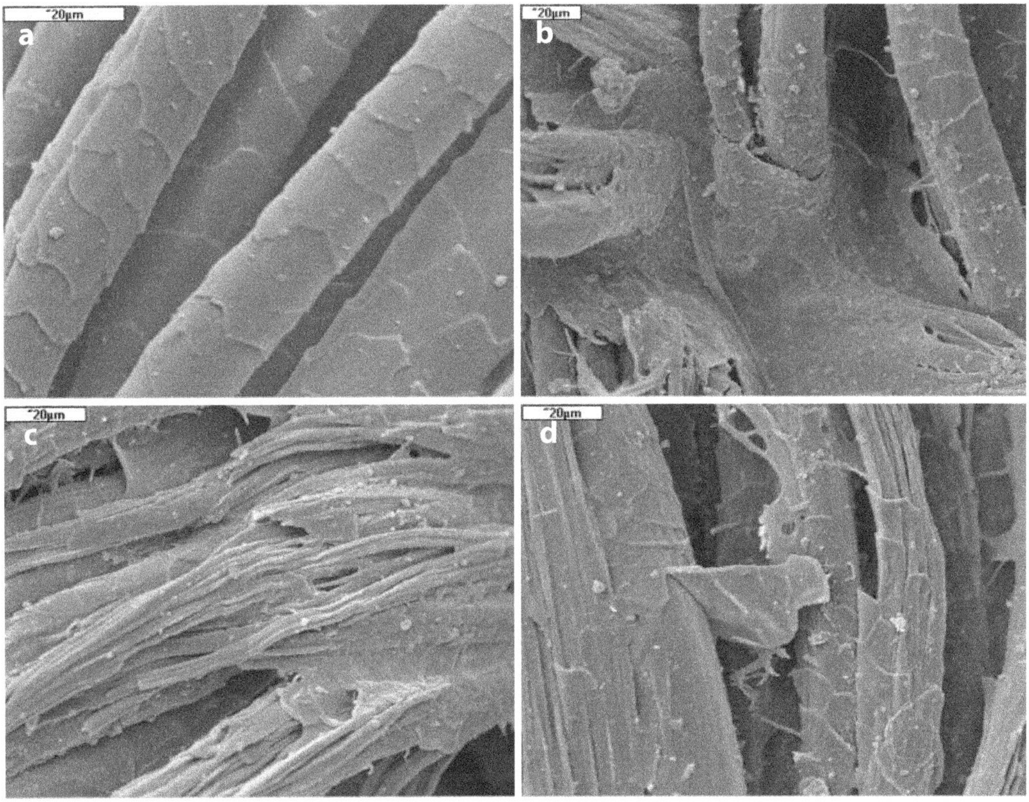

Figure 1 SEM images of synthetic yellow woollen textile fibres treated with *Acremonium camptosporum* for four weeks: (a) control ×1500, (b) ×750 and (c), (d) ×1000.

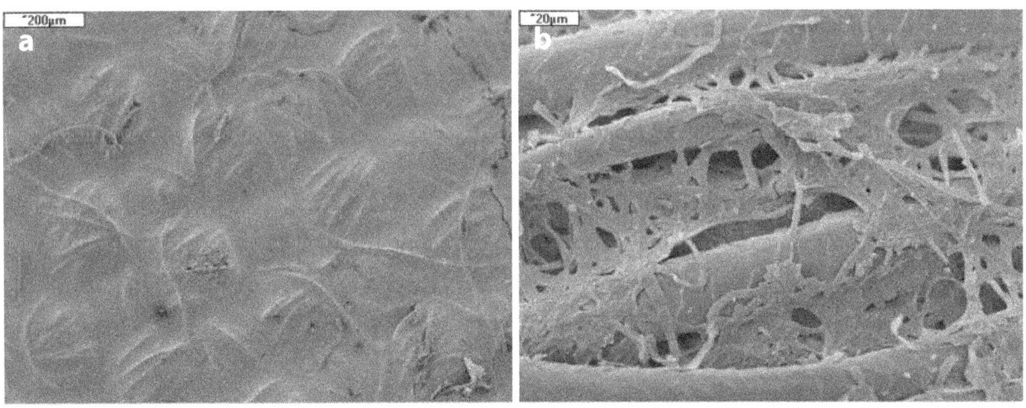

Figure 2 SEM images of fabric dyed with weld treated with *Penicillium chrysogenum* for four weeks: (a) ×100 and (b) ×750.

deterioration are a material moisture content of 7–9%, pH of 4.0–6.5, a temperature range of 25–30 °C, and RH of 65–96%. However, xerophilic fungi with high water content (52–75%) in their conidia can germinate when the water content is below average with a RH of less than 60%. It is necessary to pay special attention to such organisms when storing or exhibiting museum collections, as xerophilic fungi often induce changes in materials that

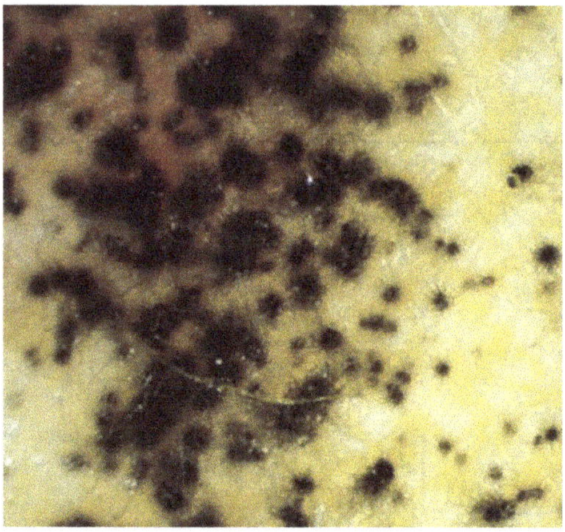

Figure 3 Stereoscopic microscope image of *Chaetomium globosum* growth and its staining of a weld-dyed fabric (×60).

are difficult to remedy. The alterations that an object may display in response to microbial activity include the obliteration of the original appearance of the material (leading to difficulties in the recognition and documentation of the primary characteristics), changes in the optical properties of fibres (gloss and colour), unsightly and difficult-to-remove stains, and structural decomposition (chemical changes in textile fibres leading to a decrease in textile strength characteristics or complete loss).

The microbial mode of action as well as the extent and severity of damage depends on the species of microorganism present. Such differences were demonstrated by the current author's studies on the biodeterioration of dyed fabrics. The growth of fungi on textiles may take the form of fabric entwinement as shown by the colonization of a woollen textile by *Acremonium camptosporum* (Fig. 1). Figure 1b shows the fabric fibres welded together with *A. camptosporum* mycelium and substances secreted by the fungus. In Figure 1c strong delamination is visible, and in Figure 1d, apart from the observed fibrillation of the fibres, the loose cuticle cells detaching from the fibres are apparent.

Upon invasion and colonization of certain fungi, increases, instead of the expected decreases, in the tensile strength and elongation of a fabric can be observed. As shown in Figure 2a, the surface of the weld-dyed fabric is completely covered with a compact mat created by *Penicillium chrysogenum*. As illustrated in Figure 2b, the fibres are covered with a dense growth of the fungus, and the resulting net around and between the fibres may contribute to the discussed increase in the strength of the material. Studies of the fibres' morphology showed no damage to their surface layers – they remained intact (see the control sample in Fig. 1a). It is possible that the organism did not have the physiological ability to utilize the fibre components, but derived nutrients from, for example, external lipids present on the fibre surface.

Aesthetic changes in the form of textile staining are another serious problem resulting from fungal colonization. Colour changes may be the result of the proliferation of pigmented fungal mycelium, conidia and sclerotia that penetrate the fibre interiors and the spaces between them. They may also be caused by the release of pigments produced by fungi and their impregnation into fibres (Fig. 3). Many fungal species generate a wide range of pigments (yellow, red, brown, green, blue, pink and violet) of various chemical compositions during their metabolic processes. In addition, the hyphal cell walls of some fungi contain melanin that may result in brown or black staining of objects. It is also possible that colour change may be due to interaction by the pigment with the material's components or by the accelerated oxidation of fatty acids due to their exposure to a high moisture level. The type of object discoloration depends on the chemical composition of the pigments and original materials, as well as the content of nutrients and metal ions (particularly those of copper, iron, manganese and zinc) in the substrate, the pH value, the co-presence of other microbes and the availability of light. Common textile discoloration consists of black, brown, green, olive, yellow, orange, ginger, pink and purple colours which considerably lower a product's value, sometimes making its use impossible. Removing stains from fabrics, although a difficult task, is important, not just for aesthetic reasons: stains contain enzymes that can continue to damage fibres even after the cells producing them have been destroyed.

There are also factors that inhibit microbial growth, especially on plant-based fibres, such as acidic conditions, high salt concentrations and the presence of copper. This may be the reason why linen fragments have been found in prehistoric salt mines, and textiles made of plant-based fibres are often preserved in contact with archaeological bronze or copper objects, or recovered from the sea (Tímár-Balázsy and Eastop 1998).

SUSCEPTIBILITY TO MICROBIAL DETERIORATION OF A WOOLLEN TEXTILE DYED WITH NATURAL AND SYNTHETIC DYES: A CASE STUDY

An important objective of conservation work is to act in a way that will ensure the longest possible survival of relics from the past. In the restoration of antique textiles, in order to preserve the original character of the object and maintain its value as a historic artefact, a material as close as possible to that from which an object is made should be used. For over four millennia, the vast majority of antique fabrics contained dyes that were made exclusively from natural ingredients. Thus, natural dyes are inherently related to historic objects.

During the 19th century, synthetic dyes became widely available and prompted a decline in the dominance of natural dyes in world markets. Nowadays, synthetic dyes are used in many ways, including in the conservation of artworks. However, a decision has to be made as to whether natural or synthetic dyes are a better choice for the protection of historic fabrics. Since this dilemma has long remained unaddressed, the current author undertook research to assess which group of dyes would render a woollen textile less susceptible to microbial damage. To achieve this, a textile made of Australian merino wool, with a plain weave and chemically untreated (raw fabric), was dyed with naturally derived yellow (weld), coral (madder), crimson (cochineal) and navy blue (indigo), and with synthetically derived yellow, coral, crimson and navy blue (Błyskal 2014).

Since fungi pose the biggest threat to historic materials of natural origin, this group of microorganisms was used in the experiments. Fungal strains determined to be particularly harmful to wool fibres were isolated from biodeteriorated antique and contemporary woollen textiles (from areas with stains, spots of brighter colour or thinning). The fungi used included: *A. camptosporum* Gams, *Microascus cirrosus* Curzi, *P. chrysogenum* Thom, *Eurotium chevalieri* Mangin, *Myceliophthora* sp. (sp. nov.), *Gymnoascus arxii* Cano and Guarro, *Chaetomium globosum* Kunze: Fries and *Alternaria alternata* (Fries: Fries) von Keissler. Strains such as *A. alternata*, *P. chrysogenum* and *C. globosum* are known for their keratinolytic activity and evoking strong destruction of keratin substrates (Błyskal 2009). These species are often also isolated from dust, and together with *E. chevalieri* are commonly found in indoor air. Because of their widespread occurrence, these organisms may pose a threat to many materials. The remaining species were isolated directly from antique textiles (except *G. arxii*), and were previously shown to be highly destructive towards keratin fibres (Błyskal 2009, 2016).

The studies investigated the intensity of fungal growth on dyed fabrics and the effect of the microorganisms on the mechanical properties of textiles as well as their colour. More intense growth of fungi was observed on fabrics dyed with synthetic dyes, of which the strongest growth appeared on the textile dyed with synthetic crimson. Of all the fabric types, the least fungal growth occurred on fabrics dyed with two natural dyes: weld and madder. Comparable intensity of mycelium growth on textiles and generated changes in the mechanical strength of dyed woollen fabrics were observed in the action of *G. arxii*, *A. alternata* and *E. chevalieri*. In the remaining cases, even though the fabrics were equally affected by the growth of fungi, their degree of decomposition varied.

The species of fungi that induced the most intense colour changes of the textiles were *M. cirrosus* and *A. alternata*. The studies

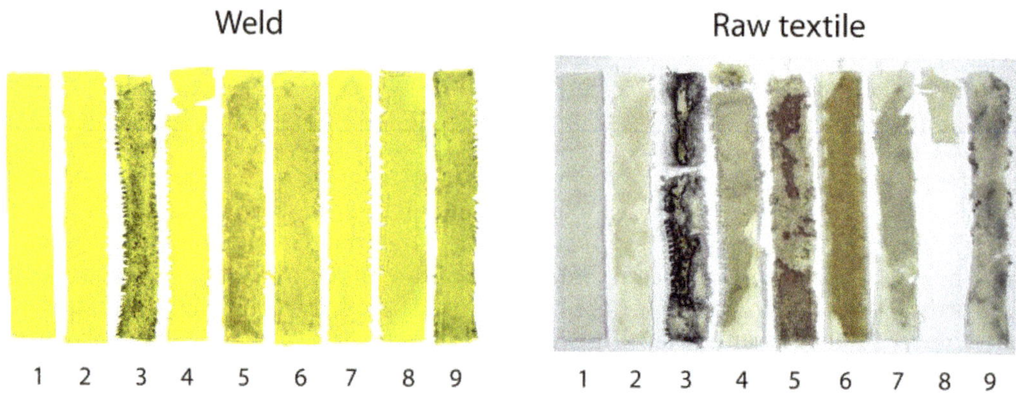

Figure 4 Images of woollen textile samples treated with fungi for a period of four weeks: (1) control, (2) *Myceliophthora* sp., (3) *Microascus cirrosus*, (4) *Acremonium camptosporum*, (5) *Chaetomium globosum*, (6) *Eurotium chevalieri*, (7) *Penicillium chrysogenum*, (8) *Gymnoascus arxii*, (9) *Alternaria alternata*.

also revealed that the nature of the changes in the aesthetic qualities of textiles depended upon the initial colour of the fabric on which the microorganisms acted. For example, *E. chevalieri* had little effect on the colour change of dark coloured fabrics, but caused severe staining on those lighter in colour. On the other hand, *Myceliophthora* sp. and *G. arxii* caused slight colour changes in light coloured fabrics, while on darker varieties they induced intense changes in the original colour of the fabrics, namely by lightening. The reason for the colour change is the alteration of the chromophoric structure of the dye. Under the influence of various factors, certain bonds of the dye molecule may undergo fragmentation, and the shorter chromophoric system shifts the absorbance of the dye to a shorter wavelength of light or even to the ultraviolet region. The colour is usually lighter and leads to fading (Tímár-Balázsy and Eastop 1998). However, can the severity of textile structural damage be predicted on the basis of changes in its colour? The data obtained from measuring the degree of fabric biodeterioration using strength tests (textile integrity) and spectrophotometric methods (colour changes) indicated that the results do not always overlap. *G. arxii*, a species with strong keratinolytic properties, caused the greatest decrease in mechanical properties of the tested fabrics and affected their aesthetic properties to a lesser extent when compared with the other strains. The opposite was observed in the case of the *M. cirrosus* strain, which was responsible for the most significant changes in fabric colour, but in terms of its influence on textile mechanical properties was not the most damaging. *A. alternata*, *C. globosum* and *E. chevalieri* demonstrated a consistent ability to destroy both the mechanical and aesthetic properties of textiles.

Studies of the dyed textiles' susceptibility to microbial attack and subsequent mechanical damage revealed that the strength of the fabrics, from the weakest to the strongest, was as follows: raw fabric, synthetic crimson, indigo, madder, synthetic coral, synthetic navy blue, cochineal, synthetic yellow and weld. Thus, the samples most distant from the remaining fabrics in terms of mechanical strength are weld and raw fabric, which constitute extreme cases of textile strength (Fig. 4).

Significant differences in resistance between fabrics were obtained for the raw fabric against all the dyed fabrics except synthetic crimson and indigo, for weld against all fabrics, and for indigo against all fabrics except the raw textile. No difference in strength was observed between fabrics dyed with synthetic dyes. Significant differences in mechanical strength between pairs of fabrics of a given colour (obtained with naturally derived and, corresponding in colour, synthetically derived dyes) occurred with reference to navy blue and yellow fabrics.

The susceptibility of dyed fabrics to biodeterioration is related to several factors: the changes in the fibre that occur in the dyeing process, the chemical structure of the dye, and the concentration of dye in the fabric. The effect of the dyeing process on the textile's initial strength was apparent in the case of the indigo-dyed fabric, where an alkaline environment was used to allow absorption of the dye. The treatment resulted in a significant lowering of strength parameters of the textile. However, even though the indigo-dyed textile was initially the weakest of all the tested fabrics, it was not the most susceptible to microbial damage. This suggested that the dye itself modulates the interaction with microbes (Błyskal 2016). The remaining natural dyes (weld, madder and cochineal) are mordant dyes, and in all cases alum was used as mordant. Therefore, it seems justified to assign the different resistances to biodeterioration of these textiles to the structure of the dyes.

The dyeing of fabrics with synthetic dyes was carried out using acid dyes. The exception was the synthetic yellow textile where additionally reactive dye was used, but like the other synthetic dyes it was applied in an acidic environment. The differences in the mean values of the mechanical parameters between fabrics dyed with synthetic dyes were statistically insignificant. For this reason, only a tendency towards greater (synthetic yellow) or weaker (synthetic crimson) resistance to microbial decomposition can be indicated. By way of explanation, it is possible that in the case of the fabric dyed with synthetic yellow, a permanent covalent bond was formed in addition to the ionic bond that occurs during processing with acid dyes. Similarly, the synthetic crimson fabric contained the highest percentages of synthetic red, and perhaps this dye was responsible for the lower resistance of this fabric to microbial attack. Given the above, it could be concluded that of textiles dyed with synthetic dyes, the susceptibility of the fabrics to biodeterioration depended on the chemical structure and concentration of the dye in the dyeing mixture. The results indicate that the use of dyes can be considered a potential method to control biodeterioration and may find an application in the restoration of historic fabrics. The use of substitute materials coloured with the appropriate dyes can reduce the negative impact of microbes on valuable works of art.

PLANT PRODUCTS AS PROTECTION AGAINST BIODETERIORATION

Many chemical or physical methods that are effective in inhibiting or inactivating microorganisms are called into question when historic textiles are considered. The least harmful method of protection against biodeterioration is monitoring of the storage or exhibition conditions. Physical protection of the material surface from contamination, management of the moisture content in the material, together with controlled temperature and RH will hinder microbial growth. However, objects with severe signs of the deteriorative action of microorganisms may require special treatment. In such cases, application of antimicrobial compounds of natural origin may prove useful to retard growth or kill microorganisms. The use of plant products for their functional properties is as old as humankind. The chemicals in dye plants that have been exploited for their colour also have medicinal as well as antimicrobial properties. Plant antimicrobials have a wide spectrum of activity and several different modes of action, including inhibition of cellular respiration, the ability to form complexes with the sterols present in the microbial cell membrane, and inhibition of nucleic acid and protein synthesis. Antimicrobial inhibition of specific enzymes that otherwise convert a bioactive component to an inactive form, or the induction prevention of a metabolizing enzyme, can also take place (Osbourn and Lanzotti 2009).

Several factors influence the activity of antimicrobials. Contributing microbial traits include those that provide inherent resistance, such as spore formation or specific strain differences, the initial population size and subsequent growth rate, interaction with other microorganisms, cellular composition and health and the ability to form biofilms. Other factors, such as material composition and microstructure, pH status and moisture content as well as temperature, RH and time of storage also affect antimicrobial activity.

Plant-derived antimicrobials are a promising tool for protecting objects from the effects of microorganisms and should play an important role in material protection in the future. Continued research on the sources and activities of such antimicrobials is needed to further our understanding of activity spectra, action mechanisms, interactions with fabric components and environmental conditions. An ideal antimicrobial would be effective when added as a whole plant-part extract. However, few antimicrobials are present at concentrations sufficiently high to act efficiently without some purification. The future of antimicrobials lies not only in producing combinations based on modes of action in order to obtain a wide spectrum of activity against microorganisms, but also on developing systems for the efficient delivery of antimicrobials to their targets. The challenge is to isolate, purify, stabilize and incorporate plant-derived antimicrobials into fabric without adversely affecting the material characteristics. Application of antimicrobial compounds to historic objects should always be preceded by a series of laboratory tests. Material protection is best achieved when the microorganisms to be inhibited, the antimicrobial type and concentration, storage conditions and the presence of other agents that may affect the process are known and taken into account.

CONCLUSIONS

The biodeterioration of historic textiles is a complex and multidimensional phenomenon. In order to fully understand and prevent it, scientists should take a collaborative, interdisciplinary research approach. Cooperative efforts should be undertaken to make plant-derived natural product research an essential contributor in future attempts to treat and restore historic textiles damaged by microbes. Biotechnology may be employed in the process of plant-derived antimicrobial manufacturing. Natural products can be produced via cell culturing of the source organism or through genetic engineering in a heterologous host. These biotechnological methods have already been shown to produce selected plant products in a relatively controlled manner, providing a supply of plant substances that are not limited by sourcing problems, such as environmental, seasonal, geographical and political factors (Osbourn and Lanzotti 2009).

Objects constituting cultural heritage are not protected exclusively for their original form and physical beauty – they are needed by society and for future generations to maintain personal contact with national history and aid in the search for world knowledge. The starting point for protecting historical artefacts is the past and the destination point is the future, for which the remnants of our cultural heritage should be protected and preserved.

REFERENCES

Balfour-Paul, J. 2006. *Indigo*. London: Archetype Publications.

Bird, C.L. 1972. *The Theory and Practice of Wool Dyeing*, 4th edn. Bradford: Society of Dyers and Colourists.

Błyskal, B. 2009. 'Fungi utilizing keratinous substrates', *International Biodeterioration and Biodegradation* 63(6): 631–53.

Błyskal, B. 2012. 'Aspects of fungal deterioration of wool', in *Technology and Innovation for a Sustainable Future: A Commodity Science Perspective*, R. Merli (ed.), 598–603. Rome: IGWT AISME.

Błyskal, B. 2014. '*Gymnoascus arxii's* potential in deteriorating woollen textiles dyed with natural and synthetic dyes', *International Biodeterioration and Biodegradation* 86(C): 349–57.

Błyskal, B. 2015. 'Fungal deterioration of a woollen textile dyed with cochineal', *Journal of Cultural Heritage* 16: 32–9.

Błyskal, B. 2016. 'Indigo dyeing and a microorganism–polymer interaction', *Journal of Cultural Heritage* 22: 974–83.

Böhmer, H. 2002. *Koekboya: Natural Dyes and Textiles. A Colour Journey from Turkey to India and Beyond*. Schweinfurt: Weppert.

Feughelman, M. and James, V. 1998. 'Hexagonal packing of intermediate filaments (microfibrils) in α-keratin fibers', *Textile Research Journal* 68(2): 110–14.

Hofenk de Graaff, J.H. 2004. *The Colourful Past: Origins, Chemistry and Identification of Natural Dyestuffs*. London: Archetype Publications.

Jones, L.N., Rivett, D.E. and Tucker, D.J. 1998. 'Wool and related mammalian fibers', in *Handbook of Fiber Chemistry*, 2nd edn, M. Lewin and E.M. Pearce (eds), 355–413. New York: Marcel Dekker, Inc.

Kent, K.P. 1983. *Prehistoric Textiles from the Southwest*. Albuquerque: University of New Mexico Press.

Liles, J.N. 1990. *The Art and Craft of Natural Dyeing: Traditional Recipes for Modern Use*. Knoxville: University of Tennessee Press.

Osbourn, A.E. and Lanzotti, V. 2009. *Plant-Derived Natural Products*. New York: Springer.

Schoeser, M. 2003. *World Textiles: A Concise History*. London: Thames & Hudson.

Teague, L.S. 1998. *Textiles in Southwestern Prehistory*. Albuquerque: University of New Mexico Press.

Tímár-Balázsy, A. and Eastop, D. 1998. *Chemical Principles of Textile Conservation*. Oxford: Butterworth-Heinemann.

Vigo, T.L. 1977. 'Preservation of natural textile fibers: historical perspectives', in *Preservation of Paper and Textiles of Historic and Artistic Value*, J.C. Williams (ed.), 189–207. Washington DC: American Chemical Society.

Vigo, T.L. 1994. *Textile Processing and Properties, Preparation, Dyeing, Finishing and Performance*. Amsterdam: Elsevier.

Warner, S.B. 1995. *Fiber Science*. New Jersey: Prentice-Hall, Inc.

AUTHOR'S ADDRESS

Barbara Błyskal, Krakow, Poland (blyskalb@poczta.onet.pl)

BIODETERIORATION OF PAINTINGS IN CAVES, CATACOMBS AND OTHER HYPOGEAN SITES

Clara Urzì, Laura Bruno and Filomena De Leo

INTRODUCTION

Caves, catacombs and similar hypogean environments (crypts, tombs and mines) have in common the fact that they are located under the terrestrial surface with highly variable extension and depth. They present a unique microhabitat and microbial communities that are in a very delicate equilibrium. Any microclimate change influences the existing equilibrium among the microbial populations and their growth on and inside the rock surface. Other contributory factors include the above-ground climate, namely the amount of annual rainfall, the density of the plant coverage, humidity, pollutants and air circulation. Conservation of these sites can be very difficult, especially if they are to be accessible to a large number of visitors with consequent changes of microclimatic parameters such as increased light, CO_2 release and the risk of introducing and transmitting pathogenic microorganisms.

CAVES

Caves are naturally formed subterranean environments excavated by constant natural water seepage or water flow through limestone, dolomite, gypsum or sandstone, or created by volcanic activity forming lava tubes or lava caves. Their length can vary from a few metres to several kilometres. Most caves are fascinating because the mineralogical formations, called speleothems, offer a magnificent exhibition of the power of biogeochemical reactions (Fig. 1). Since prehistoric times, most caves were used by humans as shelters to protect against wild animals and inclement weather conditions or as secluded places for religious or ceremonial purposes. Very often prehistoric inhabited caves harbour painted drawings on the walls or ceilings, the most ancient of which date to around 40,000 years ago (Pike *et al.* 2012). Regardless of their location, the paintings are quite similar, with animals being common subjects and humans mainly appearing as hand stencils.

Prehistoric painters utilized pigments that were available in the vicinity, such as minerals (limonite, hematite red ochre, yellow ochre and umber), charcoal, burnt bones and lime, mixed with binders such as saliva or animal fat. In most cases, these pigments were still in good condition when the caves were first discovered but by opening them up to the public, the delicate equilibrium of these ecosystems has been affected. Uncontrolled numbers of visitors and their side effects (expulsion of CO_2, need for light, etc.) irreversibly changes the microclimate, resulting in biodeterioration of speleothems and cave surfaces, including rock paintings.

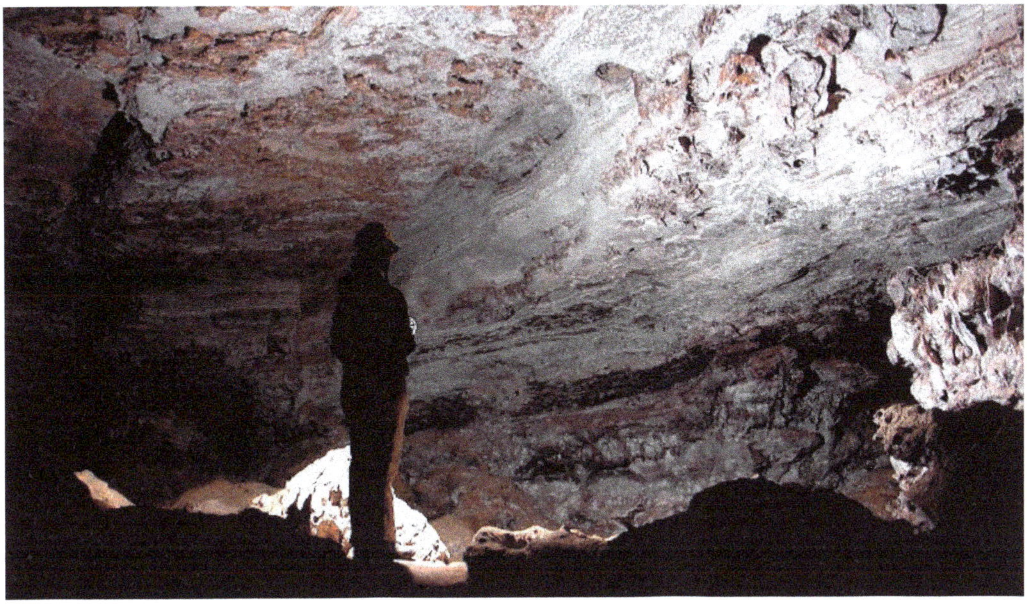

Figure 1 Wind cave in South Dakota, USA. (Photo: C. Urzì.)

One example is the Lascaux cave in France (Fig. 2). Discovered in 1940, it is famous worldwide due to the beautiful paintings that are considered to be some of the finest examples of rock art paintings (Bastian *et al.* 2009, 2010). The combination of the large number of visitors (up to 1800 per day) and the artificial lights that have been installed to allow the paintings to be viewed disturbed the microclimate and had a strong impact on the cave ecosystem. The so-called *maladie verte* was caused by a biofilm-forming green alga, the first of the various outbreaks or 'microbial outbursts' suffered by the cave, which led to its closure in 1963 (Bastian *et al.* 2010). A similar problem afflicted the Altamira cave paintings in Spain discovered in 1879 and dated around 14,000 years old. The paintings have attracted huge numbers of visitors – 175,000 in 1973, the busiest year on record. The cave was closed to the public in 2002 after photosynthetic bacteria and fungi were found biodeteriorating pigments at alarming rates (Saiz-Jimenez 2014). Other examples were reported by Jurado *et al.* (2010a) in the cave of Castañar de Ibor in Spain where the outbreak of *Mucor* and *Fusarium* species was caused by an input of organic matter from human vomit. As reported by Moñino *et al.* (2014), chewing gum found on the walls of Cueva del Tesoro, Rincón de la Victoria (Malaga, Spain) served as the carbon source for a wide microbial community. The bacteria colonizing the chewing gum were represented by members of Actinobacteria, Alphaproteobacteria and Firmicutes, specifically *Brachybacterium fresconis*, *Microbacterium aerolatum*, *Sphingopyxis italica* and *Bacillus litoralis*, while fungi belonged to the Ascomycota, order Eurotiales and Hypocreales (*Aspergillus*, *Fusarium* and *Penicillium* species).

Several exclusive microbial habitats and niches can be found in caves. A cave connected with the surface has three distinct major habitats: entrance, twilight zone and dark area (Bonacci *et al.* 2009). These three areas have different geochemical gradients that influence the nutrient availability and the demand for energy source of living microorganisms, either photo- or chemotrophic. Another important factor is the porosity of the rocks allowing the permeation of allochthonous nutrients and other energy sources suitable for autotrophy in absence of light (Engel 2007, 2010).

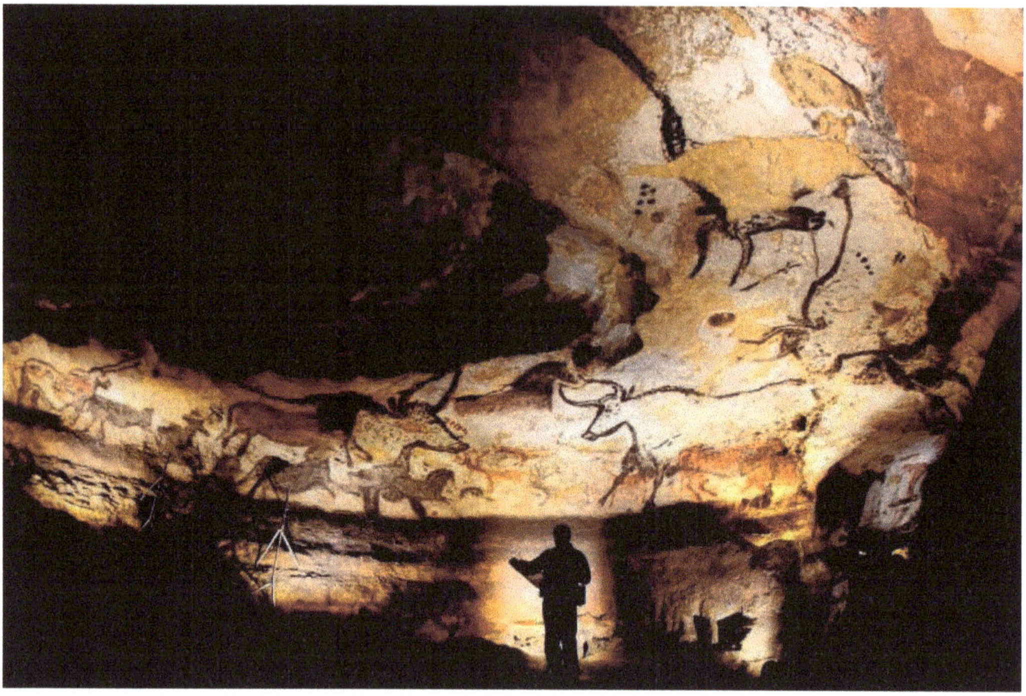

Figure 2 Detail of paintings found in the Lascaux cave. (Photo: Sisse Brimberg/National Geographic.)

CATACOMBS

The word 'catacomb' means an underground Christian cemetery of large dimensions, characterized by an elaborate network of galleries and cubicula used intensively for burial (Fiocchi Nicolai 1999). Early Christians were forced by Roman rule to bury their dead outside the city limits. Since land was expensive, they went underground, digging about 600 km of tunnels into the soft volcanic tuff present in Rome. The catacombs consist of networks of rooms lined with rectangular niches called loculi, small rooms named cubicula and arcosolia that served as a family tomb, and large niches with an arch over the opening, also used for family burials (Figs 3a and 3b). Often decorated with paintings (Figs 4a and 4b) and objects, these were places to share funeral meals and mark death anniversaries (the idea that persecuted Christians secretly worshipped there, however, is a Romantic-era legend). The Roman catacombs are the most famous, but catacombs are found worldwide, although they are mostly located in the Mediterranean area such as the Rabat catacombs in Malta, the Milos catacombs in Greece, the catacombs of Kom el Shoqafa in Alexandria, Egypt, the catacombs of Paris, France, and the catacombs of Cappuccini in Palermo, Italy.

The environmental conditions inside catacombs play a crucial role in the establishment of microbial communities on the precious surfaces. The lithic substrata present in the Roman catacombs consist of calcareous material, lime-pozzolana mortars and bricks sometimes painted with frescoes (Sánchez-Moral *et al.* 2005a). The European project 'Cyanobacteria attack rocks (CATS): control and preventive strategies to avoid damage caused by Cyanobacteria and associated microorganisms in Roman hypogean monuments' (Albertano *et al.* 2003), showed a preferential colonization by microorganisms of the different stone materials present in catacombs to be strictly correlated with the porosity of the substrates (Sánchez-Moral *et al.* 2005a, b).

Due to their archaeological, historical and religious value, catacombs attract thousands of

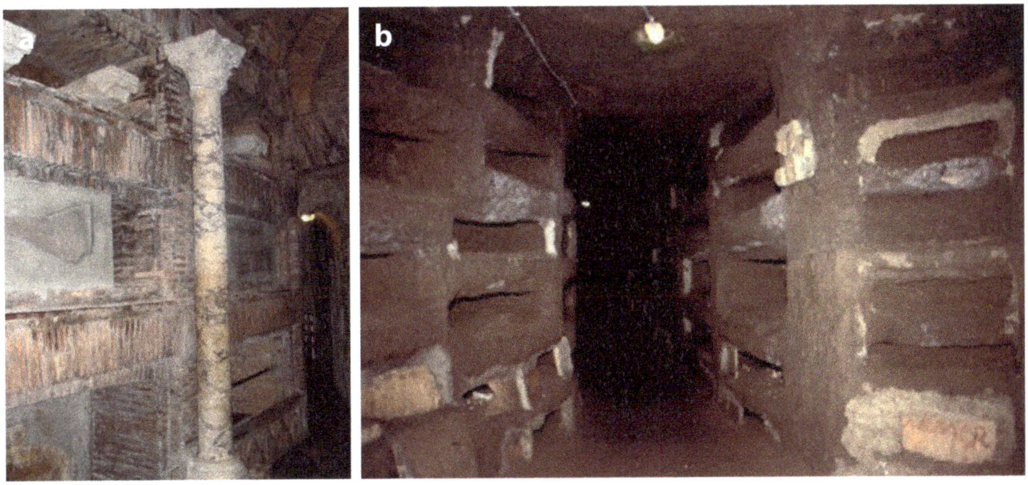

Figure 3 Catacombs of St Callixtus in Rome: (a) particular of the Cripta dei Papi, the official burial place of nine popes and, probably, of eight dignitaries of Rome's 3rd-century church and (b) a corridor along the tourist path full of loculi. (Photos courtesy of the Pontificia Commissione di Archaeologica Sacra, Rome, Italy.)

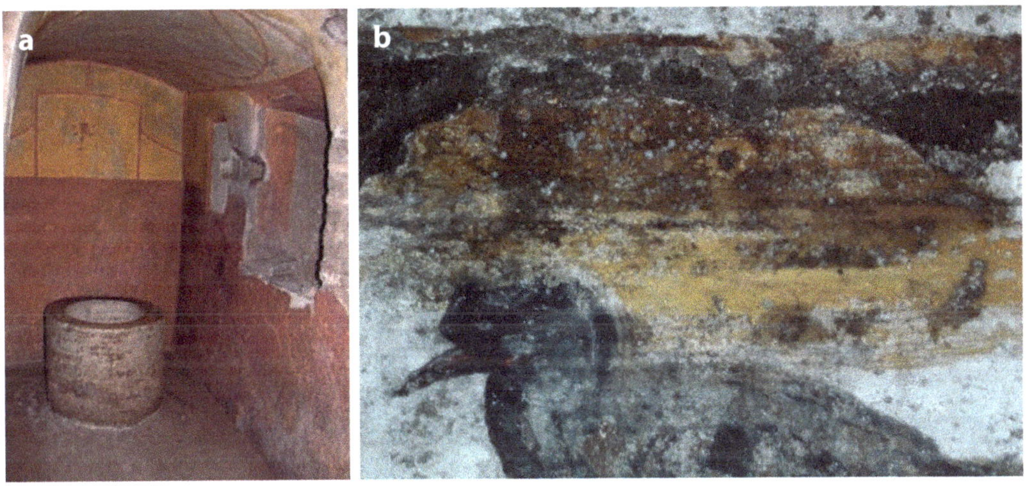

Figure 4 Frescoes in the catacombs of Domitilla in Rome: (a) paintings on the walls near the Corridor of Flavi and (b) detail from paintings in the Cubicolo of Apostoli piccoli. (Photos courtesy of the Pontificia Commissione di Archaeologica Sacra, Rome, Italy.)

visitors every year, requiring them to be illuminated with artificial light, which has deleterious effects. The unique environmental conditions inside catacombs are similar to those present in natural caves and are characterized by relative humidity (RH) higher than 87%, due either to the moisture in the ground diffusing by capillary action from the interior or to the infiltration of water from above. Furthermore, there is very little air circulation and temperature is almost constant (between 18 and 22 °C) throughout the year (Groth and Saiz-Jimenez 1999; Albertano 2012; Bruno and Valle 2017). These environmental conditions, coupled with the presence of light, at very low values of less than 5 μmol photons m^{-2} s^{-1} (Albertano and Bruno 2003), are responsible for the biodeterioration of these sites due to the development of phototrophic biofilms on the illuminated lithic faces (Fig. 5) (Albertano and Urzì 1999; Albertano 2012; Bruno *et al.* 2014).

Previous research highlighted how the presence of visitors influenced the growth of biofilms by inducing a sharp increase of temperature

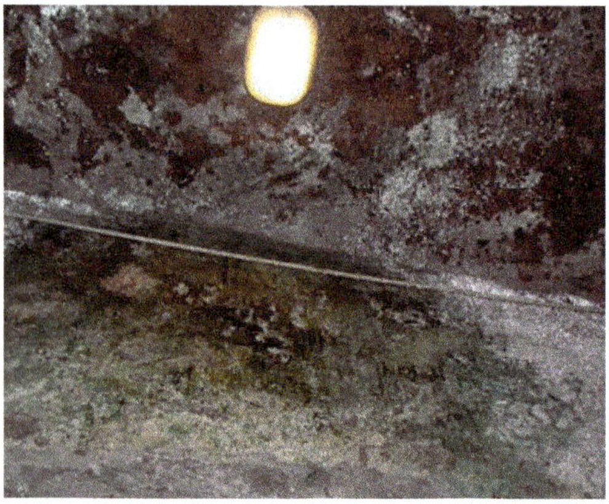

Figure 5 Phototrophic biofilms on the wall near the lamp in the catacombs of St Callixtus (Rome, Italy).

and CO_2 inside catacombs (Sánchez-Moral *et al.* 2005b). This dramatic change to the environment enhanced the water condensation on material surfaces and sustained the metabolic activity of microorganisms. The bioprecipitation of minerals was observed due to the activity of Cyanobacteria and associated bacteria constituting biofilms (Sánchez-Moral *et al.* 2003a, b). The influence of visitors on biodeterioration phenomena is also confirmed by the fact that the distribution of biofilms is restricted to the mural paintings, walls and ceilings in areas that are open to the public (Sánchez-Moral *et al.* 2005b). Some catacombs are not illuminated by artificial light but by natural light penetrating inside the hypogea from a natural opening (Figs 6a and 6b), as in the Abbatija tad-Dejr catacombs, Malta (Zammit *et al.* 2011), which is able to sustain the growth of biofilms on the illuminated stone surfaces. Chemoorganotrophic microorganisms are often associated with the cyanobacterial and algal biofilm and can also grow in the twilight zone where there is insufficient light for photosynthesis (Bruno *et al.* 2006; Urzì *et al.* 2014; Kraková *et al.* 2015).

OTHER HYPOGEAN SITES

Etruscan tombs are hypogean habitats of cultural heritage interest that present biodeterioration problems similar to those observed in caves and catacombs. These alteration patterns are often observed as the formation of whitish veils associated with the presence of *Streptomyces* as first described by Agarossi *et al.* (1985, 1988). Diaz-Herraiz *et al.* (2013) reported that painted surfaces of the Tomb of the Monkey (Tomba della Scimmia) near Chiusi (Italy) were heavily colonized by members of Actinobacteria, primarily *Nocardia* and *Pseudonocardia* species. More recently, the same authors (Diaz-Herraiz *et al.* 2014) described in another tomb in Chiusi (Tomba del Colle), white, circular to irregular aggregates or even biofilms, which extended throughout the paintings regardless of the pigment used, that were caused by the colonization of *Rhizobiales* species. This microbial growth was associated with $CaCO_3$ deposits in the shape of rosette or nest-like fabrics.

Another important hypogean site is the Necropolis of Tarquinia, included by UNESCO as a world heritage site in 2004. A survey on the Etruscan Mercareccia Tomb showed the presence of 142 strains of heterotrophic microorganisms belonging mostly to the bacterial phyla Actinobacteria and Firmicutes, while fungi isolated from this site belonged to the taxa Sordariomycetes, Eurotiomycetes, Ascomycota and Dothideomycetes (Sprocati *et al.* 2008).

MICROBIAL PROCESSES

As reported by several authors (Barton and Jurado 2007; Urzì *et al.* 2010; Yantian *et al.* 2015), caves and catacombs contain a broad diversity among members of the Alphaproteobacteria, Betaproteobacteria, Gammaproteobacteria, Firmicutes and Actinobacteria. Those results were confirmed by both culture-based techniques and non-culture-based methodologies. Strains isolated from caves or catacombs are able to carry out different metabolic activities: they can fix nitrogen, degrade complex aromatic compounds to be utilized as a carbon source, turn over lipids and scavenge scarce nutrients (Barton and Jurado 2007) as well as consume other cells that are either living or debris (Urzì *et al.* 2016).

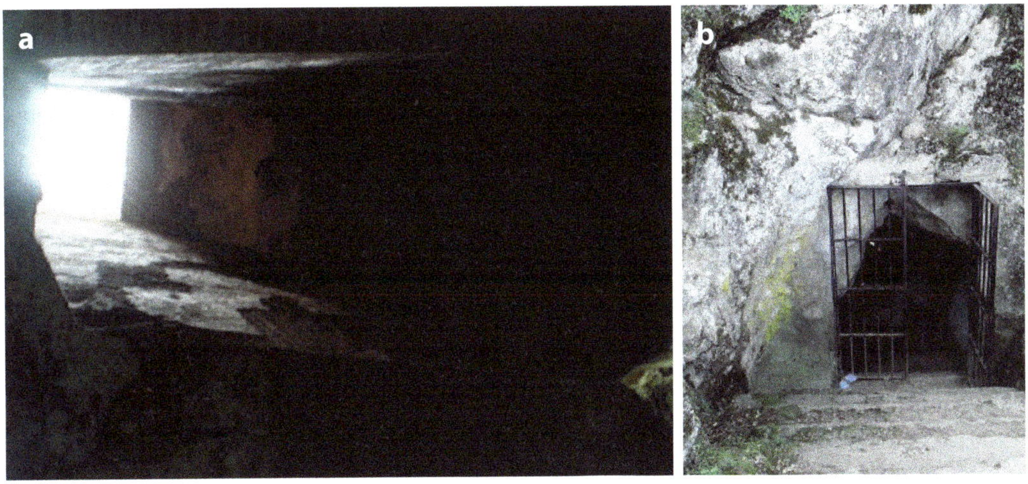

Figure 6 Light sources available inside the Roman catacombs: (a) natural light from the opening called 'skylight' and (b) at the entrance of the hypogea.

According to Kraková *et al.* (2015), the ongoing detection and isolation, in hypogean samples, of bacterial members belonging to the genera *Streptomyces*, *Pseudonocardia*, *Stenotrophomonas* and *Phyllobacterium*, and to the species *Lysobacter enzymogenes*, provided evidence for the important role of these strains in such habitats. In fact, strains belonging to the genus *Phyllobacterium* have demonstrated their ability for phosphate solubilization, N2-fixing and mineral weathering (Chen *et al.* 2009). Their presence can often be hindered by the more evident growth of photosynthetic organisms in the biofilm and they become dominant only after biocide treatment (Urzì *et al.* 2016). They can also grow in characteristic alteration patterns such as whitish veils or patinas in areas where Cyanobacteria and algae cannot reproduce, sometimes forming peculiar alterations associated with another specific microorganism (algae, Cyanobacteria, bacteria or fungi) or as an organized microbial mat (biofilm) as described below.

Microorganisms and mineral precipitation

Caves can be considered a favourable environment for chemically and microbially mediated mineral precipitation (Ehrlich 1996; Castanier *et al.* 1999; Urzì *et al.* 1999; Sánchez-Moral *et al.* 2003; Sánchez-Roman *et al.* 2007). The process of $CaCO_3$ bioprecipitation is self-induced and bacterial cells serve as nucleation centres. At the end of this process, the microbial cell becomes entrapped by its own crystal production. Under laboratory conditions more than 80% of isolated bacteria from soils were able to precipitate carbonatic minerals (Boquet *et al.* 1973; Urzì *et al.* 1999). Some microbiologists have hypothesized that by being encased, a mineral shell is one of the most primitive forms of defence against a harsh environment such as ultraviolet (UV) rays (Phoenix *et al.* 2001) and it is an efficient way to immobilize atmospheric CO_2. Several different shapes of bio-induced mineral precipitation have been described on the surfaces of caves as well as in catacombs (Miller *et al.* 2012, 2013; Banerjee and Joshi 2015). For example, in the Altamira cave, grey round spots as described by Cuezva *et al.* (2012) were associated with the presence of *Streptomyces* and *Bacillus* spp., both of which were able to produce $CaCO_3$ under laboratory conditions. Additionally, aggregates of calcite crystals (2–4 μm in size) in rosette-like shapes were previously described in the cave (Cañaveras *et al.* 1999; Laiz *et al.* 1999; Baskar *et al.* 2006).

Recently, it was proposed that different Actinobacteria strains, perhaps members of the genera *Streptomyces* and *Pseudonocardia*, are able to capture CO_2 (which is abundant in the

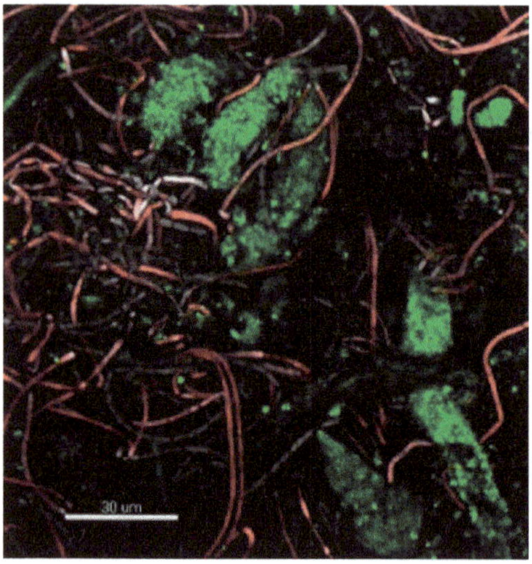

Figure 7 Mineral precipitation (green) on the sheaths of Cyanobacteria belonging to the order Nostocales. Filaments of Oscillatoriales are also visible (pink).

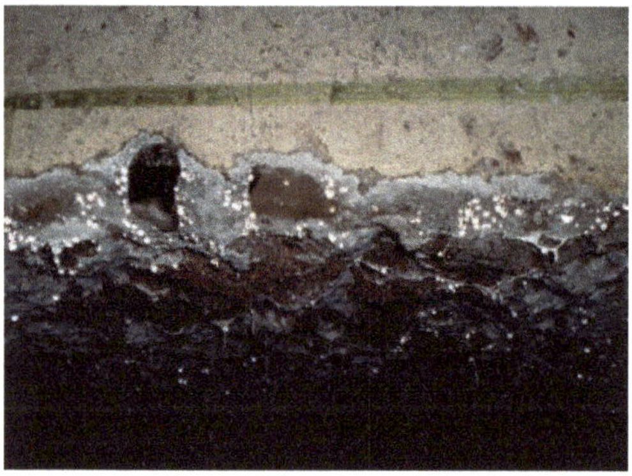

Figure 8 White biofilm formed inside the Cubicolo di Oceano in the catacombs of St Callixtus in Rome.

hypogean environment) from the air and form calcium carbonate polymorphs. This leads to dissolution of the rock substratum and the generation of $CaCO_3$ crystals (Cuezva *et al.* 2012). *Stenotrophomonas* and *Lysobacter* strains were associated with the formation of witherite in the Roman catacombs (Sánchez-Moral *et al.* 2004). Different forms of calcium carbonate crystals are associated with Cyanobacteria such as the production of mineral coats around cyanobacterial sheaths and the occurrence of calcite fibres with different morphologies on the surface of the biofilms (Fig. 7) (Zammit *et al.* 2011).

White to greyish patinas

Hypogean surfaces far from the artificial lights are often covered by white biofilms of different extension, consistency and composition (Fig. 8) (Albertano *et al.* 2003; Sánchez-Moral *et al.* 2003a; Urzì *et al.* 2008). In all cases, a high diversity of heterotrophic microflora, especially bacteria, was found. There is evidence that these patterns can be due to a diffuse mineral precipitation mediated by the activity of several groups of bacteria (Sánchez-Moral *et al.* 2003a, b), or to a network of filamentous microorganisms (Zimmermann *et al.* 2006; Zammit *et al.* 2011; De Leo *et al.* 2012).

PHOTOSYNTHETIC BIOFILMS

Algae and Cyanobacteria on hypogean surfaces can be found only near the entrance or in proximity to artificial light. Their growth and occurrence is often organized to form biofilms whose extension and complexity depends on various factors (Roldán and Hernández-Mariné 2009; Urzì *et al.* 2010), the most important of which is the light intensity. The extension of phototrophic biofilms is, in fact, strictly related to the light gradient. In the inner part of show caves and catacombs, the light available for photosynthesis is usually restricted to a limited time during the day, when visitors are present. In the Roman catacombs this area extends to 1 m from the artificial light source (Fig. 9) (Bruno *et al.* 2001; Albertano and Bruno 2003).

The microbial consortia that constitute the biofilms inside caves and catacombs are well-studied. These biofilms are comprised primarily of filamentous and coccoid Cyanobacteria (Figs 10a and 10b) associated with non-photosynthetic bacteria and occasionally a few eukaryotic microalgae, mosses, fungi and lichens (Albertano and Urzì 1999; Saarela *et al.* 2004; Bruno *et al.* 2006, 2009; Roldán and Hernández-Mariné 2009; Zammit *et al.* 2011; De Leo *et al.* 2012; Everest *et al.* 2015; Kraková *et al.* 2015; Bruno and Valle 2017). In the Roman catacombs, some strains of heterocystous Cyanobacteria have been isolated in culture, such as *Symphyonemopsis* sp.

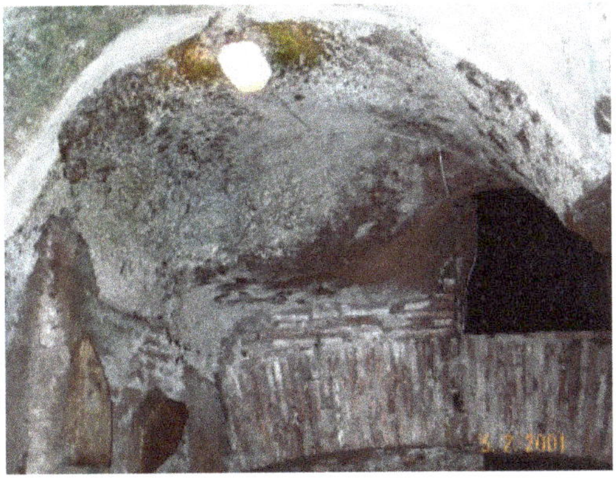

Figure 9 Diffused pattern of alterations as green and white biofilms on illuminated lithic faces in Roman hypogea.

(Fig. 11), *Scytonema julianum* and *Brasilonema* sp., and these are always associated with non-heterocystous strains of *Leptolyngbya* sp. and *Oculatella subterranea* (Bruno *et al.* 2009; Zammit *et al.* 2012; Komárek 2013).

In the Nerja cave (Malaga, Spain), some illuminated sites are covered by a greenish patina caused by Cyanobacteria and microalgae associated with heterotrophic bacteria (del Rosal *et al.* 2014). The most common colonizer was the Cyanobacterium *Chroococcidiopsis* sp., which is characterized by the ability to grow in harsh arid conditions and even in deserts (Billi and Potts 2002; Albertano 2012). It can incorporate water in its sheath and by colonizing cracks and crevices within rock it is able to provoke mechanical damage to the rocks (Billi *et al.* 2011). *Chroococcidiopsis* does not form a large biofilm but grows in spots. Other Cyanobacteria, including *Desmococcus* cf. *endolithicus*, *Geitlerinema* sp., *Leptolyngbya* spp., *Aphanothece saxicola*, *Chalicogloea cavernícola* and *Gloeocapsa atrata*, isolated from the same cave, were able to form a green biofilm in the presence of light.

In Cueva de los Murciélagos (Zuheros, Spain), green biofilms were characterized by a spatial structure very heterogeneous in thickness, density and organism composition (Fig. 12) (Roldán and Hernández-Mariné 2009). Of 17 cyanobacterial taxa identified from this site, six belonged to the order Chroococcales (*Gloeocapsa alpina*, *Gloeocapsa* sp., *Gloeothece*

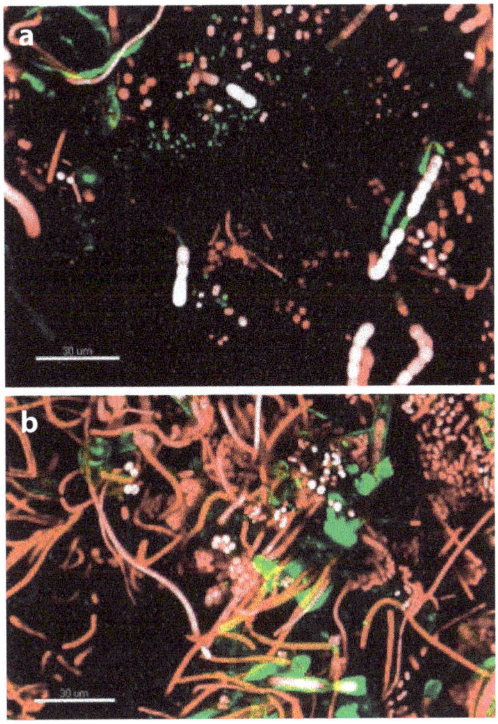

Figure 10 Images of phototrophic biofilms developing in Roman hypogea obtained with confocal laser scanning microscopy (CLSM) by using chlorophyll *a* and phycobiliproteins-based autofluorescence channels. CLSM data consisted of a set of 3D cross-sectional images in the x-y-z plane obtained with IMARIS 7.3 software (Bitplane AG, Zurich, Switzerland).

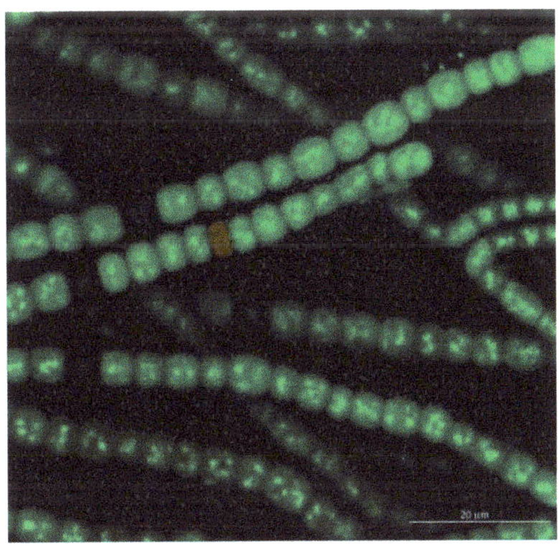

Figure 11 CLSM image of *Symphyonemopsis* sp., a heterocystous Cyanobacterium frequently isolated from phototrophic biofilms that develop on lithic faces in Roman hypogea. The sample has been stained with acridine orange for labelling nucleic acids and polysaccharides.

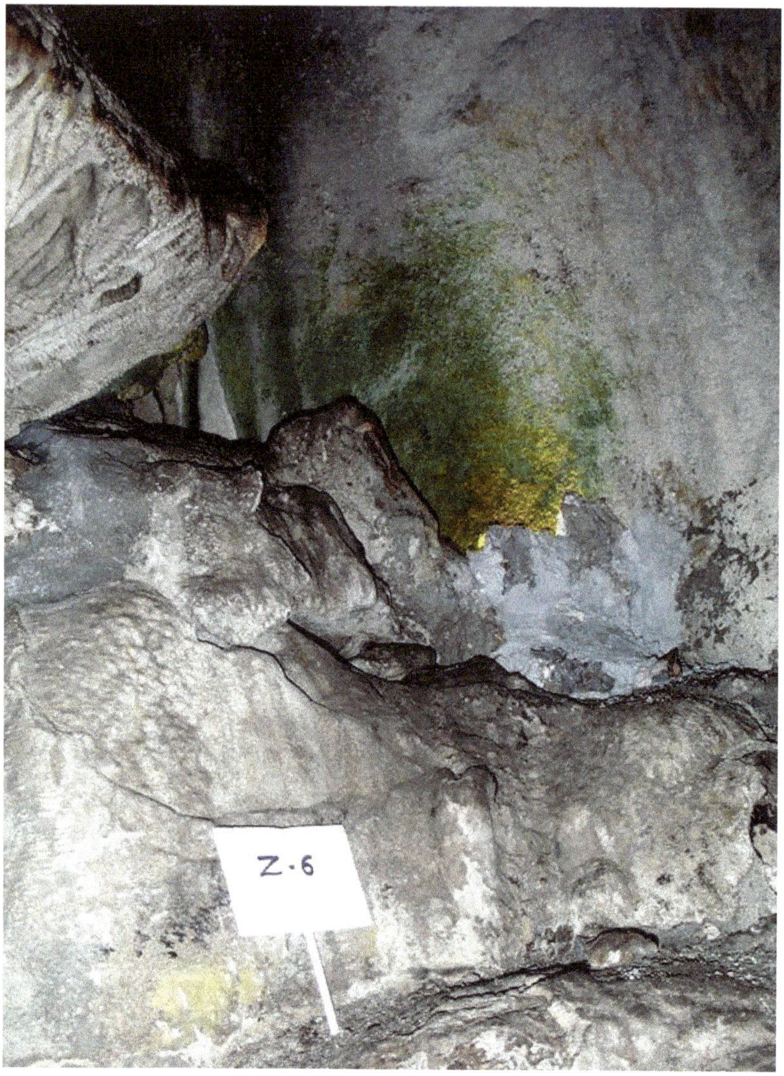

Figure 12 Green-yellow biofilm observed on illuminated rock surfaces in one area of the Cueva de los Murciélagos in Spain.

cf. *rupestris, Cyanosarcina* cf. *parthenonensis, Gloeocapsopsis* sp. and *Aphanocapsa parietina*), eight to the Oscillatoriales (*Leptolyngbya* spp., *Leptolyngbya lurida, Phormidium tenue, Phormidium* spp.), and three to the Nostocales (*Nostoc sphaericum, Tolypothrix* sp. and *Scytonema* sp.). No members belonging to the Stigonematales were found. Among the eukaryotic microalgae, 12 taxa of green coccal algae were found at most of the sampled sites. *Chlorella* species were almost always present in these communities, while *Chlorosarcinopsis* sp., *Choricystis chodatii, Coccomyxa* sp., *Klebsormidium flaccidum, Klebsormidium* sp., *Leptosira* sp., *Muriella* cf terrestris, *Myrmecia bisecta, Stichococcus bacillaris* and *Trentepohlia* sp. were found less frequently.

Among Cyanobacteria, some were particularly harmful in these hypogean sites due to their ability to precipitate calcium carbonate on their sheaths such as *Scytonema julianum* (Fig. 13) (Bruno and Albertano 1999; Albertano 2012),

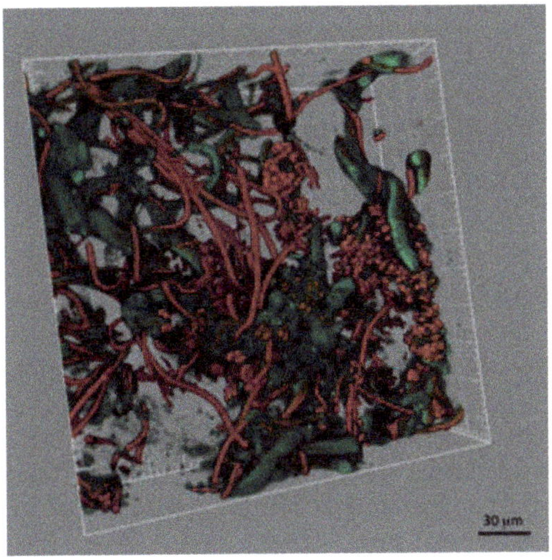

Figure 13 CLSM image of a phototrophic biofilm composed of heavy calcified filaments of *Scytonema julianum* (green) along with other Cyanobacteria belonging to the Oscillatoriales and Chroococcales. CLSM data consisted of a set of 3D cross-sectional images in the x-y-z plane obtained with IMARIS 7.3 software (Bitplane AG, Zurich, Switzerland).

Geitleria calcarea, *Herpyzonema pulverulentum* and *Loriella* sp. (Hernández-Mariné and Canals 1994; Ariño *et al.* 1997; Hernández-Mariné *et al.* 1999) and some *Leptolyngbya* and *Fischerella* spp. (Albertano 2012). An important factor for the colonization of the stone is the production and secretion of exopolymeric substances (EPS) that allows cells to stick to the substratum and to each other, forming multispecies biofilms. EPSs are polyelectrolytes of high molecular weight made by polysaccharides (up to 90%), proteins and nucleic acids that contribute to trapping water. They provide protection, are fundamental for the survival of phototrophic biofilms in a harsh environment (Rossi *et al.* 2012) and also actively participate in the weathering process of stone surfaces through the binding of solubilized minerals (De Philippis and Vincenzini 1998; Albertano and Bellezza 2001; Bellezza and Albertano 2003; Bellezza *et al.* 2005).

Among the bacteria present in Cueva de los Murciélagos there was a prevalence of Firmicutes (49.5%, mainly Bacillales) followed by Actinobacteria (41.3%) that included a variety of genera with different cell morphology, such as *Arthrobacter*, *Curtobacterium*, *Promicromonospora*, *Rhodococcus*, *Nocardia*, *Micrococcus*, *Blastococcus*, *Streptomyces* and *Kribbella*, and Proteobacteria (9.2%) with *Pseudomonas*, *Stenotrophomonas*, *Methylobacterium* and *Sphingomonas* species. Most isolated bacteria demonstrated high metabolic activity and some were attached to cyanobacterial filaments suggesting a probable role related to nitrogen assimilation following N_2 fixation by Cyanobacteria (Urzì *et al.* 2010).

FUNGI IN HYPOGEAN ENVIRONMENTS

Studies on the presence of fungi in hypogean environments demonstrated that they are relatively scarce although always present as spores (Saiz-Jimenez 2014; Docampo *et al.* 2011. Vanderwolf *et al.* (2013) reviewed research performed on the fungi, yeasts and slime moulds found in different hypogean environments around the world and listed 1029 fungal species in 518 genera mainly belonging to the Ascomycota. The authors emphasized that the origin and ecological role of fungi in caves remain relatively unknown. These fungi are mainly ubiquitous oligotrophic and psychrotrophic fungi and should always be considered harmful due to the high rate of spore dissemination and increased potential to cause serious phenomena of deterioration. Stable temperature and humidity conditions do seem favourable for the growth of fungi. Stress due to the fluctuation of microenvironmental parameters and the application of biocidal compounds may cause spore germination and rapid, intense and diffuse fungal colonization on plaster and frescoes (Caneva *et al.* 2005).

These events of dramatic fungal colonization were observed several times in Roman hypogea due to the opening and closing of the doors to allow the entrance of visitors and/or restoration activities (Florian 1996; Albertano *et al.* 2005). Similar events, described as 'fungal outbreak', were observed in the Lascaux cave in France and the Castañar Ibor cave in Spain. Another important factor limiting the growth of fungi in caves and catacombs is the natural oligotrophy of these environments. Therefore, their spread inside them depends on the allochthonous input of organic matter from outside, including visitors, or other sources such as guano, insects and roots (Saiz-Jimenez 2014).

The main type of deterioration described in caves and catacombs that are associated with fungal colonization are white and grey alterations and black stains. White alterations, formed from long, white mycelia with a fluffy appearance, have been observed in the Neolithic wall paintings of the Lascaux cave since 2001 (Dupont *et al.* 2007; Bastian *et al.* 2010) and in the Castañar Ibor cave (Jurado *et al.* 2010a). This alteration was attributed to the overgrowth of fungi identified as belonging to the *Fusarium solani* complex in the first case and *Mucor* and *Fusarium* in the second. Black stains were described as a main deterioration problem in the Lascaux cave four months after a biocide treatment of the surfaces affected by the white alteration. Their presence was attributed to an outbreak of melanized fungal species identified as *Ochroconis lascauxensis* and *O. anomala* (Martin-Sanchez *et al.* 2012).

Fungal colonization associated with grey and black alterations was observed in Cueva de los Murciélagos (also known as the Bat Cave). The few fungal species isolated were ubiquitous pigmented Hyphomycetes (*Cladosporium* spp. and *Alternaria* spp.), which is characteristic of entrance sites, while hyaline species unable to produce conidia (the so-called Mycelia Sterilia) were found mostly in the innermost area of the cave. Species in the genera *Penicillium* and *Aspergillus* had a casual distribution. Some yeasts, such as *Trichosporon* sp., were also isolated, possibly due to the presence of bat guano in the cave (Sugita *et al.* 2005). The occasional presence of fungi in the biofilms was apparently unrelated to the presence of other members of the microbial community. The organic matter available in the biological patinas may support saprophytic growth of the fungi, or they may contribute to weathering, thereby increasing the deterioration of stone surfaces (Urzì *et al.* 2010). Recently, it was hypothesized that bacteria present in the biofilm may control the growth of fungi if the conditions are more favourable to bacteria than to fungi (Urzì *et al.* 2016).

PATHOGENS

Caves are reservoirs of pathogenic microorganisms (Jurado *et al.* 2010b). A number of bacteria, fungi and viruses inhabiting caves are reported to be pathogenic to humans and animals. Similar problems could be faced in other hypogean habitats such as catacombs and tombs due to the fact that they are visited by an increasing number of tourists. Some bacterial isolates in those sites are related to pathogenic species although there is no evidence that they themselves are pathogenic. Assessment of the health risks for humans inside the catacombs (Urzì *et al.* 2014) remains unknown. In 2006, white nose syndrome (WNS) was first described as a lethal disease of bats caused by the fungus *Geomyces destructans* (Lorch *et al.* 2011). Šimonovičová *et al.* (2011) found that the same fungus was responsible for bat illness in Slovakia. For this reason, steps are being taken to avoid cross contamination due to tourists visiting, and moving between, different caves.

CONCLUSIONS

A rule among cavers is that if they see anything unusual that cannot be easily explained by geologic phenomena, it may be microbiological. Knowing that microbes can be easily and involuntarily transported from one cave to another is common knowledge among cave lovers (Barton 2006). Using common sense controls the introduction of microbes into the caves. Several recommendations are usually given to visitors, such as to stay on the flagged trails, do not touch the walls, gather up long hair during the visit, and do not leave anything in the caves (Barton 2006).

Saiz-Jimenez (2014) reported that in Europe there is a great debate concerning the conservation and management of hypogean environments. Each country has its own policy and there are no common laws at European Community level. Politics and science are often in conflict when considering these unique environments from opposite points of view. Local politicians believe that allowing tourists to visit caves and other hypogean sites can boost the local economy. On the other hand, evidence provided by scientific research shows that the uncontrolled opening of these delicate sites to a large number of visitors creates favourable conditions for the growth of numerous and varied microorganisms leading, through different mechanisms (aesthetic, mechanical, chemical), to the deterioration of valuable lithic surfaces and paintings, if present. The question that has to be raised is whether the process of biodeterioration of cultural heritage can or should be stopped under all circumstances, or whether we have to accept biodeterioration as a natural and implicit consequence of its creation (Sterflinger and Piñar 2013).

REFERENCES

Agarossi, G., Ferrari, R. and Monte, M. 1985. 'Microbial biodeterioration in the hypogea: the subterranean neopythagorean basilica of Porta Maggiore in Rome', in *5th International Congress on Deterioration and Conservation of Stone*, vol, 2, V. Furlan (ed.), 597–605. Lausanne: Presses Polytechniques Romandes.

Agarossi, G., Ferrari, R., Monte, M., Gugliandolo, C. and Maugeri, T. 1988. 'Changes of microbial system in an Etruscan tomb after biocidal treatments', in *6th International Congress on Deterioration and Conservation of Stone, Supplement*, 82–91. Torun: Nicholas Copernicus University.

Albertano, P. 2012. 'Cyanobacterial biofilms in monuments and caves', in *Ecology of Cyanobacteria II: Their Diversity in Space and Time*, B.A Whitton (ed.), 317–43. Dordrecht: Springer.

Albertano, P. and Bellezza, S. 2001. 'Cytochemistry of cyanobacterial exopolymers in biofilms from Roman hypogea', *Nova Hedwigia* 123: 501–8.

Albertano, P. and Bruno, L. 2003. 'The importance of light in the conservation of hypogean monuments', in *Molecular Biology and Cultural Heritage*, C. Saiz-Jimenez (ed.), 171–7. Lisse: Sweets and Zeitlinger.

Albertano, P. and Urzì, C. 1999. 'Structural interactions among epilithic Cyanobacteria and heterotrophic microorganisms in Roman hypogea', *Microbial Ecology* 38: 244–52.

Albertano, P., Moscone, D., Palleschi, G., Hermosin, B., Saiz-Jimenez, C., Sánchez-Moral, S. *et al.* 2003. 'Cyanobacteria attack rocks (CATS): control and preventive strategies to avoid damage caused by Cyanobacteria and associated microorganisms in Roman hypogean monuments', in *Molecular Biology and Cultural Heritage*, C. Saiz-Jimenez (ed.), 151–62. Lisse: Sweets and Zeitlinger.

Albertano, P., Urzì, C. and Caneva, G. 2005. 'Tombe, catacombe e altri ipogei', in *La biologia vegetale per i beni culturali*, vol. I, G. Caneva, M.P. Nugari and O. Salvadori (eds), 184–90. Florence: Nardini Editore.

Ariño, X., Hernández-Mariné, M. and Saiz-Jiménez, C. 1997. 'Colonization of Roman tombs by calcifying Cyanobacteria', *Phycologia* 36: 366–73.

Banerjee, S. and Joshi, S.R. 2015. 'Culturable bacteria associated with the caves of Meghalaya in India contribute to speleogenesis', *Journal of Cave and Karst Studies* 78(3):144–57.

Barton H.A. 2006. 'Introduction to cave microbiology: a review for the non-specialist', *Journal of Cave and Karst Studies* 68: 43–54.

Barton, H.A. and Jurado, V. 2007. 'What's up down there? Microbial diversity in caves', *Microbe* 2(3): 132–8.

Baskar, S., Baskar, R., Mauclaire, L. and McKenzie, J.A. 2006. 'Microbially induced calcite precipitation in culture experiments: possible origin for stalactites in Sahastradhara caves, Dehadrun, India', *Current Science India* 90: 58–64.

Bastian, F., Alabouvette, C., Jurado, V. and Saiz-Jimenez, C. 2009. 'Impact of biocide treatments on the bacterial communities of the Lascaux cave', *Naturwissenschaften* 96: 863–8.

Bastian, F., Jurado, V., Nováková, A., Alabouvette, C. and Saiz-Jimenez, C. 2010. 'The microbiology of Lascaux cave', *Microbiology* 156: 644–52.

Bellezza, S. and Albertano, P. 2003. 'A Chroococcalean species from Roman hypogean sites: characterisation of *Gloeothece membranacea* (Cyanobacteria, Synechoccaceae)', *Archiv für Hydrobiologie – Supplementbände: Algological Studies* 109: 103–12.

Bellezza, S., Albertano, P., De Philippis, R. and Paradossi, G. 2005. 'Exopolysaccharides in cyanobacterial biofilms from Roman catacombs', *Archiv für Hydrobiologie – Supplementbände: Algological Studies* 117: 117–32.

Billi, D. and Potts, M. 2002. 'Life and death of dried prokaryotes', *Research in Microbiology* 153: 7–12.

Billi, D., Viaggiu, E., Cockell, C.S., Rabbow, E., Horneck, G. and Onofri, S. 2011. 'Damage escape and repair in dried *Chroococcidiopsis* spp. from hot and cold deserts exposed to simulated space and Martian conditions', *Astrobiology* 11: 65–73.

Bonacci, O., Pipan, T. and Culver, D.C. 2009. 'A framework for karst ecohydrology', *Environmental Geology* 56: 891–900.

Bouquet, E., Boronat, A. and Ramos-Cormenzana, A. 1973. 'Production of calcite (calcium carbonate) crystals by soil bacteria is a general phenomenon', *Nature* 246: 527–9.

Bruno, L. and Albertano, P. 1999. 'Photoacclimation of sciaphilous epilithic Cyanobacteria isolated from Roman hypogea', *Archiv für Hydrobiologie – Supplementbände: Algological Studies* 94: 89–103.

Bruno, L. and Valle, V. 2017. 'Effect of white and monochromatic lights on Cyanobacteria and biofilms from Roman catacombs', *International Biodeterioration and Biodegradation* 123: 286–95.

Bruno, L., Piermarini, S. and Albertano, P. 2001. 'Characterisation of spectral emission by cyanobacterial biofilms in the Roman catacombs of Priscilla in Rome (Italy)', *Nova Hedwigia* 123: 229–36.

Bruno, L., Billi, D., Urzì, C. and Albertano, P. 2006. 'Genetic characterization of epilithic Cyanobacteria and their associated bacteria', *Geomicrobiology Journal* 23: 293–9.

Bruno, L., Billi, D., Bellezza, S. and Albertano, P. 2009. 'Cytomorphological and genetic characterization of troglophilic *Leptolyngbya* strains isolated from Roman hypogea', *Applied and Environmental Microbiology* 75: 608–17.

Bruno, L., Bellezza, S., De Leo, F. and Urzì, C. 2014. 'A study for monitoring and conservation in the Roman catacombs of St. Callistus and Domitilla, Rome (Italy)', in *The Conservation of Subterranean Cultural Heritage*, C. Saiz-Jimenez (ed.), 37–44. London: CRC Press.

Cañaveras, J.C., Sánchez-Moral, S., Sanz-Rubio, E., Bedoya, J., Soler, V., Groth, I., Schumann, P., Laiz, L., Gonzalez, I. and Saiz-Jimenez, C., 1999. 'Microbial communities associated to hydromagnesite and needle fiber aragonite deposits in a karstic cave (Altamira, northern Spain)', *Geomicrobiology Journal* 16: 9–25.

Cañaveras, J.C., Sánchez-Moral, S., Soler, V. and Saiz-Jimenez, C. 2001. 'Microorganisms and microbially induced fabrics in cave walls', *Geomicrobiology Journal* 18: 223–40.

Caneva, G., Nugari, M.P. and Salvadori, O. 2005. *La biologia vegetale per i beni culturali*, vol I. Florence: Nardini Editore.

Castanier, S., Le Metayer-Levrel, G. and Perthuisot, J.P. 1999. 'Ca-carbonates precipitation and limestone genesis: the microbiogeologist point of view', *Sedimentary Geology* 126: 9–23.

Chen, Y., Wu, L., Boden, R., Hillebrand, A., Kumaresan, D., Moussard, H., Baciu, M., Lu, Y. and Murrell, J.C. 2009. 'Life without light: microbial diversity and evidence of sulfur- and ammonium-based chemolithotrophy in Movile cave', *ISME Journal* 3: 1093–104.

Cuezva, S., Fernandez-Cortes, A., Porca, E., Pašić, L., Jurado, V., Hernández-Mariné, M., Serrano-Ortiz, P., Hermosin, B., Cañaveras, J.C., Sánchez-Moral, S. and Saiz-Jimenez, C. 2012. 'The biogeochemical role of Actinobacteria in Altamira cave, Spain', *FEMS Microbiology Ecology* 81: 281–90.

De Leo, F., Iero, A., Zammit, G. and Urzì, C.E. 2012. 'Chemoorganotrophic bacteria isolated from biodeteriorated surfaces in cave and catacombs', *International Journal of Speleology* 41(2): 125–36.

De Philippis R. and Vincenzini, M. 1998. 'Exocellular polysaccharides from Cyanobacteria and their possible applications', *FEMS Microbiology Reviews* 22: 151–75.

del Rosal, Y., Liñan, C. and Hernández-Mariné, M. 2014. 'The conservation of the Nerja cave: preserving anthropogenic impact in a tourist cave', in *The Conservation of Subterranean Cultural Heritage*, C. Saiz-Jimenez (ed.), 193–206. London: CRC Press.

Diaz-Herraiz, M., Jurado, V., Cuezva, S., Laiz, L., Pallecchi, P., Tiano, P., Sánchez-Moral, S. and Saiz-Jimenez, C. 2013. 'The actinobacterial colonization of Etruscan paintings', *Scientific Reports* 3:1440, doi: 10.1038/srep01440.

Diaz-Herraiz, M., Jurado, V., Cuezva, S., Laiz, L., Pallecchi, P., Tiano, P., Sánchez-Moral, S. and Saiz-Jimenez, C. 2014. 'Deterioration of an Etruscan tomb by bacteria from the order Rhizobiales', *Scientific Reports* 4: 3610, doi: 10.1038/srep03610.

Docampo, S., Trigo, M.M., Recio, M., Melgar, M., García-Sánchez, J. and Cabezudo, B. 2011. 'Fungal spore content of the atmosphere of the cave of Nerja (southern Spain): diversity and origin', *Science of the Total Environment* 409(4): 835–43.

Dupont, J., Jacquet, C., Dennetiere, B., Lacoste, S., Bousta, F., Orial, G., Cruaud, C., Couloux, A. and Roquebert, M.F. 2007. 'Invasion of the French paleolithic painted cave of Lascaux by members of the *Fusarium solani* species complex', *Mycologia* 99(4): 526–33.

Ehrlich, H.L. 1996. 'How microbes influence mineral growth and dissolution', *Chemical Geology* 132: 5–9.

Engel, A.S. 2007. 'Observations on the biodiversity of sulfidic karst habitats', *Journal of Cave and Karst Studies* 69: 187–206.

Engel, A.S. 2010 'Microbial diversity of cave ecosystems', in *Geomicrobiology: Molecular and Environmental Perspective*, L. Barton, M. Mandl and A. Loy (eds), 437. Dordrecht: Springer.

Everest, E.J., Curtis, S.M., De Leo, F. and Urzì, C. 2015. 'Description of *Kribbella italica* sp. nov. isolated from a Roman catacomb', *International Journal of Systematic and Evolutionary Microbiology* 65: 491–6.

Fiocchi Nicolai, V. 1999. 'The origin and development of Roman catacombs', in *The Christian Catacombs of Rome: History, Decoration, Inscriptions*, V. Fiocchi Nicolai, F. Bisconti and D. Mazzoleni (eds), 208. Regensburg: Verlag Schnell and Steiner GmbH.

Florian, M.L.H. 1996. 'The role of the conidia of fungi in fox spots', *Studies in Conservation* 41(2): 65–75.

Groth, I. and Saiz-Jimenez, C. 1999. 'Actinomycetes in hypogean environments', *Geomicrobiology Journal* 16: 1–8.

Hernández-Mariné, M. and Canals, T. 1994. '*Herpyzonema pulverulentum* (Mastigocladaceae): a new cavernicolous atmophytic and lime-incrusted cyanophyte', *Archiv für Hydrobiologie – Supplementbände: Algological Studies* 75: 123–36.

Hernández-Mariné, M., Asencio, A.D., Canals, A., Ariño, X., Aboal, M. and Hoffmann, L. 1999. 'Discovery of populations of the lime-encrusting genus *Loriella* (Stigonematales) in Spanish caves', *Archiv für Hydrobiologie – Supplementbände: Algological Studies* 94: 121–38.

Hurst, J.F. 1897. *History of the Christian Church. 1. Eaton and Mains*. New York.

Jurado, V., Porca, E., Cuezva, S., Fernandez-Cortes, A., Sánchez-Moral, S. and Saiz-Jimenez, C. 2010a 'Fungal outbreak in a show cave', *Science of the Total Environment* 408: 3632–8.

Jurado, V., Laiz, L., Rodriguez-Nava, V., Boiron, P., Hermosin, H., Sánchez-Moral, S. and Saiz-Jimenez, C. 2010b. 'Pathogenic and opportunistic microorganisms in caves', *International Journal of Speleology* 39(1): 15–24.

Komárek, J. 2013. *Cyanoprokaryota 3. Teil/3rd part: Heterocytous Genera*. Süßwasserflora von Mitteleuropa, Springer SpeKtrum.

Kraková, L., De Leo, F., Bruno, L., Pangallo, D. and Urzì, C. 2015. 'Complex bacterial diversity in the white biofilms of St. Callistus catacombs in Rome evidenced by different investigation strategies', *Environmental Microbiology* 17(5): 1738–52.

Laiz, L., Groth I., Gonzalez, I. and Saiz-Jimenez, C.1999. 'Microbiological study of the dripping waters in Altamira cave (Santillana del Mar, Spain)', *Journal of Microbiological Methods* 36(1): 129–38.

Laiz, L., Miller, A.Z., Jurado, V., Akatova, E., Sánchez-Moral, S., Gonzalez, J.M., Dionísio, A., Macedo, M.F. and Saiz-Jimenez, C. 2009. 'Isolation of five strains from biodeteriorated monuments', *Naturwissenschaften* 96(1): 71–9.

Lorch, J.M., Meteyer, C.U., Behr, M.J., Boyles, J.G., Cryan, P.M., Hicks, A.C., Ballmann, A.E., Coleman, J.T.H., Redel, D.N., Reeder, D.M. and Blehert, D.S. 2011. 'Experimental infection of bats with *Geomyces destructans* causes white-nose syndrome', *Nature* 480(7377): 376–8.

Martin-Sanchez, P.M., Nováková, A., Bastian, F., Alabouvette, C. and Saiz-Jimenez, C. 2012. 'The case of the Lascaux cave in France', *Environmental Science & Technology* 46: 3762–70.

Miller, A.Z., Hernández-Mariné, M., Jurado, V., Dionísio, A., Barquinha, P., Fortunato, E., Afonso, M.J., Chaminé, H.I. and Saiz-Jimenez, C. 2012. 'Enigmatic reticulated filaments in subsurface granite', *Environmental Microbiology Reports* 4(6): 596–603.

Miller, A.Z., Dionísio, A., Jurado, V., Cuezva, S., Sánchez-Moral, S., Cañaveras, J.C. and Saiz-Jimenez, C. 2013. 'Biomineralization by cave dwelling microorganisms', in *Advances in Geochemistry Research*, J. Sanjurjo Sanchéz (ed.), 77–105. New York: Nova Science Publishers.

Moñino, D.I., Jurado, V., Candelera, R. and Hermosin, B. 2014. 'Human impact on show caves: chewing gum stuck to the walls', in *The Conservation of Subterranean Cultural Heritage*, C. Saiz-Jimenez (ed.), 247–52. London: CRC Press.

Phoenix, V.R., Konhauser, K.O., Adams, D.G. and Bottrell, S.H. 2001. 'Role of biomineralization as an ultraviolet shield: implications for archean life', *Geology* 29: 823–6.

Pike, A.W.G., Hoffmann, D.L., García Diez, M., Pettitt, P.B., Alcolea, J., De Balbín, R., González-Sainz, C., de las Heras, C., Lasheras, J.A., Montes, R. and Zilhão, J. 2012. 'U-series dating of Paleolithic art in 11 caves in Spain', *Science* 336(6087): 1409–13.

Roldán, M. and Hernández-Mariné, M. 2009. 'Exploring the secrets of the three-dimensional architecture of phototrophic biofilms in caves', *International Journal of Speleology* 38: 41–53.

Rossi, F., Micheletti, E., Bruno, L., Adhikary, S.P., Albertano, P. and De Philippis, R. 2012. 'Characteristics and role of the exocellular polysaccharides produced by five Cyanobacteria isolated from phototrophic biofilms growing on stone monuments', *Biofouling* 28(2): 215–24.

Saarela, M., Alakomi, H.L., Suihko, M.L., Maunuksela, L., Raaska, L. and Mattila-Sandholm, T. 2004. 'Heterotrophic microorganisms in air and biofilm samples from Roman catacombs, with special emphasis on Actinobacteria and fungi', *International Biodeterioration and Biodegradation* 54: 27–37.

Saiz-Jimenez, C. 2014. *The Conservation of Subterranean Cultural Heritage*. London: CRC Press.

Sánchez-Moral, S., Bedoya, J., Luque, L., Cañaveras, J.C., Jurado, V., Laiz, L. and Saiz- Jimenez, C. 2003a. 'Biomineralization of different crystalline phases by bacteria isolated from catacombs', in *Molecular Biology and Cultural Heritage*, C. Saiz-Jimenez (ed.), 179–85. Lisse: Sweets and Zeitlinger.

Sánchez-Moral, S., Cañaveras, J.C., Laiz, L., Saiz-Jimenez, C., Bedoya, J. and Luque, L. 2003b. 'Biomediated precipitation of calcium carbonate metastable phases in hypogean environments', *Geomicrobiology Journal* 20: 491–500.

Sánchez-Moral, S., Luque, L., Cañaveras, J.C., Laiz, L., Jurado, V., Hermosin, B. and Saiz-Jimenez C. 2004. 'Bioinduced barium precipitation in St. Callixtus and Domitilla catacombs', *Annals in Microbiology* 54(1): 1–12.

Sánchez-Moral, S., Luque, L., Cañaveras, J.C., Soler, V., Garcia-Guinea, J. and Aparicio, A. 2005a. 'Lime pozzolana mortars in Roman catacombs: composition, structures and restoration', *Concrete and Cement Research* 35: 1555–65.

Sánchez-Moral, S., Luque, L., Cuezva, S., Soler, V., Benavente, D., Laiz, L., Gonzalez, J.M. and Saiz-Jimenez, C. 2005b. 'Deterioration of building materials in Roman catacombs: the influence of visitors', *Science of the Total Environment* 349: 260–76.

Sánchez-Roman, M., Rivadeneyra, M.A., Vasconcelos, C. and McKenzie, J.A. 2007. 'Biomineralization of carbonate and phosphate by moderately halophilic bacteria', *FEMS Microbiology Ecology* 61: 273–84.

Šimonovičová, A., Pangallo, D., Chovanová, K. and Lehotská, B. 2011. '*Geomyces destructans* associated with bat disease WNS detected in Slovakia', *Biologia* 66(3): 562–4.

Sprocati, A.R., Alisi, C., Tasso, F., Vedovato, E., Barbabietola, N. and Cremisini, C. 2008. 'A microbiological survey of the Etruscan Mercareccia Tomb (Italy): contribution of microorganisms to deterioration and restoration', in *The 9th International Conference on NDT of Art, Jerusalem, Israel.*

Sterflinger, K. and Piñar, G. 2013. 'Microbial deterioration of cultural heritage and works of art: tilting at windmills?', *Applied Microbiology and Biotechnology* 97: 9637–46.

Sugita, T., Kikuchi, K., Makimura, K., Urata, K., Someya, T., Kamei, K., Niimi, M. and Uehara, Y. 2005. '*Trichosporon* species isolated from guano samples obtained from bat-inhabited caves in Japan', *Applied and Environmental Microbiology* 71: 7626–9.

Urzì, C., Garcia-Valles, M., Vendrell, M. and Pernice, A. 1999. 'Biomineralization processes on rock and monument surfaces observed in field and in laboratory conditions', *Geomicrobiology Journal* 16: 39–54.

Urzì, C., De Leo, F. and Schumann, P. 2008. '*Kribbella catacumbae* sp. nov. and *Kribbella sancticallisti* sp. nov., isolated from whitish-grey patinas in the catacombs of St. Callistus in Rome, Italy', *International Journal of Systematic and Evolutionary Microbiology* 58: 2090–97.

Urzì, C., De Leo, F., Bruno, L. and Albertano, P. 2010. 'Microbial diversity in Paleolithic caves: a study case on the phototrophic biofilms of the Cave of Bats (Zuheros, Spain)', *Microbial Ecology* 60: 116–29.

Urzì, C., De Leo, F., Bruno, L., Pangallo, D. and Kraková, L. 2014. 'New species description, biomineralization processes and biocleaning applications of Roman catacombs-living bacteria', in *The Conservation of Subterranean Cultural Heritage*, C. Saiz-Jimenez (ed.), 65–72. London: CRC Press.

Urzì, C., De Leo, F., Kraková, L., Pangallo, D. and Bruno, L. 2016. 'Effects of biocide treatments on the biofilm community in Domitilla's catacombs in Rome', *Science of the Total Environment* 572: 1252–62.

Vanderwolf, K.J., Malloch, D., McAlpine, D.F. and Forbes, G.J. 2013. 'A world review of fungi, yeasts, and slime molds in caves', *International Journal of Speleology* 42(1): 77–96.

Yantian, M., Zhang, H., Du, Y., Tian, T., Xiang, T., Liu, X., Wu, F. An, L., Wang, W., Gu, J. and Feng, H. 2015. 'The community distribution of bacteria and fungi on ancient wall paintings of the Mogao Grottoes', *Scientific Reports* 5: 7752, doi: 10.1038/srep07752.

Zammit, G., Sánchez-Moral, S. and Albertano, P. 2011. 'Bacterially mediated mineralisation processes lead to biodeterioration of artworks in Maltese catacombs', *Science of the Total Environment* 409: 2773–82.

Zammit, G., Billi, D. and Albertano, P. 2012. 'The subaerophytic Cyanobacterium *Oculatella subterranea* (Oscillatoriales, Cyanophyceae) gen. et sp. nov.: a cytomorphological and molecular description', *European Journal of Phycology* 47(4): 341e354.

Zimmermann, J., Gonzalez, J.M. and Saiz-Jimenez, C. 2006. 'Epilithic biofilms in Saint Callixtus catacombs (Rome) harbour a broad spectrum of Acidobacteria', *Antonie Van Leeuwenhoek* 89: 203–8.

AUTHORS' ADDRESSES

- Clara Urzì, Department ChiBioFarAM, University of Messina, Viale F. Stagno d'Alcontres 98166 Messina, Italy (urzicl@unime.it)
- Laura Bruno, Department of Biology, University of Rome 'Tor Vergata', Via Cracovia 1 00133, Rome, Italy
- Filomena De Leo, Department ChiBioFarAM, Viale F. Stagno d'Alcontres 98166, University of Messina, Messina, Italy (fdeleo@unime.it)

LIMESTONE BIODETERIORATION: EXAMPLES FROM PORTUGAL[1]

A. Catarina Pinheiro, Nuno Mesquita and António Portugal

INTRODUCTION

The history of men, art and stone is forever intertwined. For millennia, stone has been used not only to protect us but also as a support for our initial artistic impressions of the world – first as a rudimentary canvas and later as an almost pliable material providing the basis for our sites of worship. Despite its eminence and apparent sturdiness, one of stone's biggest threats comes from the microscopic world. Susceptible to the actions of algae, Cyanobacteria, bacteria, fungi and lichens, each with their own set of adaptive traits, stone does not escape the changes imposed by these small invaders and, in time, these may render it difficult for future generations to study, learn and appreciate the works of art and built heritage made from it.

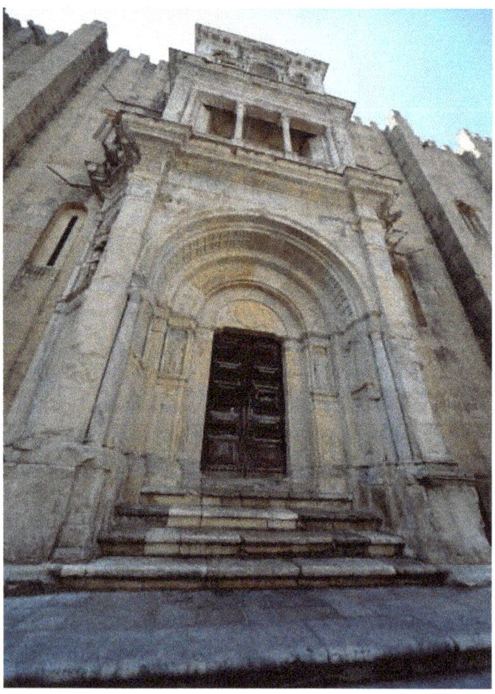

Figure 1 Limestone and dolostone were both used for the Porta Especiosa, a Renaissance-style portal on the north facade of the Old Cathedral in Coimbra.

From all the possible types of stone – and there are many – carbonate rock such as limestone is more prone to biodeterioration (Miller *et al.* 2006, 2010) revealing a higher intrinsic bioreceptivity. In order to actively protect this rock, it is vital to understand not only the specific characteristics which make limestone particularly vulnerable to biodeterioration but also the microbial communities that thrive in them.

LIMESTONE: TYPES, FEATURES AND FRAILTIES

Different types of stone display different inherent stone bioreceptivities (i.e. susceptibility to colonization by living organisms) (Guillitte 1995). Composed mostly of calcite and aragonite, limestone can also contain quartz, sand, chert and clay (Anthony *et al.* 1990–2003). Dolomitic limestone (also known as dolomite and dolostone) depends on the relative presence of the mineral dolomite (Fig. 1).

Limestone has long been favoured for both building and ornamenting cultural heritage structures. Its softness and ease of handling is accompanied by a brightness that makes it irresistible to artists and architects. However, these same attributes also render it more susceptible to biodeterioration agents. In an extensive review of monuments, statues and historic buildings subjected to biological contamination assessment (Macedo *et al.* 2009), it was reported that most were built of limestone (32%) and marble (30%), with the lithotypes travertine (7%) and dolomite (2%) less represented. In Portugal, this type of stone is abundant and can be found in different colours according to the region from where it is extracted (Fig. 2).

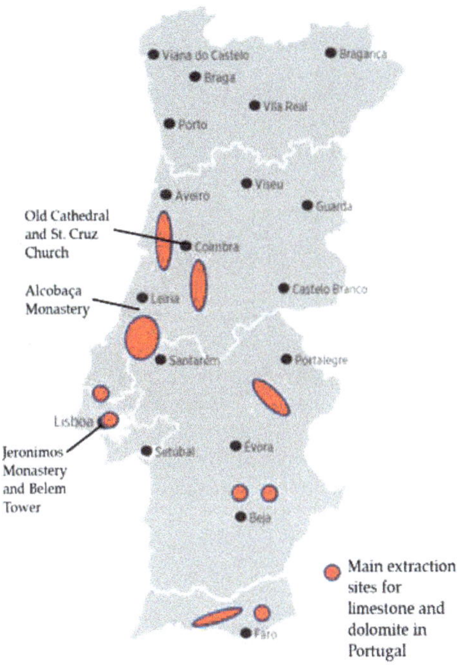

Figure 2 Map showing the main extraction sites for limestone and dolostone in Portugal together with the locations of some of the monuments for which these stones were used (Manupella *et al.* 1981; Carvalho *et al.* 2001).

The focus here are the three main limestones found in Portugal: Portunhos (from which Ançã is a very characteristic subtype), Lioz and the aforementioned dolomite. Portunhos, named after the region from where it is extracted, is pure, homogenous and mostly composed of $CaCO_3$ (96.5%). The colour ranges from white to grey and due to its low compressive strength and hardness, it is ideal for highly detailed carvings. It was used extensively for Portuguese monuments (Miller *et al.* 2010) and its quality was also appreciated abroad – the facade of the Royal Hospital in Santiago de Compostela in Spain features a magnificent portal made from Ançã limestone. Lioz limestone is extracted from the northern region of Lisbon and can vary in colour between whitish-yellow and bluish-grey (Carvalho *et al.* 2001). It is comprised of pure calcite and has a coarser grain. Due to its inherent hardness, it is frequently used for outdoor applications. The fossil content that is characteristic of this limestone makes it aesthetically relevant for decorative purposes.

In addition to $CaCO_3$, dolomitic limestone or dolostone is also composed of the mineral dolomite, a calcium magnesium carbonate ($CaMg(CO_3)_2$) mineral. This stone's colours range from greyish-light brown to a yellowish tone (Manupella *et al.* 1981; Carvalho *et al.* 2001) according to the proportion of magnesium and iron contents. A yellow-toned dolomitic stone (low magnesium and high iron content) has been used extensively in the region of Coimbra (Manupella *et al.* 1981; Aires-Barros and Picciochi Alves 1987). Figure 3 shows examples of the use of these types of limestone for Portuguese-built cultural heritage.

The physical and chemical features of Portuguese limestones have been studied extensively (Manupella *et al.* 1981; Carvalho *et al.* 2001; Figueiredo *et al.* 2007, 2010; Miller *et al.* 2012). In terms of resistance to compression and/or flexion and water absorption rate, Lioz limestone presents the best characteristics and hence is more able to withstand harsh meteorological conditions, followed by dolomite and Portunhos (Ançã) limestone (Carvalho *et al.* 2001). Regarding bioreceptivity, however, other factors must be taken into consideration such as pH, porosity, texture, roughness, moisture, density, and the chemical composition of the surface layer (Guillitte 1995; Miller *et al.* 2012). Given these parameters, Portunhos stone seems

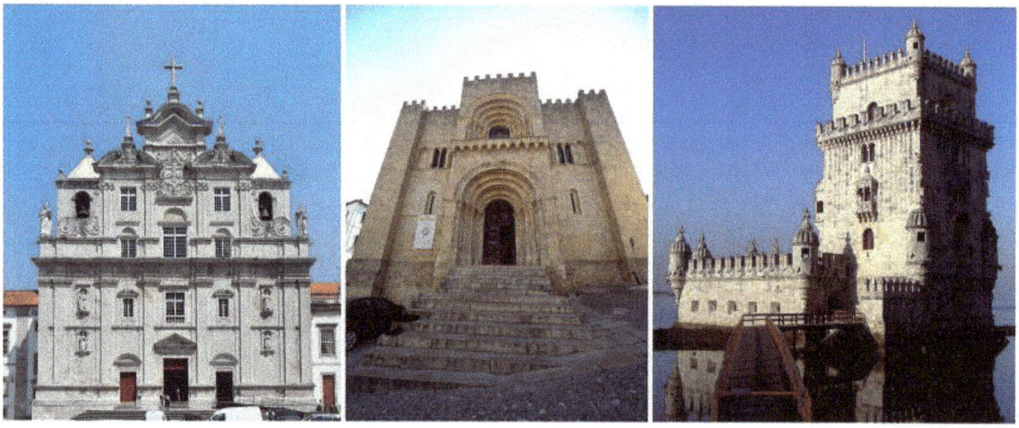

Figure 3 Ançã-Portunhos and dolomitic limestone were frequently used in the Coimbra area. Examples include the New Cathedral, constructed of Ançã-Portunhos (left) and the Old Cathedral, built using Ançã-Portunhos and dolomite (centre). Lioz limestone was used for the Belém Tower (right) in Lisbon.

to be the most susceptible to biodeterioration (Miller *et al.* 2006).

These intrinsic factors, coupled with extrinsic environmental factors, determine the establishment of a biological community. In fact, having described the stone, the next step is to assess the biological contribution to stone decay by determining the exposure conditions for the entire object or building, including water and nutrient presence (Warscheid and Braams 2000). Prior conservation/restoration treatments and the current conservation efforts are relevant factors that influence the type of communities found in limestone (and other stone types). This strong dependence on external factors, such as exposure to light and water availability, explains why (on the same monument and lythotype) it is possible for different locations to display different colonizing populations (Hallmann *et al.* 2013; Li *et al.* 2016). Moderate climates with regular rainfall support a mixed consortium of microorganisms on exposed stone surfaces, whereas in semi-arid areas that experience less rain and higher temperatures, the growth of more specialized microorganisms such as Cyanobacteria, black yeasts and lichens is more common (Scheerer 2008). Knowledge of the stone type and its location are good indicators for predicting susceptibility to colonization by microorganisms.

COLONIZING AGENTS IN PORTUGUESE LIMESTONES

Microalgae, bacteria, Cyanobacteria and fungi colonize worldwide. The different metabolisms and survival strategies of microbes translate into different damage inflicted on the stone on which they thrive: biofilm formation, discoloration, salting, physical damage and the production of osmolytes and organic acids (Scheerer *et al.* 2009) are just some of the effects and consequences. Several reviews have elaborated on the relationship between microorganisms and stone monuments (Griffin *et al.* 1991; Warscheid and Braams 2000; Gaylarde *et al.* 2003; Crispim and Gaylarde 2005; McNamara and Mitchell 2005; Macedo *et al.* 2009; Gaylarde *et al.* 2012; Keshari and Adhikary 2014) and the available studies on limestone report on the consistent populations comprising some of these groups (frequently fungi, algae and lichens).

In Portugal, there is little information on the relationship between stone and microbes; there are no studies addressing the total biodiversity on limestone-built heritage. Figure 4 summarizes the results gathered from published case studies concerning the biological contamination of Portuguese limestone monuments and/or limestone-based architectural features placed inside monuments (Ascaso *et al.* 1998, 2002;

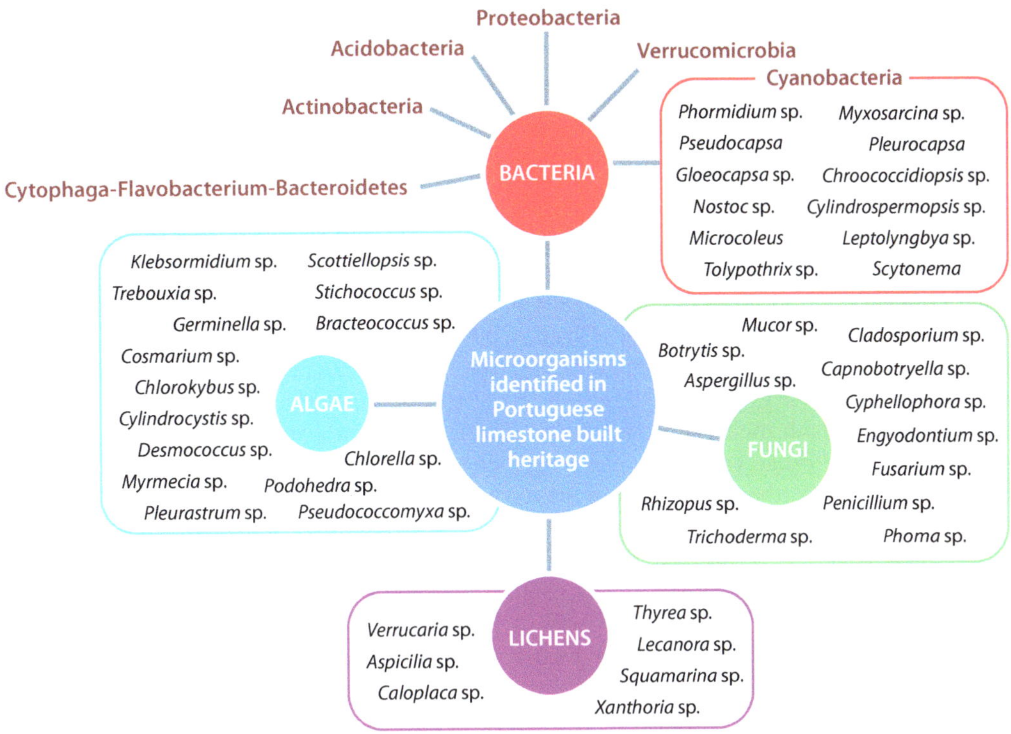

Figure 4 Microorganisms identified in Portuguese limestone built heritage. These data were collected from Santa Cruz church, the altarpiece from Sé de Guarda, Santa Clara-a-Velha monastery, Ajuda National Palace, Belém Tower, Jerónimos monastery and the Convent of Christ, Tomar.

Girbal *et al.* 2002; Miller *et al.* 2006, 2008, 2009; Macedo *et al.* 2009; Miller 2010; Mateus *et al.* 2013).

All microorganisms depicted in Figure 4 exert some sort of mechanical and chemical stress on the stone material (Garcia-Rowe and Saiz-Jimenez 1991; Gaylarde *et al.* 2003; Crispim and Gaylarde 2005; Dakal and Cameotra 2012) and they typically thrive as a community. Photolithoautotrophic organisms, such as algae and Cyanobacteria, may play an important role as primary colonizers, providing support for the posterior settlement of other organisms. However, despite their synergic nature, it is worth examining each group individually and focusing on the main agents of deterioration and their effects on limestone.

Bacteria

Generally, this large domain of prokaryotic microorganisms can discolour, disrupt, weaken and dissolve a wide variety of materials (Gaylarde *et al.* 2003). There are many ways to classify this large group of microorganisms, one of the most popular being the means by which they obtain their energy and carbon. Some are photolithoautotrophic (e.g. Cyanobacteria) and use carbon from the atmosphere and sunlight for energy while releasing oxygen, whereas chemolithoautotrophic bacteria utilize the oxidation of inorganic compounds for energy and carbon from the atmosphere. Nitrous acids (e.g. *Nitrosomonas* spp.), nitric acid (e.g. *Nitrobacter* spp.) and sulfuric acid (e.g. *Thiobacillus* spp.) are released as a result. This group also includes elements that are capable of growing mixotrophically using organic nutrients for the synthesis of cell components (chemolitho-mixotroph). Chemoorganotrophic bacteria (and fungi) use organic substrates and release complexing corrosive organic acids or weaken the mineral lattice by the oxidation of metal cations such as iron or manganese (Warscheid and Braams 2000). More recent studies have stressed the importance of these bacteria and fungi, together with photoautotrophs, as the primary microbial colonizers of archaeological stone (Warscheid and Braams 2000). Given the above, it is clear that one of the main risks bacteria pose to stone in general, and limestone in particular, relates to their acidic metabolic byproducts, which cause corrosion. The pH decrease induced by these metabolites has already been identified as a tool to predict and prevent biodeterioration (Skipper and Skipper 2014).

Most studies focusing on Portuguese limestones have pinpointed the general presence of bacteria, but lack a thorough phylogenetic identification. The phyla or groups found in Portuguese limestone are presented in Figure 4: Cyanobacteria, Proteobacteria, Acidobacteria, Actinobacteria, Bacteroidetes (aka Cytophaga-Flavobacterium-Bacteroides groups) (McNamara and Mitchell 2005) and Verrucomicrobia, identified in Portunhos limestone (Miller *et al.* 2008, 2009, 2010).

Within the comprehensive group of bacteria are the Cyanobacteria, present in almost every study performed on biodiversity in Portuguese limestone. The most frequently identified Cyanobacteria belong to the genera *Gloeocapsa* and *Phormidium*, which are considered ubiquitous in the Mediterranean basin (Macedo *et al.* 2009). Among the survival skills displayed by Cyanobacteria is the ability to survive under alternating cycles of drying and rehydration by means of a gelatinous sheath that acts as a water reservoir and also increases their adhesion to the substrate (Miller 2010). Many of the species from *Gloeocapsa, Phormidium, Scytonema* and *Microcoleus* share this ability (Miller 2010). This sheath is sometimes pigmented which protects them from ultraviolet radiation and causes colour alterations (Dakal and Cameotra 2012). The genera *Nostoc, Tolypothrix* and *Gloeocapsa* are known for their ability to produce extracellular polymeric substances that facilitate their attachment and survival in an inhospitable environment (Coutinho *et al.* 2013). This polymeric matrix also traps pollutants from the surrounding environment (Warscheid and Braams 2000).

Scytonema, Gloeocapsa and *Pleurocapsa* are capable of causing discoloration and degradation of the stone structure (Crispim *et al.* 2006). The presence of *Chroococcidiopsis* in the Ajuda National Palace located in Lisbon is relevant since this is the most desiccation-resistant Cyanobacterium known (Scheerer *et al.* 2009) and may indicate endolithic growth (Miller 2010). Moreover, the presence of nitrogen-fixing Cyanobacteria such as *Cylindrospermopsis, Nostoc* and *Pleurocapsa* is relevant for the establishment and development of other organisms such as heterotrophic bacteria (Miller 2010). More data are needed to support the results obtained by the presented studies. Understanding the relationship between Cyanobacteria in this particular setting and whether some of them are specific to this stone type is necessary since this correlation has not yet been determined (Macedo *et al.* 2009).

The Proteobacteria encompass a large number of organisms with different nutritional pathways. Alphaproteobacteria are generally phototrophic while Beta- and Deltaproteobacteria are either nitrifying or sulfur- and iron-reducing bacteria. Sulfur-reducing bacteria and Actinomycetes are frequently associated with stone deterioration (Scheerer *et al.* 2009; Dakal and Cameotra 2012). Due to the important deterioration role of Proteobacteria, further and more in-depth studies are necessary on Portuguese stone and limestone in particular. Also in need of further investigation is the Acidobacteria group, which has been identified on stone both at a national site (Ajuda National Palace) and throughout the world (Scheerer *et al.* 2009).

Members of the Actinobacteria inhabit stone more effectively than most of the single-celled bacteria because of their filamentous growth and effective use of various nitrogen and carbon sources (Saarela *et al.* 2004). Being heat resistant, Actinobacteria are to be expected in the temperate climate of Portugal (Scheerer *et al.* 2009) and have been identified as colonizers in Portuguese dolomite (in the Santa Clara-a-Velha monastery in Coimbra for example). Actinobacteria of the genera *Streptomyces*, *Micromonospora* and *Micropolyspora* exhibit a high prevalence on decayed stone (Scheerer *et al.* 2009) and *Rubrobacter* was shown to promote biofilm formation and hyphae penetration, discoloration and stone deterioration (Laiz *et al.* 2009; Scheerer *et al.* 2009; Dakal and Cameotra 2012; Jurado *et al.* 2012). Recently, a new species of *Rubrobacter* was identified in a Portuguese stone monument (Jurado *et al.* 2012), reinforcing the need to perform such studies in limestone and other stones used in cultural heritage. *Rubrobacter* has been held accountable for the rosy discoloration of wall paintings and stones, especially in monuments where the formation of salt efflorescence is common (Schabereiter-Gurtner *et al.* 2001; Ettenauer *et al.* 2010).

Spore-forming bacteria are also expected in dry/temperate locations because of their increased resistance to certain environmental conditions. This is the case of the genus *Bacillus*, a common colonizer of stone buildings that has demonstrated an ability to hasten stone degradation (Laiz *et al.* 2003; Scheerer *et al.* 2009; Piñar *et al.* 2014a; Skipper and Skipper 2014). However, no reference exists on the presence of this genus in Portuguese limestones, highlighting the need to further address this subject.

The phylum Bacteroidetes is a comprehensive group of bacteria and its presence in stone has been associated with halophilic species (Piñar *et al.* 2014b). In Portugal, this phylum was present in the Santa Clara-a-Velha monastery (Miller *et al.* 2008, 2009; Miller 2010). Not represented in Figure 4 is the Archaea, probably due to the difficulty in isolating them in pure culture. Archaea have been largely overlooked (Ettenauer *et al.* 2010) and underestimated in most molecular studies that make use of universal primers commonly used in molecular biology protocols optimized to study bacteria (Piñar *et al.* 2001; Ettenauer *et al.* 2010; Piñar *et al.* 2014b). In this domain, however, there are potentially relevant halophilic and alkaliphilic groups (McNamara and Mitchell 2005), which might find their perfect habitat in stones with abundant salt efflorescences.

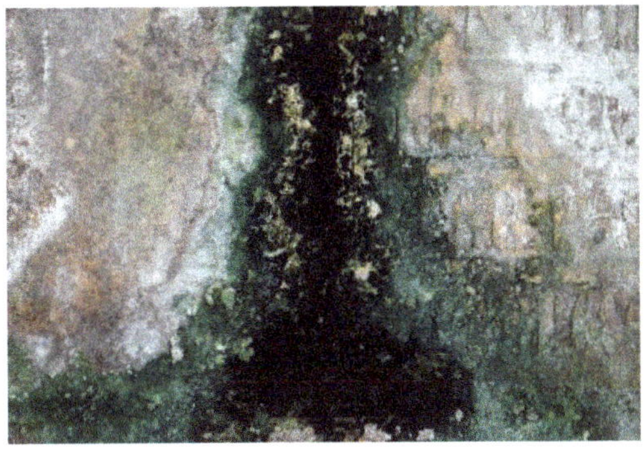

Figure 5 Algae growing on a stone substrate where water is readily available. They form powdery or viscous layers and, as seen here, are responsible for dramatic chromatic changes and eventually stone corrosion.

Algae

Algae are a diverse group of eukaryotic organisms that contain pigments such as chlorophyll, carotenoids and xanthophyll (Jayakumar 2009) and are responsible for the highly visible green and black stains as shown in Figure 5 (Macedo *et al.* 2009; Sterflinger and Piñar 2013; Dakal and Cameotra 2012).

Algae share common ground with Cyanobacteria in that their acidic metabolites corrode limestone and marble. In addition, algae accumulate sulfur, calcium and water inside their cells, which causes them to enlarge and place added pressure on existing pores (Vijayakumar 2014). They also share the capacity for photosynthesis, which means they can establish a pioneer community on limestone (Dakal and Cameotra 2012). Algae are able to form an extracellular matrix of proteins and sugar polymers (biofilm), increasing self-adherence and protecting themselves from desiccation (Wakefield and Jones 1998; Scheerer *et al.* 2009; Skipper and Skipper 2014) while easing the settlement of heterotrophic organisms.

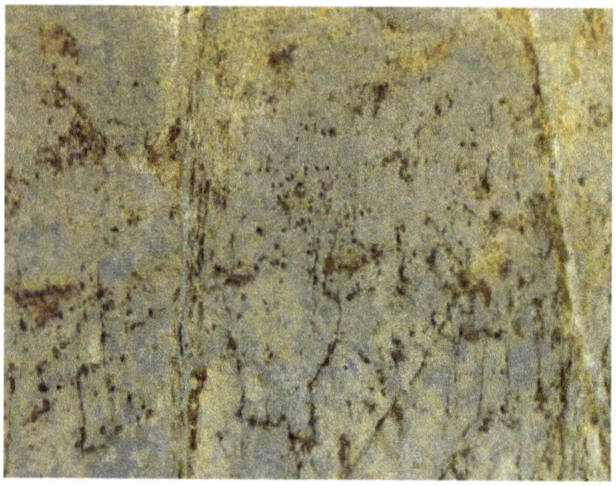

Figure 6 Biopitting results in the formation of point-like millimetric cavities (image taken in the Old Cathedral in Coimbra, Portugal).

In line with international results (Macedo *et al.* 2009), the most commonly encountered algae are *Chlorella* sp., *Myrmecia* sp., *Stichococcus* sp. and *Trebouxia* sp. Considered a major threat in other stone monuments (Dakal and Cameotra 2012), *Cosmarium* sp. was identified in the studies performed at Santa Cruz church in Coimbra. *Klebsormidium flaccidum*, considered responsible for the formation of green and black sulfated crusts and biofilm in the cathedral of Seville, was also present in Portuguese limestone (Macedo *et al.* 2009). *Trebouxia* sp. should also be mentioned since these microalgae could be involved in the lichenization process as it occurs in almost 20% of all lichens and has rarely been found in free-living form (Miller 2010).

Fungi

Among the heterotrophs present in a stone-associated community are fungi, a group characterized by filamentous hyphae. Fungi display erosive abilities capable of causing chemical, physical and mechanical deterioration (Sterflinger and Piñar 2013). As with other microbes, organic and inorganic acid production is the suspect mechanism since it allows fungi to weaken and dissolve the stone matrix. This effect may, however, vary between fungal species and different mineral substrates (Gadd 2007). Chelating agents bind reactive mineral constituents, such as manganese and iron (Burford *et al.* 2003; Gadd 2007), and further promote biodeterioration (Gaylarde *et al.* 2003; Dakal and Cameotra 2012). Species in the genera *Botrytis*, along with *Mucor*, *Penicillium* and *Trichoderm*, have demonstrated an ability to solubilize calcium, magnesium and zinc silicates (Gadd 2007).

From the data collected internationally (Gadd 2007), limestone appears to display less diversity in terms of fungal colonization, at least when compared to granite, marble or sandstone. *Aspergillus*, *Engyodontium*, *Fusarium* and *Penicillium* species are among the most commonly identified. Gadd points out that these records are probably underestimated, highlighting the need for further studies. The type of fungus present in stone is related to the surrounding climate. The common airborne fungi (*Alternaria*, *Aspergillus*, *Cladosporium*, *Epicoccum*, *Aureobasidium*, *Phoma* and *Ulocladium* species) are found in humid and moderate climates and almost all are also encountered in Portuguese limestone (see Fig. 4). When water and nutrient availability is reduced, black yeasts and microcolonial fungi, such as *Capnobotryella* spp., start to appear (Sterflinger and Piñar 2013). While Hyphomycetes establish their hyphal network through the stone pores (Scheerer *et al.* 2009; Sterflinger and Piñar 2013) black yeasts and microcolonial fungi become deeply embedded in the stone matrix, resulting in biopitting (Fig. 6) (Sterflinger and Piñar 2013).

In recent studies, dematiaceous fungi with meristematic and yeast-like growth patterns have been sharing the spotlight with lichens and Cyanobacteria as the most successful inhabitants of marble, limestone, granite and other rock types in arid and semi-arid environments (Harutyunyan *et al.* 2008). In the past these fungi were overlooked but are now raising concern since they can cover an entire stone facade with a thick black layer (Cuzman and Olmi 2013; Sterflinger and Piñar 2013; Onofri *et al.* 2014).

Aspergillus and *Penicillium* include numerous species, and their taxonomy and chemo-diversity is currently under intense study. For example, *A. fumigatus* produces a green pigment, oxalic acid, and can be damaging to limestone, while *A. niger*, also an oxalic acid producer, is well known to degrade olivine, dunite, serpentine, muscovite, feldspar, spodumene, kaolin and nepheline (Sterflinger 2000). *A. niger* however, like many rock-dwelling dark pigmented fungi (such as *P. simplissimum* and *Scopulariopsis brevicaulis*), can actively penetrate limestone and produce pits of up to 2 cm in diameter on rock surfaces (Sterflinger 2000; Gadd 2007; Scheerer *et al.* 2009). Furthermore, the genus *Cladosporium* has been associated with damage in stone monuments as well as on the reinforcement substances used to protect them (Dakal and Cameotra 2012). Together with *Penicillium*, *Fusarium*, *Phoma* and *Trichoderma*, this genus is also associated with acid secretion and bioweathering (Dakal and Cameotra 2012). *C. sphaerospermum* is a relevant melanin producer (Sterflinger 2000), as are *Alternaria alternata*, *Aureobasidium pullulans*, *Phoma glomerulata* and *Ulocladium chartarum* among others (Sterflinger 2000). Besides inducing discoloration, melanin is believed to play a major role in the capacity of fungi as biosorbents (Sterflinger 2000). Due to these intrinsic characteristics, it is relevant to fully identify the fungal community present in order to determine how best to manage degradation.

Lichens

Until recently, lichens were believed to be the result of two organisms combined: alga (mainly green alga) or a Cyanobacterium, and a fungus (mainly an Ascomycete), coexisting in a symbiotic relationship. However, recent studies indicate the possible presence of a third element in this relationship, as non-filamentous Ascomycota (yeasts) have been identified and contribute to some of the features displayed by the lichens (Spribille *et al.* 2016). Since this is a very recent discovery, there are still no data on whether this particular trio may also be present in limestone. Lichens are especially resilient in situations of extreme temperatures and lack of available water, a protection guaranteed by the heterotrophic (fungal) element, which also delivers minerals it can leach from the stone. In return, the fungus uses the organic nutrients produced by the alga or Cyanobacterium through photosynthesis (Warscheid and Braams 2000).

To get inside the stone and guarantee better adherence, the hyphae from the fungal component of lichens promote the formation of cracks and fissures that lead to structural and physical damage. Ultimately, this can result in the peeling and detachment of the upper mineral layer (Scheerer *et al.* 2009; Dakal and Cameotra 2012). The effects are particularly apparent on limestone where the lichen thallus may even fully integrate into the stone matrix (Lisci *et al.* 2003); these are termed endolithic lichens. The epilithic lichens are classified as crustose, foliose or fruticose, depending on the way these attach to the stone (Garcia-Rowe and Saiz-Jimenez 1991). Most of the lichens found in Portuguese limestone monuments are crustose – these, along with the endolithic type, are also very capable of chemical and physical weathering (Garcia-Rowe and Saiz-Jimenez 1991) as their removal is particularly difficult due to the strong physical connection to the substrate (Lisci *et al.* 2003; Scheerer *et al.* 2009; Pinna 2014). Lichens also produce organic acids and chelating compounds through which they form complexes with the mineral cations of the substratum (Lisci *et al.* 2003), further increasing stone susceptibility to biodeterioration. This mineral uptake perpetuates the pitting effects of lichens on stone long after they die (Dakal and Cameotra 2012). Microcolonial fungi and black yeasts are often found in close physical proximity with lichens (Sterflinger and Piñar 2013).

In Portugal, the study of colonization by lichens has been performed on Lioz limestone at the Jerónimos monastery and Belém Tower heritage sites and on unspecified limestone at the Convent of Christ in Tomar. In this latter location, microbial identification was not pursued.

Different conditions stimulate the presence of different lichens, even on the same monument and for the same genera (Lisci *et al.* 2003). Various *Caloplaca* spp. for instance, prefer either shaded or sun-exposed areas as well as locations where nitrates are readily available (Lisci *et al.* 2003). Nitrogen-fixation abilities further enhance the lichen's weathering potential (Garcia-Rowe and Saiz-Jimenez 1991; Lisci *et al.* 2003; Dakal and Cameotra 2012). *Caloplaca* sp., generally of the crustose type (Garcia-Rowe and Saiz-Jimenez 1991), was identified in both locations studied for the presence of lichens: Belém Tower and Jerónimos monastery. *Xanthoria* sp. was also found in the Belém Tower where the species *X. parietina* displayed a foliose-type thallus and a preference for sunlit surfaces (Garcia-Rowe and Saiz-Jimenez 1991). As mentioned for *Caloplaca*, the genus *Lecanora* also shows differences according to species: *L. muralis* has been demonstrated to induce mechanical decay while *L. dispersa* appeared harmless (Lisci *et al.* 2003). The genera *Aspicilia*, *Thyrea* and *Verrucaria*, all predominantly crustose in nature, have been identified on stone (Garcia-Rowe and Saiz-Jimenez 1991; Scheerer *et al.* 2009).

CONCLUSIONS

Numerous agents, some more complex than others, can cause biodeterioration of Portuguese limestone. The phototrophic algae and Cyanobacteria are likely to initiate a complex community where heterotrophs can establish due to the conditions maintained by their predecessors: the build-up of photosynthetic biomass provides an excellent organic nutrient base for subsequent heterotrophic microflora (Warscheid and Braams 2000). The establishment of a heterotrophic community, however, is possible even without the participation of phototrophs (Scheerer *et al.* 2009) since byproducts of natural or synthetic substances applied during restoration treatments can act as a carbon source (Wakefield and Jones 1998; Crispim and Gaylarde 2005; Scheerer *et al.* 2009; Skipper and Skipper 2014).

In siliceous sandstones, granite, basalt, slate, limestone and metamorphic rocks (gneiss, quartzite, marble), when the size of the most abundant grain is less than 0.1 mm, there is an Hg porosimetry of ~14% and poor water penetration. As a result, the formed biofilm is usually superficial or found along cracks and fissures in the stone, and is mostly dominated by phototrophic agents and fungi. The initial effects of biodeterioration are the discoloration caused by biogenic pigments and the oxidation of iron and manganese. The extracellular polymeric substances contained in this biofilm accumulate atmospheric particles, causing the formation of thin-skinned scales which is followed by biopitting from the presence of organic acids (Warscheid and Braams 2000).

The potential for damage increases even further as communities evolve and become ever more complex. The presence of lichens, and after these, Bryophytes, leads to mechanical and chemical damage to stone, which then gives rise to invasion by vascular plants, whose roots further increase the damage to the structure of buildings (Garcia-Rowe and Saiz-Jimenez 1991). Studying, monitoring and acting (when necessary and advisable) on the microbiological communities present in our stone legacy is vital to better address this natural succession and its impending threats.

Data obtained across the globe, especially when achieved with modern molecular biology protocols, coupled with a wide selection of microbiological media, point to biological diversity that has still not been encountered in Portuguese limestone monuments. The Portuguese panorama is, in this sense, still obscure. Very few monuments have been studied and the results do not provide a full picture of the biological communities that are present. Further and more thorough analysis is warranted.

The actions that microbial communities employ when colonizing stone can be very harmful for the conservation of historic monuments. To effectively repair or prevent the damage potentially inflicted by these microorganisms it is first necessary to identify and understand the biological organisms, their modus operandi, the interactions between them and other established communities, and the stone on which they thrive. After this comprehensive task has been performed, countermeasures can be initiated to reduce the impact and prevent further damage. Reducing the water exposure and nutrient availability is a key aspect. Biocides can be used, but with caution, and never before testing the compatibility between the biocide and the stone (Warscheid and Braams 2000). In addition, drastically reducing the population of a given microorganism may have a severe impact on the existing community. Artificial selection can clear the way for one genera or species to take over and in some cases, this is especially damaging (Martin-Sanchez *et al.* 2012). Finally, the nutritive function of organic biocides demands a careful use of these chemicals in the conservation practice (Warscheid and Braams 2000).

ACKNOWLEDGEMENTS

The authors acknowledge financial support by FEDER - Fundo Europeu de Desenvolvimento Regional funds through the COMPETE 2020 - Operational Programme for Competitiveness and Internationalisation (POCI), and by Portuguese funds through FCT - Fundação para a Ciência e a Tecnologia in the framework of the project POCI-01-0145-FEDER-PTDC/EPH-PAT/3345/2014, POCH - Programa Operacional Capital Humano (co-funding by the European Social Fund and national funding by MCTES), Post-Doc Research grant (SFRH/BPD/112830/2015), Project IN0756 - INV.EXPLORATORIA - IF/01061/2014 and contract reference IF/01061/2014. Photographs were kindly provided by Miguel Mesquita.

NOTE

1. Other contributors include: J. Trovão, F. Soares, I. Tiago, C. Coelho and H. Paiva de Carvalho (all from the Centre for Functional Ecology, University of Coimbra, Coimbra, Portugal), F. Gil (from the Centre for Physics of the University of Coimbra and the UI&D Molecular Chemical-Physics UC QFM-UC), L. Catarino (from the Geosciences Centre, University of Coimbra) and G. Piñar (University of Natural Resources and Life Sciences (BOKU), Vienna, Austria.

REFERENCES

Aires-Barros, L. and Picciochi Alves, L.M. 1987. 'Estudo sobre o estado de alteração de alguns monumentos portugueses', *Boletim da Sociedade Portuguesa de Química* 28(II): 59–63.

Anthony, J.W., Bideaux, R.A., Kenneth, W.B. and Nichols, M.C. (eds) 1990–2003. *Handbook of Mineralogy*, 5 vols. Chantilly VA: Mineralogical Society of America. Available at: http://www.handbookofmineralogy.org/ (accessed 7 June 2018).

Ascaso, C., Wierzchos, J., Rodrigues, J.D., Aires-Barros, L., Henriques, F.M.A. and Charola, A.E. 1998. 'Endolithic microorganisms in the biodeterioration of the Tower of Belem', *Internationale Zeitschrift für Bauinstandsetzen* 4: 627–40.

Ascaso, C., Wierzchos J., Souza-Egipsy, V., de los Rios, A. and Rodrigues, J.D. 2002. '*In situ* evaluation of the biodeteriorating action of microorganisms and the effects of biocides on carbonate rock of the Jerónimos monastery (Lisbon)', *International Biodeterioration and Biodegradation* 49(1): 1–12.

Burford, E.P., Kierans, M. and Gadd, G.M. 2003. 'Geomycology: fungi in mineral substrata', *Mycologist* 17(3): 98–107.

Carvalho, J., Manuppella, G. and Moura, A.C. 2001. 'Calcários ornamentais portugueses', *Boletim de Minas do Instituto Geológico e Mineiro* 37(4): 223–32.

Coutinho, M.L., Miller, A.Z., Gutierrez-Patricio, S., Hernández-Mariné, M., Gomez-Bolea, A., Rogerio-Candelera, M.A., Philips, A.J.L., Jurado, V., Saiz-Jimenez, C. and Macedo, M.F. 2013. 'Microbial communities on deteriorated artistic tiles from Pena National Palace (Sintra, Portugal)', *International Biodeterioration and Biodegradation* 84: 322–32.

Crispim, C.A. and Gaylarde, C.C. 2005. 'Cyanobacteria and biodeterioration of cultural heritage: a review', *Microbial Ecology* 49(1): 1–9.

Crispim, C.A., Gaylarde, P.M., Gaylarde, C.C. and Neilan, B.A. 2006. 'Deteriogenic Cyanobacteria on historic buildings in Brazil detected by culture and molecular techniques', *International Biodeterioration and Biodegradation* 57(4): 239–43.

Cuzman, O. and Olmi, R. 2013. 'Preliminary study on controlling black fungi dwelling on stone monuments by using a microwave heating system', *International Journal of Conservation Science* 4(2): 133–44.

Dakal, T. and Cameotra, S. 2012. 'Microbially induced deterioration of architectural heritages: routes and mechanisms involved', *Environmental Sciences Europe* 24: 36.

Ettenauer, J., Sterflinger, K. and Piñar, G. 2010. 'Cultivation and molecular monitoring of halophilic microorganisms inhabiting an extreme environment presented by a salt-attacked monument', *International Journal of Astrobiology* 9(1): 59–72.

Figueiredo, C.A.M., Aires-Barros, L., Basto, M.J., Graca, R.C. and Mauricio, A. 2007. 'The weathering and weatherability of Basilica da Estrela stone, Lisbon, Portugal', in *Building Stone Decay: From Diagnosis to Conservation*, R. Přikryl and B.J. Smith (eds), Special Publication 271, 99–107. London: Geological Society.

Figueiredo, C., Aires-Barros, L. and Neto, M.J. 2010. 'The church of Santa Engracia (the National Pantheon, Lisbon, Portugal): building campaigns, conservation works, stones and pathologies', in *Limestone in the Built Environment: Present-Day Challenge for the Preservation of the Past*, B.J. Smith, M. Gomez-Heras, H.A. Viles and J. Cassar (eds), Special Publication 331, 183–93. London: Geological Society.

Gadd, G.M. 2007. 'Geomycology: biogeochemical transformations of rocks, minerals, metals and radionuclides by fungi, bioweathering and bioremediation', *Mycological Research* 111(1): 3–49.

Garcia-Rowe, J. and Saiz-Jimenez, C. 1991. 'Lichens and Bryophytes as agents of deterioration of building materials in Spanish cathedrals', *International Biodeterioration* 28(1–4): 151–63.

Gaylarde, C.C., Ribas Silva, M. and Warscheid, T. 2003. 'Microbial impact on building materials: an overview', *Materials and Structures* 36(5): 342–52.

Gaylarde, C.C., Gaylarde, P.M. and Neilan, B.A. 2012. 'Endolithic phototrophs in built and natural stone', *Current Microbiology* 65(2): 183–8.

Girbal, J., Iglesias, M., Prada, J.L., Pugés, M. and Rocabayera, R. 2002. 'Estudio de las patologias del retablo del altar mayor de la Sé da Guarda', *Estudos do Patrimonio* 3: 70–78.

Griffin, P.S., Indictor, N. and Koestler, R.J. 1991. 'The biodeterioration of stone: a review of deterioration mechanisms, conservation case histories, and treatment', *International Biodeterioration* 28(1–4): 187–207.

Guillitte, O. 1995. 'Bioreceptivity: a new concept for building ecology studies', *Science of the Total Environment* 167(1–3): 215–20.

Hallmann, C., Stannek, L., Fritzlar, D., Hause-Reitner, D., Friedl, T. and Hoppert, M. 2013. 'Molecular diversity of phototrophic biofilms on building stone', *FEMS Microbiology Ecology* 84(2): 355–72.

Harutyunyan, S., Muggia, L. and Grube, M. 2008. 'Black fungi in lichens from seasonally arid habitats', *Studies in Mycology* 61: 83–90.

Jayakumar, S. 2009. *Studies on the Biodeterioration of Concrete by Marine Algae*, PhD dissertation, School of Engineering, Pondicherry University, India.

Jurado, V., Miller, A.Z., Alias-Villegas, C., Laiz, L. and Saiz-Jimenez, C. 2012. '*Rubrobacter bracarensis* sp. nov.: a novel member of the genus *Rubrobacter* isolated from a biodeteriorated monument', *Systematic and Applied Microbiology* 35(5): 306–9.

Keshari, N. and Adhikary, S.P. 2014. 'Diversity of Cyanobacteria on stone monuments and building facades of India and their phylogenetic analysis', *International Biodeterioration and Biodegradation* 90: 45–51.

Laiz, L., Piñar, G., Lubitz, W. and Saiz-Jimenez, C. 2003. 'Monitoring the colonization of monuments by bacteria: cultivation versus molecular methods', *Environmental Microbiology* 5(1): 72–4.

Laiz, L., Miller, A.Z., Jurado, V., Akatova, E., Sanchez-Moral, S., Gonzalez, J.M., Dionísio, A., Macedo, M.F. and Saiz-Jimenez, C. 2009. 'Isolation of five *Rubrobacter* strains from biodeteriorated monuments', *Naturwissenschaften* 96(1): 71–9.

Li, Q., Zhang, B., He, Z. and Yang, X. 2016. 'Distribution and diversity of bacteria and fungi colonization in stone monuments analyzed by high-throughput sequencing', *PLoS One* 11(9): 1–17.

Lisci, M., Monte, M. and Pacini, E. 2003. 'Lichens and higher plants on stone: a review', *International Biodeterioration and Biodegradation* 51: 1–17.

Macedo, M.F., Miller, A.Z., Dionísio, A. and Saiz-Jimenez, C. 2009. 'Biodiversity of Cyanobacteria and green algae on monuments in the Mediterranean Basin: an overview', *Microbiology* 155(11): 3476–90.

Manupella, G., Moreira, J.C. and Romão, M.L. 1981. 'Panorama dos dolomitos e calcários dolomíticos Portugueses', *Boletim de Minas do Instituto Geológico e Mineiro* 4: 3–15.

Martin-Sanchez, P.M., Nováková, A., Bastian, F., Alabouvette, C. and Saiz-Jimenez, C. 2012. 'Use of biocides for the control of fungal outbreaks in subterranean environments: the case of the Lascaux cave in France', *Environmental Science and Technology* 46: 3762–70.

Mateus, D.M.R., Silva, R.B., Costa, F.M.C. and Coroado, J.P.F. 2013. 'Diversidade microbiológica do edifício da Sacristia Incompleta do Convento de Cristo, em Tomar, e avaliação do seu controlo por biocidas', *Conservar Património* 17: 11–20.

McNamara, C.J. and Mitchell, R. 2005. 'Microbial deterioration of historic stone', *Frontiers in Ecology and the Environment* 3(8): 445–51.

Miller, A.Z. 2010. *Primary Bioreceptivity of Limestones from the Mediterranean Basin to Phototrophic Microorganisms*, PhD dissertation, New University of Lisbon.

Miller, A., Dionísio, A. and Macedo, M.F. 2006. 'Primary bioreceptivity: a comparative study of different Portuguese lithotypes', *International Biodeterioration and Biodegradation* 57(2): 136–42.

Miller, A.Z., Laiz, L., Gonzalez, J.M., Dionísio, A., Macedo, M.F. and Saiz-Jimenez, C. 2008. 'Reproducing stone monument photosynthetic-based colonization under laboratory conditions', *Science of the Total Environment* 405(1–3): 278–85.

Miller, A.Z., Laiz, L., Dionísio, A., Macedo, M.F. and Saiz-Jimenez, C. 2009. 'Growth of phototrophic biofilms from limestone monuments under laboratory conditions', *International Biodeterioration and Biodegradation* 63(7): 860–67.

Miller, A.Z., Leal, N., Laiz, L., Rogerio-Candelera, M.A., Silva, R.J.C., Dionísio, A., Macedo, M.F. and Saiz-Jimenez, C. 2010. 'Primary bioreceptivity of limestones used in southern European monuments', in *Limestone in the Built Environment: Present-Day Challenge for the Preservation of the Past*, B.J. Smith, M. Gomez-Heras, H.A. Viles and J. Cassar (eds), Special Publication 331, 79–92. London: Geological Society.

Miller, A.Z., Sanmartín, P., Pereira-Pardo, L., Dionísio, A., Saiz-Jimenez, C., Macedo, M.F. and Prieto, B. 2012. 'Bioreceptivity of building stones: a review', *Science of the Total Environment* 426: 1–12.

Onofri, S., Zucconi, L., Isola, D. and Selbmann L. 2014. 'Rock-inhabiting fungi and their role in deterioration of stone monuments in the Mediterranean area', *Plant Biosystems – An International Journal Dealing with all Aspects of Plant Biology: Official Journal of the Societa Botanica Italiana* 148(2): 384–91.

Piñar, G., Saiz-Jimenez, C., Schabereiter-Gurtner, Blanco-Varela, M.T., Lubitz, W. and Rölleke, S. 2001. 'Archaeal communities in two disparate deteriorated ancient wall paintings: detection, identification and temporal monitoring by denaturing gradient gel electrophoresis', *FEMS Microbiology Ecology* 37(1): 45–54.

Piñar, G., Sterflinger, K. and Ettenauer, J. 2014a. 'La vie en rose: a review of the rosy discoloration of subsurface monuments', in *The Conservation of Subterranean Cultural Heritage*, C. Saiz-Jimenez (ed.), 113–24. London: CRC Press.

Piñar, G., Kraková, L., Pangallo, D., Piombino-Mascali, D., Maixner, F., Zink, A. and Sterflinger, K. 2014b. 'Halophilic bacteria are colonizing the exhibition areas of the Capuchin catacombs in Palermo, Italy', *Extremophiles* 18: 677–91.

Pinna, D. 2014. 'Biofilms and lichens on stone monuments: do they damage or protect?', *Frontiers in Microbiology* 5: 133–6.

Saarela, M., Alakomi, H.L., Suihko, M.L., Maunuksela, L., Raaska, L. and Mattila-Sandholm, T. 2004. 'Heterotrophic microorganisms in air and biofilm samples from Roman catacombs, with special emphasis on Actinobacteria and fungi', *International Biodeterioration and Biodegradation* 54: 27–37.

Schabereiter-Gurtner, C., Piñar, G., Vybiral, D., Lubitz, W. and Rölleke, S. 2001. '*Rubrobacter*-related bacteria associated with rosy discolouration of masonry and lime wall paintings', *Archives of Microbiology* 176(5): 347–54.

Scheerer, S. 2008. *Microbial Biodeterioration of Outdoor Stone Monuments: Assessment Methods and Control Strategies*, PhD dissertation, University of Cardiff.

Scheerer, S., Ortega Morales, B.O. and Gaylarde, C. 2009. 'Microbial deterioration of stone monuments: an updated overview', *Advances in Applied Microbiology* 66: 97–139.

Skipper, P.J.A. and Skipper, L.K. 2014. 'A survey of bacterial colonisation of historic limestone buildings: Lincoln Cathedral and St. Peter at Gowts, United Kingdom', in *Proceedings of the International Conference on Preservation, Maintenance and Rehabilitation of Historic Buildings and Structures*, R. Amoêda, S. Lira and C. Pinheiro (eds), 1003–12. Tomar: Green Lines Institute.

Spribille, T., Tuovinen, V., Resl, P., Vanderpool, D., Wolinski, H., Aime, M.C., Schneider, K., Stabentheiner, E., Toome-Heller, M., Thor G., Mayrhofer, H., Johannesson, H. and McCutcheon, J.P. 2016. 'Basidiomycete yeasts in the cortex of ascomycete macrolichens', *Science* 353(6298): 488–92.

Sterflinger, K. 2000. 'Fungi as geologic agents', *Geomicrobiology Journal* 17(2): 97–124.

Sterflinger, K. and Piñar, G. 2013. 'Microbial deterioration of cultural heritage and works of art: tilting at windmills?', *Applied Microbiology and Biotechnology* 97(22): 9637–46.

Vijayakumar, S. 2014. 'Role of Cyanobacteria in the biodeterioration of historical monuments: a review', *BMR Microbiology* 1(1): 1–13.

Wakefield, R.D. and Jones, M.S. 1998. 'An introduction to stone colonizing micro-organisms and biodeterioration of building stone', *Journal of Engineering Geology* 31(4): 369–73.

Warscheid, T. and Braams, J. 2000. 'Biodeterioration of stone: a review', *International Biodeterioration and Biodegradation* 46(4): 343–68.

AUTHORS' ADDRESSES

- Catarina Pinheiro, Centre for Functional Ecology (CFE), Department of Life Sciences, University of Coimbra, Calçada Martim de Freitas, 3000-456 Coimbra, Portugal (catmarpin@gmail.com)
- Nuno Mesquita, Centre for Functional Ecology (CFE), Department of Life Sciences, University of Coimbra, Calçada Martim de Freitas, 3000-456 Coimbra, Portugal (nuno@plasticsounds.com)
- António Portugal, Centre for Functional Ecology (CFE), Department of Life Sciences, University of Coimbra, Calçada Martim de Freitas, 3000-456 Coimbra, Portugal (aportuga@bot.uc.pt)

THE REMOVAL OF BIOLOGICAL MATERIAL FROM CALCAREOUS STONE MONUMENTS[1]

Margaret Breuker and Joannie Bottkol

INTRODUCTION

> The American Institute for Conservation of Historic and Artistic Works (AIC) is dedicated to the preservation of the material evidence of our past so that we can learn from it today and understand it in the future. Paragraph III of the AIC Code of Ethics states, 'While recognizing the right of society to make appropriate and respectful use of cultural property, the conservation professional shall serve as an advocate for the preservation of cultural property.'[2]

As conservators, we are bound by our code of ethics not only to preserve cultural property, but also to educate ourselves, our colleagues and our communities as to the most effective ways of pursuing preservation. Although preservation frequently demands active interventions, such as repairing a damaged object or cleaning a monument on-site, occasionally it calls for passive intervention, which might entail changing the environment or conditions around an object or site in order to prevent further deterioration. Any intervention, no matter the type, requires investigation. This almost always includes research into degradation processes, appropriate materials for cleaning or repair, along with asking ourselves for which reasons, or to whom or how, the object or site in question is significant. An important way in which conservators fulfill their mission of preserving the material evidence of the past and cultural resources for the future is through the careful and thoughtful maintenance of historic monuments and sites. This often includes cleaning marble and limestone monuments suffering from the disfiguring and destabilizing effects of biological growth.

Certainly there are circumstances under which preservation or conservation interventions are not appropriate. For example, many of the site-specific sculptures and land artworks by artist Andy Goldsworthy, fabricated from natural materials like rock, stones, sticks, ice and mud, are intended to be consumed by nature and to weather with time. Left to the elements, his configurations of materials come apart and break down. A 2016 book review in *Architectural Digest* states, 'his pieces are designed to disappear as nature takes its course: Ice melts, wind blows and rain falls, factors that shape how viewers experience Goldsworthy's constructions over the course of their temporary life spans' (Funderberg 2016). Goldsworthy's works explore the relationship between time and nature, lifecycles and temporality. Interventions that would hamper the decay of works designed to be altered by nature and time would be antithetical to their intended narrative. But in the case of most monuments and sacred sites, the intent of the artist or the memorialized, and the wishes of those invested in the site or sculpture, is to have them preserved.

With these goals for, and exceptions to, preservation in mind, we ask: how does the biological colonization of calcareous stone generate instability or make our monuments less readable? Where do we find value in monuments? Why do we choose to preserve them? What happens when we allow monuments and sites to go untended? Of course, the answers to these questions are myriad and nuanced, but perhaps this paper can prompt new or further discussion on these complicated issues.

SCIENTIFIC CONSIDERATIONS

Factors affecting when and how a stone develops microflora and weathers biogenically are diverse and may include pollution types and levels, exposure to light and moisture, temperature fluctuations, or geographic location. Research has shown that, in general, the main determining factor behind microbial colonization is the overall climate within which a stone is located (Scheerer *et al.* 2009). In the relatively moderate climate of the Northeastern United States, however, climatic extremes are not the main drivers of biological degradation so much as the air pollution present in this somewhat densely populated region, which supplies the vital nutrients required for the development of a mixed consortium of microflora (de la Torre *et al.* 1993). Urban environments tend to be rich in both nitrogen dioxide (NO_2) and sulfur dioxide (SO_2), and these pollutants are transported to stone surfaces through wind and rain (Saiz-Jimenez 1993). Bacteria have been shown to obtain their energy through the oxidation of sulfur and nitrogen compounds, transforming those compounds into sulfuric and nitric acids capable of degrading calcareous stones, including marble and limestone (Mitchell and Gu 2000). The sulfuric and nitric acids produced by organisms that colonize calcareous stones weaken the stones' surfaces by dissolving the binding minerals between the grains (Tiano 1998). Lichens are particular offenders, extracting minerals directly from the stone surfaces upon which they reside into their thalli for metabolic purposes, leaving crater-like areas of erosion after they die or are removed.

While ample information concerning the general mechanisms of microbial acid production and acid dissolution of minerals is available, not as much is known about the less visible but perhaps equally corrosive effects of mineral-dissolving biofilm components, specifically extracellular polymeric substances (EPS), on stones. Organisms living on limestone isolated from archaeological sites in Cyprus and Mexico have shown that these exopolysaccharides can induce the same effects as acids, causing similar mineral weathering and leaving stones weak through intergranular disintegration (Perry *et al.* 2005b).

Once established on a stone, biofilms serve as effective nutrient-catchers for additional species of microflora, encouraging new organisms to colonize the stone surface. As these colonies develop, they can form highly complex communities resulting in symbiotic relationships between both epilithic and endolithic organisms. The surface texture of a biological patina provides the ideal capture mechanism for holding moisture, hydrocarbons and airborne particulates to a stone's surface (Kemmling *et al.* 2004). These communities then work together to retain moisture and create physical avenues within the stone by way of fungal hyphae, leading to the deposition of more airborne and hydrocarbon particulates (Ciferri 2000). Subsequent physical spalling can occur, obscuring inscriptions, reliefs and decorative elements, and leading to the loss of valuable information integral to the understanding of the site, monument or object.

Another type of damage that ensues with microbial colonization is biological staining, which can be significantly disfiguring in monuments, especially those made of pale or white calcareous stone. The black or red pigments that can appear on pale calcareous stone with biofilms are likely exudates produced by fungi to protect themselves from environmental stresses (Perry *et al.* 2005a). Furthermore, the fungi appear to only produce the pigment when calcium is present, as was true in the case of the red staining on the US Naval Academy's Tripoli Monument at Annapolis (Maryland), in which the two pigment-producing fungi isolated from the monument, *Epicoccum nigrum* and *Drechslera* sp., seemed to only produce a red pigment in the presence of calcium carbonate stone (Perry *et al.* 2005b). These pigments are carried through the fungal hyphae shaft in times of stress, and are exuded and entrapped within the polysaccharides, making this type of bio-staining extremely difficult, if not impossible, to remove.

DEFINING VALUE

Having established that biological growth on calcareous stone can pose a danger to the stability of stone monuments, we return to the questions: in what ways do we find value in our monuments and why are we concerned about their preservation? Simon Schama begins his article 'The remains of that day' about the 9/11 Memorial situated in Lower Manhattan by asking:

> What are public memorials for? Are they meant to perpetuate the sorrow of loss, pay a debt of respect or set a boundary about grief by turning it to public reverence? Must their primary obligation always be to the immediately bereaved? Should such places be no more than a site where those victimized by slaughter can find consolation in a community of mourning? Or is a public memorial, by definition, created to make something more universally redeeming from atrocious ruin? ... The competing claims of lamentation and instruction were embodied in the guidelines provided for the memorial by the Lower Manhattan Development Corporation ... The document also called for a work that evoked 'the historical significance' of 9/11 (Schama 2011).

In the case of the 9/11 Memorial and the Flight 93 National Memorial in Pennsylvania, the monuments are newly erected and we expect that their meaning will change and develop over time. Their preservation and maintenance, however, must be prioritized if the evolution of meaning and education about the past are to continue into the future.

In the introduction to her book *Monuments to Shattered Myths*, Harriet F. Senie notes a number of examples of monuments whose meaning and import have changed throughout the years. One such example cited is the Bunker Hill monument, erected in 1843 to commemorate the revolutionary troops in a 1775 Revolutionary War battle. Over the years, the area in which the monument stood, Charlestown, Massachusetts, fell into socio-economic decline, and by the 1990s, 'contemporary perspectives challenged the local meaning of the memorial'. Senie describes a 1998 temporary public artwork for which voices of local people talking about loved ones lost to violence were broadcast aloud while a video projected onto the monument. 'The once-heroic national monument now valorized local victims ... the meaning of memorials is never fixed' (Senie 2016: 2). As the meaning and import of monuments and sites changes with transitions in communities and environments (political, economic, climatic, etc.), so too will approaches to maintenance and conservation change. The fact remains, however, that if we do not maintain our monuments, sites and objects, they may be lost to future generations, eliminating opportunities to reinterpret, learn, feel, consider, relate, discuss or honor what they represent.

At the still older site of the Hellenic and pre-Hellenic archaeological excavations at the Sanctuary of the Great Gods in Samothrace, Greece, marble architectural blocks are maintained every summer by visiting conservators. Student conservators have worked under senior conservators to clear plants and grasses from

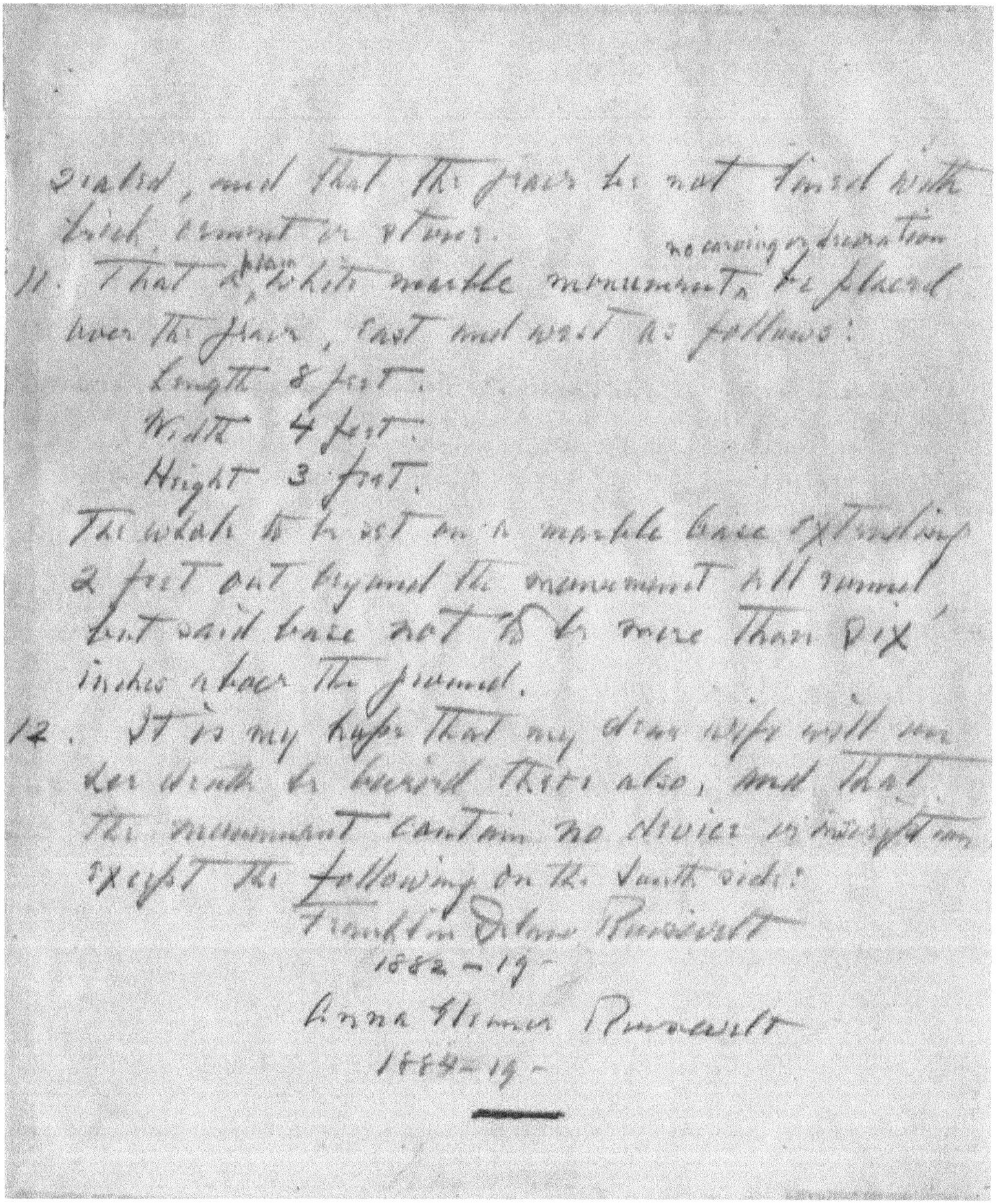
sealed, and that the grave be not lined with brick, cement or stones.

11. That a plain white marble monument, no carving or decoration, be placed over the grave, east and west as follows:

Length 8 feet.

Width 4 feet.

Height 3 feet.

The whole to be set on a marble base extending 2 feet out beyond the monument all round but said base not to be more than six inches above the ground.

12. It is my hope that my dear wife will on her death be buried there also, and that the monument contain no device or inscription except the following on the south side:

Franklin Delano Roosevelt

1882–19–

Anna Eleanor Roosevelt

1884–19–

Figure 1 Page 3 of Franklin D. Roosevelt's four-page handwritten burial instructions dated December 26, 1937. (Image from the Franklin D. Roosevelt Library, Hyde Park, New York, USA, President's Secretary's File, Subject File: 'Roosevelt, Franklin D. - Burial Instructions.')

and perform lichen abatement treatments on the ancient, inscribed marble blocks annually since 2002 (Koob 2010: 206). The readability of the inscribed blocks, which are unsheltered and outside year-round, provide critical evidence of the dates of various buildings on the site and of activities which took place within the Sanctuary in the 4th century BC. The site, although approximately 2,500 years old, is still a place of remembrance and learning, attracting visitors ranging from those wishing to practice modern spiritual meditations among the ruins to those expressing a desire to learn more about ancient cultures. Certainly the meaning of this sacred sanctuary has changed with the intentions of visitors to the site over the years, from the earliest Greek sailors who visited for the purpose of making ritual offerings at the Sanctuary, to the earliest amateur archaeologists who unearthed the Winged Victory of Samothrace in 1863 and transported it to the Musée du Louvre in Paris, to today's tourists from all around the world. Without annual maintenance and cleaning of the marble architectural and dedication blocks, they would become more and more illegible, eventually becoming indecipherable, and the site would permanently lose a non-renewable resource and important cultural heritage specimens.

The above illustrates that monuments and special sites often serve as places to reflect, remember, mourn, honor, learn, worship, and contextualize. If a quick web search for 'cleaning marble monuments' were to be conducted, a profusion of websites and YouTube videos offered up by monument manufacturers, private individuals, churches and community preservation organizations would be found. This may be interpreted as evidence that there is demand for information on maintaining memorials and monuments and, perhaps even further, that people feel compelled to clean and maintain memorials as a sign of respect, love and as a ritual of remembrance: as a way to connect. That is most certainly the case described in a 2011 *Washington Post* article by Patricia Sullivan about the maintenance of the Vietnam Memorial in Washington DC:

> The washing of the dead, with its religious resonances, arose out of frustration. In 1998 ... Jan Scruggs of the Vietnam Veterans Memorial Fund ... handed 37 toothbrushes to visiting vets from Wisconsin, who scrubbed the filth away ... A little more than a decade ago ... the organized weekly cleanings began. 'It's a show of respect. These are not just names; these are people who gave their lives for the freedom we enjoy. We have got to keep it clean,' said Mike Luftman ... 'As a veteran and an American, this is probably one of the highest honors you can have, to make sure this monument is in pristine shape.'

Schama (2011) also points out that the 9/11 Memorial's architect originally intended for names of those lost that day to be inscribed on the granite walls that form the two pools, far out of reach and below ground level with water falling endlessly over them. That incarnation of the design was rejected by the victims' families and other stakeholders, as they wanted the names to be accessible and touchable. In the final design, the names were placed on low bronze parapets in the plaza, above and around the sunken pools, where visitors can interact with them, supporting the development of connections.

In addition to honoring memorialized events or individuals, and the needs and wants of visitors to sites and monuments, there are occasions when, by keeping a monument clean and stable, we are also honoring the wishes, instructions or intent of the memorialized. Such is the case with the white marble monument at the Springwood Estate in Hyde Park, New York where Franklin Delano Roosevelt and Eleanor Roosevelt are interned. President Roosevelt left detailed instructions in a four-page, handwritten sealed letter to be opened after his death. The letter contained not only instructions for his funeral and internment services, but also detailed instructions[3] for the design of his grave monument, as shown in Figure 1:

> That a plain white marble monument, no carving or decoration, be placed over the grave, east and west as follows:
>
> Length 8 feet
> Width 4 feet
> Height 3 feet
>
> The whole to be set on a marble base extending 2 feet out beyond the monument all around, but said base not to be more than six inches above the ground.... and that the monument

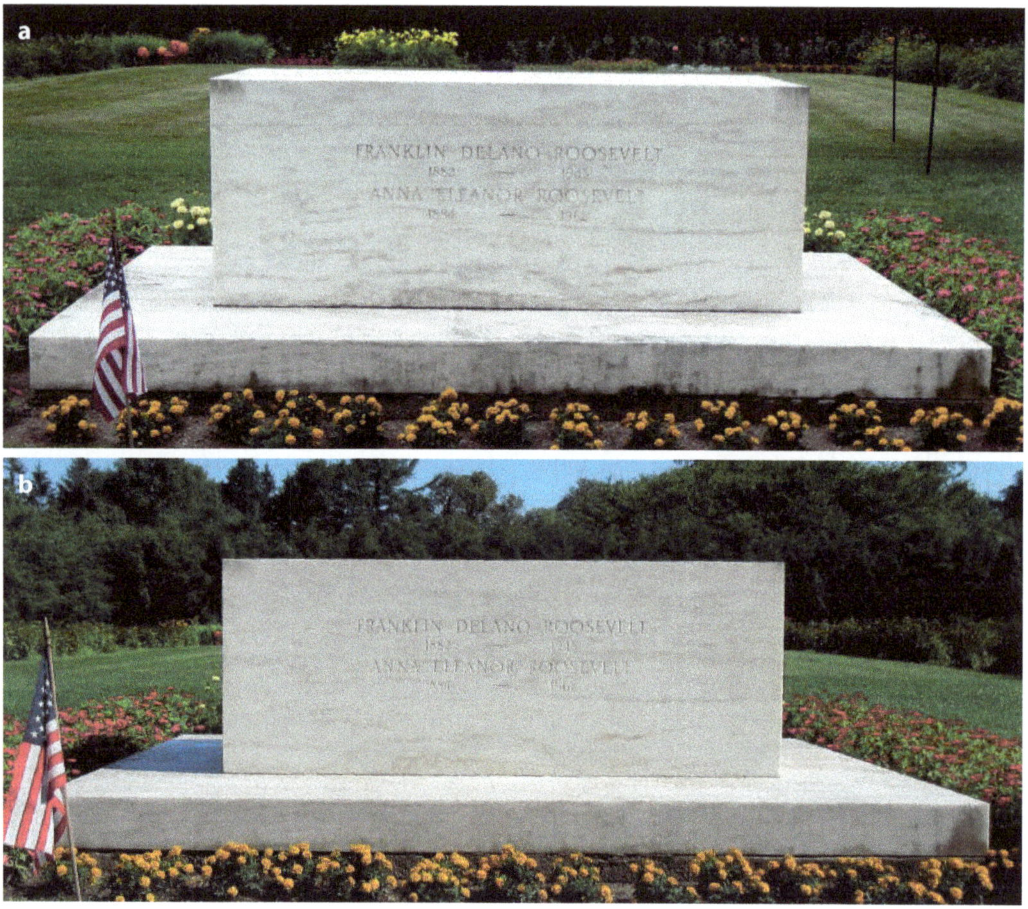

Figure 2 Before (a) and after (b) removal of biological growth from the south face of the white marble Franklin D. and Eleanor Roosevelt grave monument in Hyde Park, New York, USA. (Photo: Joannie Bottkol.)

> contain no device or inscription except the following on the South side:
>
> Franklin Delano Roosevelt
> 1882–19–
> Anna Eleanor Roosevelt
> 1884–19–

To allow this unadorned, simple and pristine white monument to become overgrown or stained with biogrowth would be an offense to the clean, modernist aesthetic so specifically described in President Roosevelt's instructions. A moldering monument in this instance would be in opposition to the intent of the memorial as well to the wishes of the memorialized as shown in Figure 2.

When we do allow for works to fall into disrepair or accumulate dirt and grime, not only might we fail to honor the visitors and the memorialized, but a site's or object's meaning can be clouded by a new and inaccurate narrative, one which may obscure history rather than building upon an ever-evolving interpretation or story that is tethered to history. When context is lost in this way, the cultural resource can also be lost. In David Lowenthal's *The Past is a Foreign Country*, he delves into what it means for cultural objects to be 'dirty' and have the 'patina of time'. He states that 'interaction with a heritage continually alters its nature and context, whether by choice or by chance' (Lowenthal 1985: 263). In some instances, the age patina of dirt, soot and biological growth has endowed cultural material with a kind of false credibility; the illusion of trueness imbued by age that separates it visually from something new or fake. Over many years, an age patina can become the aesthetically familiar and thus preferred appearance. This thinking is what we invoke when we imagine that ancient temples should appear covered in vines and moss, or that tombstones should look wasted away like physical embodiments of *memento mori*. The promotion of this aesthetic requires preservation of the biological growth that accumulates when monuments are left untended, rather than the preservation of the monument itself.

Just such a debate over the removal versus preservation of soot and age patina has been taking place of late over the conservation of the interior of Chartres Cathedral, located approximately 80 miles outside of Paris. In 2017, Benjamin Ramm wrote about this highly controversial conservation treatment and cleaning in the *New York Times*:

> In ... 800 years, the building has changed almost beyond recognition, as smoke from burning candles, oil lamps and fires darkened the walls, the statues ... and the exquisite stained glass ... The restoration aims not only to clean and maintain the structure, but also to offer an insight into what the cathedral would have looked like in the thirteenth century. Its interior was designed to be a radiant vision, as close to heaven on earth as a pilgrim might come ... Archaeological investigations beginning in the 1980s demonstrated that what appeared to be exposed stonework was in fact an accretion of dirt hiding decaying whitewash and two layers of paint ... The bony white colonettes and the multicolored ceiling keystones may seem garish, but they were aspects of the medieval cathedral ... Jeffrey F. Hamburger, a medieval art historian at Harvard, said that there is 'no reason to be nostalgic or romantic about the dirt.' The association of gothic buildings with 'dark, brooding gloom' is 'fundamentally misguided,' he said; they are 'not monuments to melancholy.' The restoration seeks to reconstitute a temple of light, to challenge the popular perception of gothic dejection.

There have most certainly been cases in which visual and structural damages have resulted from overzealous or ill-advised conservation treatments, but a poorly maintained object or site that shows little sign of stewardship or maintenance can also incur damage, just by virtue of the community around it believing that no one has a stake in its preservation. This is, of course, the scenario described in the 1982 article by George Kelling and James Wilson entitled 'Broken windows: the police and neighborhood safety':

> Disorder and crime are usually inextricably linked ... if a window in a building is broken and is left unrepaired, all the rest of the windows will soon be broken. This is as true in nice neighborhoods as in run down ones ... one unrepaired broken window is a signal that no one cares, and so breaking more windows costs nothing.

CONSEQUENCES OF INACTION

Maintaining our monuments, sites and objects demonstrates to the community and visitors that our cultural heritage sites and objects are valued, and this alone acts as a preventive conservation measure, discouraging graffiti and other forms of vandalism from ever taking place. Furthermore, just as neglect may encourage damage through vandalism, studies have shown that there are particular destructive organisms that tend to inhabit stone which is decayed preferentially over so-called 'sound' stone. In other words, poor or compromised surface conditions of stone may influence the type of microflora that inhabit stone in a way that encourages further damage and decay (May *et al.* 2000). Removing lichen hyphae that have been allowed to become entrenched in stone can also be destructive as the hyphae form a physical attachment to stones and take up mineral grains, leaving voids and fissures after their removal where original stone surface has been digested or lost (Ciferri 2000). These voids can create niches in which new organisms can attach, furthering the cycle of degradation. These processes of deterioration suggest that there are significant benefits to keeping stone surfaces clear of biological growth, especially of boring organisms such as lichen and fungi.

Generally, the more damage incurred on a monument, the more complicated and expensive the remediation or repair. A stone monument that may have once simply required that nearby plants be cut back or weeds growing around it be pulled to prevent the development of biological growth, might, if left alone, eventually require a more in-depth cleaning using biocides to remove biological staining and growth. If left to molder for even longer, the same monument may develop even greater problems of a structural nature, demanding consolidation or other more serious and costly interventions for its preservation.

There is no doubt in the scientific community that biofilms change both surface and subsurface characteristics of stones. Destabilizing changes to pore size, hardness, and water retention are just some of the ways in which the development of a microfloral community can adversely affect stone. On the other hand, research has also shown that certain types of 'biological patina' can be protective to some types of stones while they still reside in their natural, undisturbed environment. For example, EPS has been demonstrated to protect stone by consolidating and binding together stone grains that would otherwise be weak and vulnerable to dissolution, and can even increase the rate of water evaporation in some sandstones still in their original geological locations (Hammes *et al.* 2013). Biofilms can also temporarily protect the uppermost layers of certain types of stone by developing a symbiotic relationship with that stone's surface. Epilithic lichens on sandstone, for example, have been found to secure the uppermost layers of the sandstone in order to protect the underlying substratum and biofilm matrix (Hoppert *et al.* 2004). However, when a stone is removed from its original location, quarried, carved and transformed into a monument, the once-protective microflora community that established itself over many decades in the old location is not necessarily a relevant source of protection in the new environment, to the new surface or to the new functions of the stone.

The decision to clean an outdoor monument or building is often made in response to the appearance of biological growth or staining. When a stone becomes a monument, its meaning changes, it no longer retains its original appearance, and rarely its original geological location. The stone is removed and placed into a foreign ecosystem with different microflora. The symbiotic community that once inhabited the biogenically weathered and patinated surface of the stone in its original geological location is largely destroyed after the stone is quarried and carved into a monument placed into a secondary location, often far from the stone's origin. Many factors, both scientific and ethical, must be weighed when determining a sustainable conservation treatment.

CONCLUSIONS

Deciding whether to undertake a conservation intervention such as cleaning an object or monument can be complicated. The intent of the artist, the intent of the memorial, the narrative of the cultural history and the needs of the community must be considered and taken into account. The goal of the intervention needs to be determined: is it to preserve readability or usability, to prevent an imminent catastrophic damage, or to retard a gradual decline in stability? The level of intervention required needs to be determined as does a future maintenance plan. And just as the field of conservation requires knowledge of both science and art history, the preservation of stone monuments necessitates both an understanding of the roles that microorganisms can play in carbonate stone degradation, and a willingness to consider the myriad ways in which we find meaning and value in monuments: as sites for lamentation, education and connections, and as objects which can adapt to meet the changing needs of communities while continuing to illuminate old narratives alongside new ones.

NOTES

1. The views and opinions of the authors or any Federal employee interviewed for this chapter does not state or reflect those of the United States Government. Any reference within this paper to any specific commercial products, process, or service by trade name, trademark, manufacturer, or otherwise, does not constitute or imply its endorsement, recommendation, or favoring by the United States Government.
2. AIC Position Statement on Confederate and Other Historic Public Monuments, issued September 2017: https://www.conservation-us.org/docs/default-source/governance/position-paper-on-confederate-monuments-(septber-2017).pdf?sfvrsn=2 (accessed 8 September 2017).
3. Franklin D. Roosevelt Library, Hyde Park, New York. *FDR's Handwritten Burial Instructions*, p. 3. 26 December 1937 Folder: 'Roosevelt, Franklin D. - Burial Instructions' (Box 160), Series: Subject File, Collection: President's Secretary's File President's Secretary's Files.

REFERENCES

Ciferri, O. 2000. 'Conservation of cultural heritage', in *Encylopedia of Microbiology*, 2nd edn, vol. 1, J. Lederberg (editor-in-chief), 863–72. New York: Academic Press.

de la Torre, A., Gomez-Alarcon, G., Vizcaino, C., Garcia, T. and de la Torre, M.A. 1993. 'Biochemical mechanisms of stone alteration carried out by filamentatious fungi living in monuments', *Biogeochemistry* 19: 129–47.

Funderburg, L. 2015. 'Andy Goldsworthy: natural man', *Architectural Digest* 26 November 2015. Available at: https://www.architecturaldigest.com/gallery/andy-goldsworthy-book-ephemeral-works (accessed 7 June 2017).

Hammes, E., Floyd, M. and Kurtz, Jr., H.D. 2013. 'Assessment of iron(II) binding by a cryptoendolithic bacterial community: implications for iron cycling in the Jurassic Navajo Sandstone', *Journal of Arid Environments* 97: 49–55.

Hoppert, M., Flies, C., Pohl, W., Gunzl, B. and Schneider, J. 2004. 'Colonization strategies of lithobiontic microorganisms on carbonate rocks', *Environmental Geology* 46: 421–8.

Kelling, G.L. and Wilson, J.Q. 1992. 'Broken windows: the police and neighborhood safety', *The Atlantic* March 1992.

Kemmling, A., Kemper, M., Flies, C., Schieweck, O. and Hoppert, M. 2004. 'Biofilms and extracellular matrices on geomaterials', *Environmental Geology* 46: 429–35.

Koob, S.P. 2010. 'The conservation program in the Sanctuary of the Great Gods on Samothrace', in *Samothracian Connections: Essays in Honor of James R. McCredie*, O. Palagia and B. Wescoat (eds), 206. Oxford: Oxbow Books.

Lowenthal, D. 1985. *The Past is a Foreign Country*. Cambridge: Cambridge University Press.

May, E., Papida, S., Hesham, A., Taylor, S. and Dewedar, A. 2000. 'Comparative studies of microbial communities on stone monuments in temperate and semiarid climates', in *Of Microbes and Art: The Role of Microbial Communities in the Degradation and Protection of Cultural Heritage*, O. Ciferri, P. Tiano and G. Mastomei (eds), 49–62. New York: Kluwer Academic/Plenum Publishers.

Mitchell, R. and Gu, J.D. 2000. 'Changes in the biofilm microflora of limestone caused by atmospheric pollutants', *International Biodeterioration and Biodegradation* 46: 299–303.

Perry, T.D., McNamara, C.J. and Mitchell, R. 2005a. *Scientific Examination of Art: Modern Techniques in Conservation and Analysis* (Sackler NAS Colloquium). Washington DC: The National Academies Press, doi: 10.17226/11413.

Perry, T.D., Klepac-Ceraj, V., Zhang, X.V., McNamara, C.J., Polz, M.F., Matrin, S.T., Berke, N. and Mitchell, R. 2005b. 'Binding of harvested bacterial exopolymers to the surface of calcite', *Environmental Science and Technology* 39: 8770–75.

Ramm, B. 2017. 'A controversial restoration that wipes away the past', *New York Times* 1 September 2017. Available at: https://www.nytimes.com/2017/09/01/arts/design/chartres-cathedral-restoration-controversial.html?_r=1 (accessed 2 September 2017).

Saiz-Jimenez, C. 1993. 'Atmospheric environment, Part B', *Urban Atmosphere* 27: 77–85.

Schama, S. 2011. 'The remains of that day', *Financial Times* 2 September 2011. Available at: https://www.ft.com/content/34b187c2-d3d8-11e0-bc6b-00144feab49a (accessed 30 August 2017).

Scheerer, S., Ortega Morales, B.O. and Gaylarde, C. 2009. 'Microbial deterioration of stone monuments: an updated overview', *Advances in Applied Microbiology* 66(5): 103.

Senie, H.F. 2016. *Memorials to Shattered Myths: Vietnam to 9/11.* New York: Oxford University Press.

Sullivan, P. 2011. 'Washing the wall to remember Vietnam vets', *The Washington Post* 29 May 2011. Available at: https://www.washingtonpost.com/local/washing-the-wall-to-remember-vietnam-vets/2011/05/09/AGJxZsCH_story.html?utm_term=.8999ef2d9909 (accessed 8 September 2017).

Tiano, P. 1998. 'Biodeterioration of monumental rocks: decay mechanisms and control methods', *Science and Technology for Cultural Heritage* 7(2): 19–38.

AUTHORS' ADDRESSES

- Margaret Breuker, National Park Service, Northeast Region, Historic Architecture, Conservation, and Engineering Center, 115 John Street, 4th Floor, Lowell, MA 01852, USA (margaret_breuker@nps.gov)
- Joannie Bottkol, National Park Service, Northeast Region, Historic Architecture, Conservation, and Engineering Center, 115 John Street, 4th Floor, Lowell, MA 01852, USA (joannie_bottkol@nps.gov)

SUBMERGED ARCHAEOLOGICAL WOODEN STRUCTURES

Charlotte Björdal

INTRODUCTION

Wood is one of the most significant raw materials used by our ancestors. Apart from providing heat, it was an important building material for large-scale construction projects such as trackways, houses, fortifications, bridges, harbours and ships, as well as smaller products including weapons, tools, house interiors, containers and food-related articles used in daily life. It can be claimed that wood provided a basis for the development of early human cultures. Archaeological wooden objects should therefore be present in large numbers in our museums, but this is not the case: few wooden objects have 'survived' to the present time.

During the last 150 years some fantastic and unique waterlogged archaeological wooden objects have been found in Scandinavia. In Norway and Denmark, Viking ships with elegant construction and wonderfully carved ornamentations are on display in Bygdö Viking museum in Oslo and in the Viking ship museum near Copenhagen. In Sweden an impressive number of 16th- and 17th-century shipwrecks have been located on the seabed of the Baltic Sea. The most spectacular is probably the warship *Vasa*, which was raised in 1961 and is on display at the Vasa Museum in Stockholm. Many of these ships capsized and sank during war conflict, whereas others, such as the Oseberg ship and the Gokstadship in Norway, were found in deep soil on land, the result of a burial ritual.

Today, with the help of modern survey techniques, new wreck sites and wooden remains are being discovered in marine waters and on land site excavations. These wooden finds have survived hundreds of years in nature and are available for research opportunities. In order to understand why these finds are preserved, unlike other wooden objects from the past, we need to understand wood as material, the degradation process and the site and environment where the finds were left by our ancestors.

WOOD AS MATERIAL

Wood is literally the material of trees and an organic material. Trees are divided into two groups: softwoods (gymnosperms) and hardwoods (angiosperms). Living trees of softwood such pine and spruce generally have needles, whereas hardwood trees like oak, birch and maple have leaves. Just beneath the bark of the living tree, the cambium continuously produces new wood fibres during the growth season and the tree increases in size each year. Growth rings are produced continuously, especially in temperate climates, and are often possible to see with the naked eye in a cross-section of a log (Fig. 1). Most of the fibres produced by the tree are oriented along the living stem and connected to each other by small openings (pits) that secure a water flow from the root system to the top of the tree (Fig. 2). Some fibres, called rays, are oriented

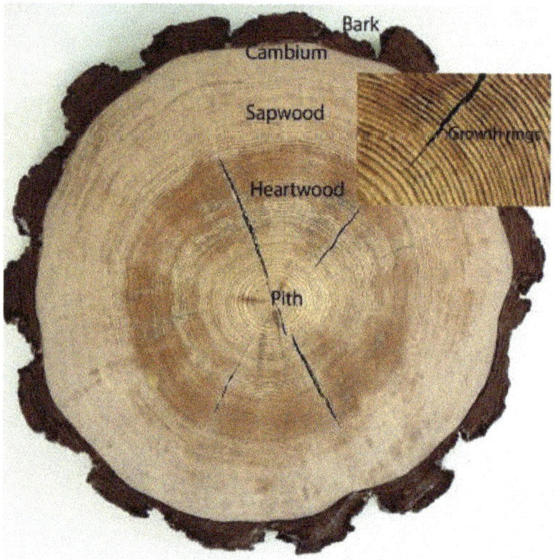

Figure 1 Cross-section of a pine disc showing important wood features and their position.

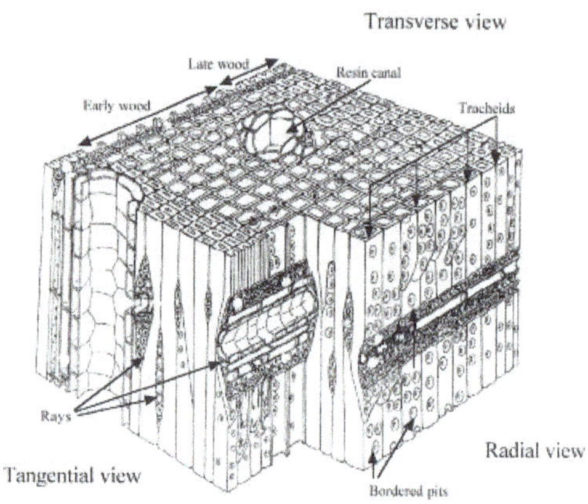

Figure 2 Illustration showing a 3D cube of softwood consisting of about 150 longitudinal fibres (tracheids). Each has channels in the centre called lumen and small openings in the cell wall called pits that connect them to adjacent fibres. Horizontal tracheids and rays (observed in groups) are shown in transversal and longitudinal directions.

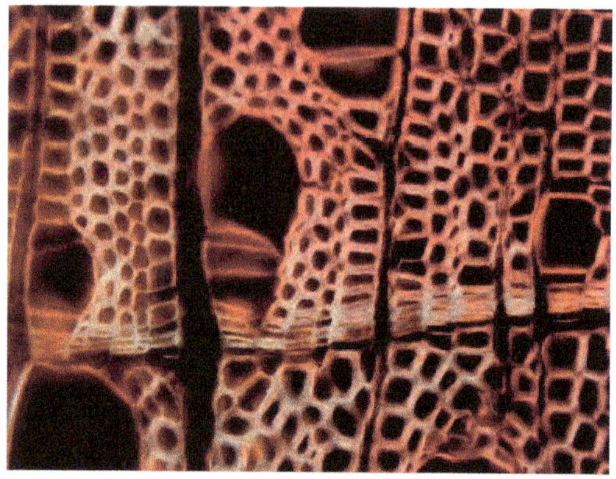

Figure 3 Cross-section of hardwood (birch: *Betula sp.*), showing fibres, rays and vessels (large openings) that have empty spaces used for water and nutrient transport (black).

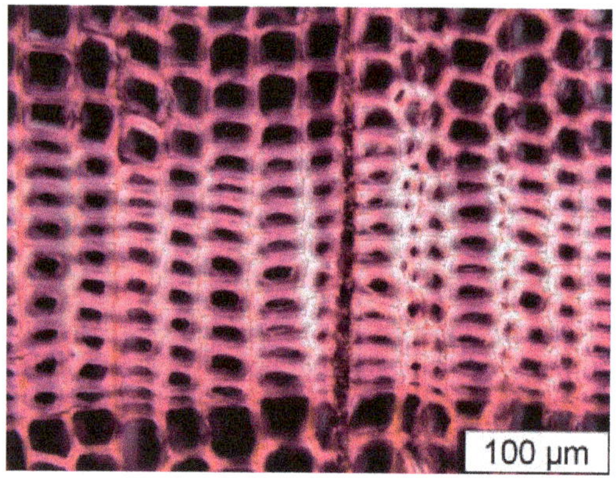

Figure 4 Cross-section of softwood (pine: *Pinus sylvestris*) showing fibres (tracheids) and one ray that have empty spaces used for water and nutrient transport. This is a more simple structure compared to hardwood seen in Figure 3.

across the stem. As the tree matures, water flow stops in the inner part of the stem. This part is then called heartwood whereas the outer part with an active flow is called sapwood (Fig. 1). The principle of growth is the same in hardwood and softwood, but the type of fibres produced in the cambium differs. As shown in Figures 3 and 4, hardwood is more 'advanced' and has a wider variation of specialized fibres compared to softwood, including water-conducting vessels. The anatomical differences can be observed either by a magnifying lens or light microscopy (Hoadley 1990).

A wood fibre can be described as a straw-like construction with a channel for water transport in the middle (a lumen) (Figs 2–4). The fibres are positioned on top of each other and connected only by small pores that allow upward water transport from fibre to fibre (Bowyer *et al.* 2007). The three-dimensional (3D) structure of wood anatomy is rather complicated when microbial decay must be interpreted.

Chemically, the wood fibre is composed of ~50% cellulose, ~ 25% hemicellulose and

~25% lignin, and is often described as a lignocellulolytic material. These three main molecules are bonded to each other in a complicated way and are arranged in microfibrils and macrofibrils. The individual fibres are 'glued' and held together in a matrix called the middle lamellae, which is rich in lignin. The wood fibre can be divided into a primary cell wall and secondary cell wall, the latter being further divided into three layers: S1, S2 and S3. The thickest layer is often the S2 layer, which also has the highest cellulose concentration. In addition, wood contains smaller quantities of extractives (fat, tannins, acids and resins) and minerals. The contents and concentrations vary greatly among the different wood species (Rowell 2012).

Wood is a hygroscopic material, i.e. it is a 'living' material that responds to the surrounding moisture in the environment by water uptake, evaporation or equilibrium. It is also an anisotropic material, meaning that its physical properties vary in the three directions, or axes, of the wood material: transversal, radial and tangential (see Fig. 2). One example of varying properties is the dimensional instability at different moisture contents since wood swells as it wets and shrinks as it dries. For example, when wood is dried it is more than twice as high in the tangential direction compared to the radial direction.

Another example is that the strength is much higher along the fibre direction (transversal) than across the grain (radial and tangential). Different wood species have different properties and therefore different values for properties such shrinkage and strength. This varies both within and between the groups of softwood and hardwoods (Rowell 2012). Some of the properties that are normally regarded as important for different building purposes are strength, elasticity and hardness. Knowledge of these variables and selection of the right wood species for a specific purpose were known by skilled engineers and craftsmen in historic times and evident to us when these structures are analysed today.

Wood durability is strongly related to the presence of extractives as well as the concentrations and type of lignin. Lignin, a phenolic compound, is a very complex molecule that is still not fully described. It is one of the hardest plant-related compounds to naturally degrade. Coniferyl-derived lignin is the only lignin component of softwood and is more resistant to biological attack than sinapyl-derived lignin, another lignin component in hardwood (Eaton and Hale 1993). Therefore, softwoods are generally more resistant to biological attack than hardwoods. However, content and concentration of the extractives may have a toxic effect, giving some hardwoods more durability than the softwoods. Furthermore, heartwood is generally more resistant to biological attack than sapwood (see Fig. 1). This is mainly related to the chemical change of the lignin-cellulose material and the increase of extractives that occurs as the tree matures.

BIODETERIORATION IN A SUBMERGED ENVIRONMENT

Biodeterioration of wood plays a major role in carbon recycling. There are organisms that, with access to oxygen, moisture and favourable temperatures, are able to rapidly degrade and decompose wood. Specialized fungi, bacteria and insects transform the wood polymers by complex processes, liberating CO_2, water and mineral elements (Zabel and Morrell 1992). On land, with full access to oxygen, this process is mainly driven by fungi, especially the Basidiomycetes (white rot and brown rot fungi), which have advanced enzymatic systems for the effective degradation of cellulose, hemicellulose and lignin. Under optimal conditions certain brown rot species are able to produce a 60% mass loss of wood material in less than one year (Blanchette *et al.* 1990), resulting in major strength loss. Due to this efficient decomposition on land and in ground contact, most historical artefacts from the past have rotted away relatively fast, if not already destroyed by fire.

However, if the wooden structures, by accident or on purpose, have been abandoned in a waterlogged environment, they might be preserved for hundreds or thousands of years.

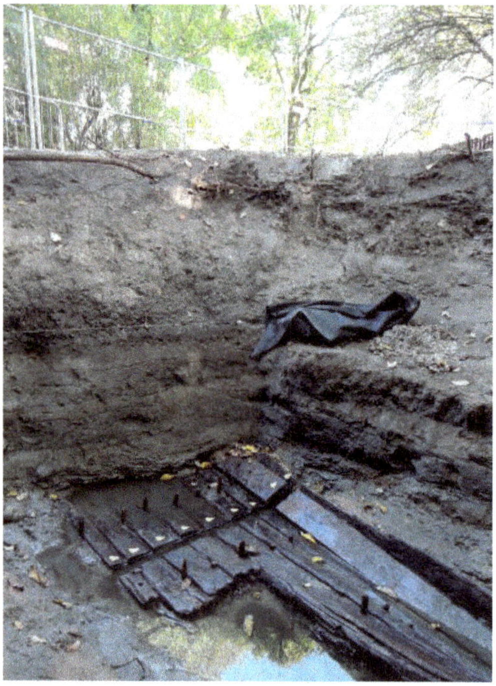

Figure 5 A terrestrial archaeological excavation site in Sweden showing exposed parts of a medieval boat. One side of the boards is exposed and it was decided to preserve the remains *in situ* within the wet and clay-rich environment.

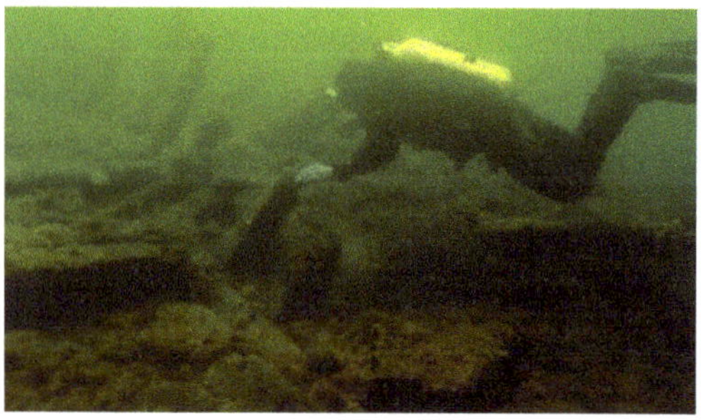

Figure 6 Marine archaeological excavation of a shipwreck in the Baltic Sea. The wreck is protected by law, but not from physical and biological deterioration. (Photo: Gripshunden by Johan Rönnby.)

This depends on the nature of the submerged environment, which can be marine, aquatic or terrestrial, as illustrated in Figures 5 and 6. Characteristic for these environments is that they are all waterlogged and contain less oxygen than above ground. Decay by the aggressive white rot and brown rot fungi is strongly repressed and eliminated, mainly as a consequence of fungal demands for high oxygen concentration. These three environments are described below.

Marine environments

A marine environment in this context is described as a saline milieu such as most of the ocean and seas that surround us where wood material is efficiently degraded by marine crustaceans (invertebrates) and bivalves (molluscs). The wood-boring 'shipworms' (molluscs of the families Teredinidae and Pholadidae) as well as gribbles (crustaceans) are generally found in all saline and oxygenated waters (Nair and Saraswarthy 1971; Eaton and Hale 1993; Cragg *et al.* 1999). Parallel to the attack by the marine borers, marine wood-degrading fungi and bacteria also colonize and degrade the wood (Pointing and Hyde 2000; Björdal 2012a). Whereas gribbles, fungi and bacteria degrade the surface layer of the wood, the shipworm penetrates the interior and forms large burrows (Figs 7 and 8). As the shipworm grows, the tunnels expand in length. Under optimal conditions they can rapidly decompose and transform solid wood structures into smaller fragments. Experiments show that in less than one year, wood materials are severely degraded or totally decomposed when infested with shipworm (Jones *et al.* 1972; Björdal and Nilsson 2008; Borges 2014), and even archaeological wood that still contains cellulose would be degraded if sediments were to be removed and the wood exposed to water again (Eriksen *et al.* 2015).

As shipworms are very aggressive and able to degrade most wood species (Şen *et al.* 2009), decay by soft rot fungi and bacteria play a minor role as degradation by these organisms is very slow. However, when it comes to degradation of wood buried in sediment, the situation is different: marine borers are excluded due to very low oxygen concentrations and microorganisms become the main degraders. Soft rot fungi and tunnelling bacteria are active in degrading the first centimetre of the sediment layer, whereas erosion bacteria are the only organisms able to degrade wood in the deeper and more anoxic layers of the sediment (Björdal and Nilsson 2008; Björdal 2012a). The inability of erosion bacteria to degrade all of the lignocellulose material makes it possible for us to discover archaeological wooden remains that appear well preserved and can be recovered from marine environments. One good example showing the protective role of sediments is the warship *Mary Rose*, which sank in the Solent near Portsmouth in the UK. One side of the ship was embedded in sediment and preserved, whereas the other half, which was exposed in the saline water column, was totally decomposed. The remaining hull was raised from the sediment, together with other objects, and conserved (Mouzouras *et al.* 1986, 1990; Jones and Rule 1990).

Non-marine environments

Non-marine environments in this context are defined as fresh and brackish waters. This includes lakes, estuaries and rivers, as well as their respective sediments. Marine borers are not found in these low-saline environments. Wood-degrading fungi and bacteria are the major cause of decay both in the water column and within the sediments. Similar to the marine environment, erosion bacteria are the only degraders of the anoxic sediment layers. Recently, a review on microbial degradation of historic shipwreck timber in the brackish Baltic Sea was published in which both soft rot fungi and bacterial degradation are described in detail (Björdal 2012a). Both groups initially attack the wood surface and proceed slowly inwards. The decay processes are faster in water than within the sediments. At sites exposed to currents and waves, decay increases as the weakened surface is continuously peeled away by streaming or turbulent water. In such locations, ornamental details on wood-carved sculptures will fade away with time. Some sculptures, such as the recently discovered griffin hound figure from the wreck of the *Gripshunden* (Eriksson 2015), and a post head from the *Ghost* wreck, exposed to the water column of the Baltic Sea, showed clear signs of this slow and irremediable damage (Eriksson and Rönnby 2012), whereas the man sculpture raised from sediments next to the wreck *Kronan* was in exceptionally good condition and is on display at Kalmar Länsmuseum, Sweden (Einarsson 1990).

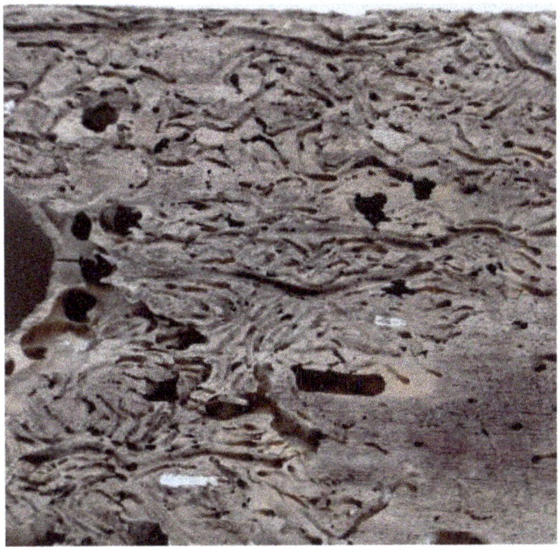

Figure 7 Attack of isopods (*Limnoria* spp.) in the wood surface where minute, irregular burrows remain after decay by these motile organisms. At the left, larger holes in the underlying layer are the result of attack by shipworm.

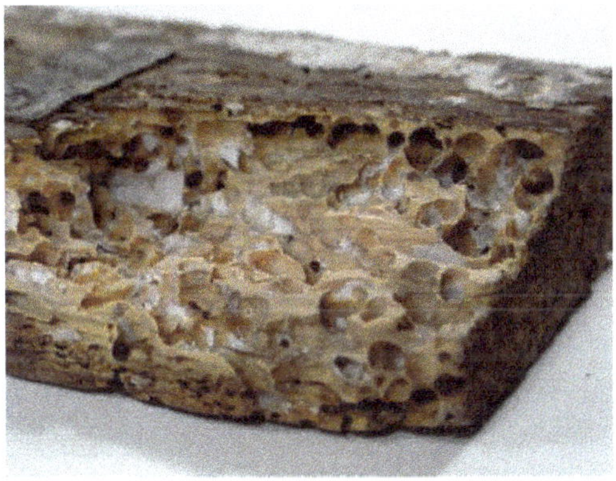

Figure 8 Severe attack by shipworms (*Teredo navalis*), which form individual tunnels covered with calcareous deposits in order to protect their soft bodies. Decay takes place inside the wood and is not visible from outside.

Terrestrial sites

In terrestrial waterlogged environments, such as wetlands, peatlands, medieval culture layers and clay soils with high water-holding capacity, wooden objects may survive for hundreds of years (Björdal *et al.* 1999; Björdal 2012b). These locations could be characterized as soils with a natural high water content or a high water level and are especially vulnerable to a decrease in ground water level (Caple 1994; Jordan 2001).

Clay soils are generally very protective for wooden objects as they are able to raise water through fine pores from lower water levels and upward, in contrast to sandy soils that do not have this capacity (Fetter 1994). Extraction of peat and drainage of wetlands, as well as infrastructural projects in urban environments, may often lead to drainage and desiccation of larger areas (Gregory *et al.* 2002; Matthiesen *et al.* 2008; Corfield 2012). From several archaeological important wetlands and peats in the UK, we have learned that drainage of water levels is a constant threat to unique archaeological remains (Brunning 1996, 2012; Cox *et al.* 2001). The archaeological wood, already degraded by erosion bacteria, may undergo final degradation by aggressive fungi when oxygen concentrations are increased (Powell *et al.* 2001; Björdal and Nilsson 2002).

From forestry research, especially pertaining to wood durability of normal compared to preservative-treated wood, it has been found that normal wood is generally quickly degraded by ground/soil contact (Eaton and Hale 1993). At terrestrial sites, fungal decay is dominant, both on the ground and within the upper soil layers. The oxygen demanding and aggressive Basidiomycetes (white rot and brown rot fungi) are repressed when the oxygen concentration decreases with the depth of the soil profile, and therefore the more low oxygen-tolerant soft rot fungi (Ascomycetes, Fungi imperfecti) and bacteria (tunnelling and erosion bacteria) are more dominant in terrestrial contexts (Blanchette *et al.* 1990; Blanchette 2000). The decay process for all wood-degrading microorganisms is strictly correlated to the surrounding environment with respect to oxygen concentrations, temperature, moisture, wood species and soil type. Deeper in the soils, where near-anaerobic conditions exist, erosion bacteria are the only active wood degraders (Kim *et al.* 1996; Björdal *et al.* 1999).

Fungal and bacterial degradation

In waterlogged environments, such as those described above, oxygen is only available, if at all, in very low concentrations as a gas dissolved in water. This favours decay by two groups of specialized wood degraders: soft rot fungi and bacteria (Björdal *et al.* 1999; Blanchette 2010; Pedersen *et al.* 2013).

Soft rot fungi are named after their characteristic way of softening the wood surface. Just as with decay by brown rot and white rot fungi, this classification is directly related to the visual appearance of the wood material after decay (Blanchette *et al.* 1990). However, the colour is in fact a result of their capacity to degrade the wood components lignin, cellulose and hemicellulose (Rowell 2012). Wood-degrading soft rot fungi belong to the group of Ascomycetes and Fungi Imperfecti and are saprophytic, which means that they feed on dead wood. The vegetative element of fungi, the hyphae, is an elongated structure with filamentous growth. It is therefore a so-called multicellular organism with continuous growth at the tip of the hyphae. It also branches, and all hyphae together form a large network called the mycelium. Using an advanced enzymatic process (Zabel and Morrell 1992), the soft rot hyphae grow and spread inside the wood material by penetrating the adjacent fibre wall and the lignin-rich middle lamellae. Inside the secondary cell wall, it utilizes the cellulose-rich material as a carbon source for growth.

Based on their visual characteristics resulting from breakdown of the wood cell wall, soft rot is divided into two types: the hyphae of the more common Type 1 forms cavities within the secondary cell wall following the microfibril orientation of the cell layer (S1, S2, S3), whereas the hyphae of Type 2 erode the secondary cell wall and are positioned in the lumen (Schwarze 2007). In Figure 9, typical decay features of Type 1 can be seen in cross-section.

The majority of soft rot fungi are found in terrestrial environments and there was a long-held belief that wood-degrading fungi could not survive in aquatic environments. However, in 1979 about 107 soft rot species were identified (Kohlmeyer and Kohlmeyer 1979), and in the year 2000 the number rose to 444 (Pointing and Hyde 2000). New DNA techniques will probably enable the identification of even more species in the future, both on land and in water. The aquatic fungi are particularly fascinating organisms that spread by dispersal of spores tailored for water transport and for fastening onto wood material (Kohlmeyer and Kohlmeyer 1979). When attached to the wood surface, the spore produces hyphae which grow and spread into the wood fibres as described previously.

Erosion bacteria are gram-negative, slime-producing and gliding rod-shaped bacteria with a length of 1–8 μm and a diameter of about 0.5–0.9 μm. The bacteria are attached to the secondary cell wall and aligned in individual channels during decay. The direction of bacterial attachment follows the microfibrillary orientation of the actual cell wall layer (Daniel and Nilsson 1986; Singh and Butcher 1991; Daniel 1994; Björdal *et al.* 2000).

Wood degraded by erosion bacteria may appear sound at the excavation site, as the ornamentations, dimensions, colour and other physical structures remain intact. Closer observation, however, reveals that the wood is very soft, spongy and fragile in the surface layers, sometimes even so weak that it is possible to easily press a needle right through. The apparently well-preserved wood is in fact heavily degraded; only the lignin-rich middle lamellae remain and as long as the wood is waterlogged, this skeleton maintains the structure of the material.

In the 1960s a non-fungal decay pattern present in fibres of submerged wood was investigated. This peculiar type of decay was found both in foundation poles supporting old buildings as well as in archaeological wooden objects. It was speculated to be either of chemical or bacterial origin, but no evidence or verifications were possible at that time (Boutelje and Bravery 1968; Greaves 1969). Beginning in the 1980s, aided by electron microscopy, it was possible to confirm that the 'new' type of degradation was of bacterial origin. It was evident that true bacterial decay took place in the cell walls of submerged wood extracted from near-anaerobic environments (Nilsson and Holt 1983; Nilsson and Daniel 1986). The bacteria were termed erosion bacteria because of the characteristic erosion of the cell wall. Later, it was found that these organisms are cosmopolitan and that bacterial degradation is the major form of decay in waterlogged archaeological wood (Kim *et al.* 1996; Björdal *et al.* 1999).

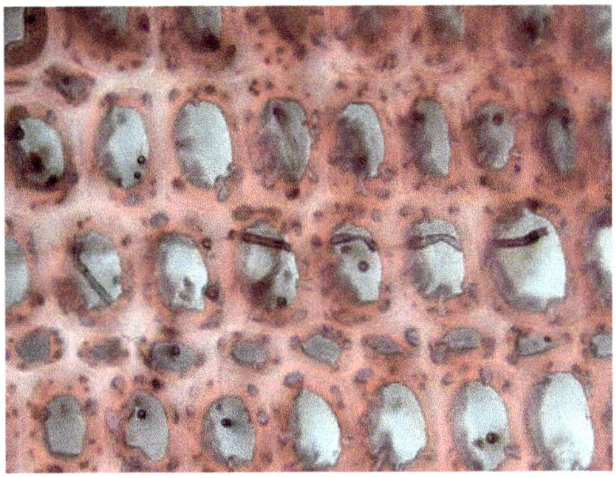

Figure 9 Cross-section of pinewood degraded by soft rot fungi. There are typical holes in the secondary cell wall after decay by hyphae. Growth and penetration through several fibres by brown hyphae can be observed (×400).

Further studies on erosion bacteria have shown that they generally have very slow decay rates (Kim and Singh 2000). Foundation poles as well as archaeological shipwrecks from terrestrial or aquatic/marine environments are not very affected by decay despite hundreds of years of exposure. Under stable, submerged and near anoxic conditions the erosion bacteria might degrade only the first two centimetres of the wood surface (Björdal 2012a). The speed of degradation relates to several factors: age of the find, wood species, whether it is sapwood or heartwood, type of soil or sediment, depth of burial (soil/sediment), thickness of the object, and the environmental conditions (Björdal *et al.* 1999). Ongoing activity by erosion bacteria was evident in marine Viking poles exposed for more than 1000 years in sediment (Björdal *et al.* 2000).

Degradation begins from the surface of the wood and proceeds slowly inwards. Bacteria enter the lumen via anatomical openings such as rays, vessels and pits where they penetrate the S3 layer in order to access the cellulose and carbon-rich S2 layer. At the final stage of decay the less carbon-rich and more lignin-rich S1 and S2 layers are also degraded (Fig. 10). As a result, the secondary cell wall is turned into slime and residual material, whereas the compound middle lamellae, rich in lignin, remain intact (Singh *et al.* 1990). This

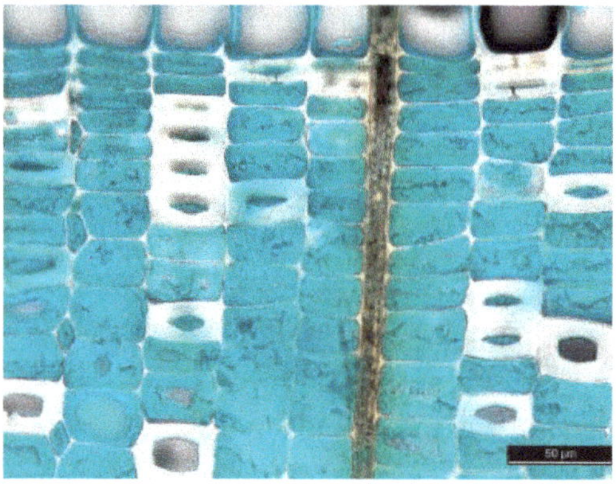

Figure 10 Cross-section showing typical decay by erosion bacteria forming a chess-like pattern. The normal, healthy fibres are white, whereas the degraded fibres are stained blue. Note the residual material occupying the lumen of many degraded fibres (×400).

remaining middle lamellae skeleton maintains the original form of each fibre providing it is kept wet and waterlogged. If the wood is allowed to dry out, water evaporation, creating capillary forces, leads to extreme shrinkage and deformation, as the fibres are extremely weak. Desiccation will ultimately destroy the integrity of the archaeological wood (Björdal 2012b). Characteristic for waterlogged archaeological wood degraded by erosion bacteria is the very high water content and a strong decrease in cellulose and hemicellulose concentrations; lignin remains seemingly unaltered (Pedersen *et al.* 2015). To date it has not been possible to identify erosion bacteria, the main degraders of archaeological submerged wood. Through the use of modern molecular DNA technology, a wide diversity of bacterial species has been found inside waterlogged archaeological wood, but their identification, role and capacities are not as yet defined (Clausen 1996; Helms *et al.* 2004; Landy *et al.* 2008).

Tunnelling bacteria are present in both terrestrial and aquatic/marine environments (Kim and Singh 2000). They are often found active together with soft rot fungi decay, but have slower decay rates. Microscopic investigations on their morphology indicate that they are similar to erosion bacteria (rod-shaped, gram negative, slime producers), but their decay capacities differ greatly. Decay starts at the wood surface where they enter the cell wall by forming fine tunnels covered with cross-walls. In contrast to erosion bacteria they have the ability to degrade both the lignin-rich and lignin-poor layers of the cell wall as well as the middle lamellae, allowing them to spread in any direction (Daniel and Nilsson 1985; Daniel *et al.* 1987). They are also able to degrade contaminated and preservative-treated wood, which makes them the ultimate wood degraders (Daniel and Nilsson 1985; Nilsson *et al.* 1992).

PRESERVATION AND CONSERVATION STRATEGIES AND METHODS

Development of conservation methods

Preservation of archaeological wood from waterlogged sites started in the mid-19th century. Unique finds from sacrificial peaks and lakes led to the development of methods and techniques for long-term preservation with the aim of displaying the finds to the public. Larger objects such as shipwrecks and other constructions demand spacious areas for storage and conservation, and may require long and complicated transport from the archaeological site to the conservation facilities.

The main challenge is to dry the waterlogged and very weak material without damaging the fragile soft surface that may contain unique ornamentation, decoration or other important archaeological information. The methods used should also prevent shrinkage, splitting and other physical disintegration of the delicate material. After treatment, the object must correspond as closely as possible to the original size and form, and visually still look like wood; in other words the integrity of the object must be maintained.

The solution is to replace the large volume of water inside the wood with another material with the ability to physically or chemically support the degraded wood cell wall so that the wood fibres are prevented from physical collapse during drying. One of the first techniques developed and used frequently by museums in the Nordic countries was the alum method.

The still wet, but cleaned, archaeological wood was submerged in a hot alum-salt solution for a number of hours allowing salt to penetrate the wood material. After treatment the wood could be air-dried with good results. Later, in the 1950s, it was found that wood treated with this method was very sensitive to changes in humidity and salt precipitations. Cracks and loose powder pockets inside the wood were also found among many precious items in museums and in storage (Rosenquist 1959; Christensen 1970). Today, re-conservation of alum-treated objects still takes place and recently an extensive research project on this topic started in Norway, where the famous Oseberg find urgently needs to be re-treated (Braovac and Kutzke 2012).

When the famous warship *Vasa* was found in the archipelago of Stockholm in 1957, serious efforts were undertaken to find a new conservation method that would be stable in the long term with the potential to treat large structures (Håfors 1990). Given the ship's huge size, it was decided to spray the conservation solution onto the hull that was attached to a cradle below, a process that took over 10 years. The treatment was based on a water-soluble synthetic wax called polyethylene glycol (PEG), which had previously been used in the wood industry to reduce shrinkage during drying. During the long period of impregnation treatment, different molecular weights of PEG were tested and used (Håfors 2010). Tests showed that a lower molecular weight of PEG (PEG 200–PEG 4000), when diluted in water, was able to penetrate the degraded surface of the waterlogged hull timbers and support the fibre cell wall during the final controlled air drying. The result was timbers with acceptable levels of shrinkage and cracking. The PEG method proved to be very efficient in combination with freeze-drying (Hoffmann and Fortuin 1991; Schnell and Jensen 2007; Babiński 2011). It was found that wood species and the density of the microbially degraded wood were important parameters for the decision on the concentration as well as the molecular weight of PEG used in the treatment. Smaller as well as larger objects have been treated successfully using this often preferred technique.

The PEG method is used frequently around the world but recently the long-term effect of the treatment was questioned when the *Vasa* and other objects in storage revealed salt precipitations with a low pH on the surface. A research project was undertaken in order to determine the chemical processes behind the salt precipitation and the consequences for the long-term preservation. It was concluded that chemical processes take place inside the wood and those that affect the pH are most likely related to the presence of sulfur and iron (Fors 2008; Almkvist and Persson 2011). There is no clear evidence that these processes reduce the strength of the wood in a way that would give rise to major concern, but we have learned that the extraction of iron and sulfur compounds before conservation would probably improve the long-term efficacy as well as the stability of the treatment.

Other conservation treatments have also been developed. Reversible methods, such as sugar and low molecular weight sugar impregnations, and irreversible methods, such as gamma irradiation, silicon oil injection and plastic embedding are also available. Nevertheless, PEG remains the most preferred option both because of the long-term experience with good results and for its reversibility should new options become available in the future (Hoffmann 2013). Under the umbrella of the International Council of Museums Committee for Conservation (ICOM-CC), triennial meetings of the Wet Organic Archaeological Materials (WOAM) working group take place. Conservation methods and preservation issues on waterlogged wood have been discussed and presented in these meetings since the late 1970s by conservators, archaeologists and wood scientists.[1]

PRESERVATION *IN SITU* AND REBURIAL

In parallel with the development of active conservation treatments for waterlogged archaeological wood, another approach for the preservation of unique cultural heritage objects has been explored. In the late

1970s, the Netherlands and Denmark started testing storage of excavated and documented shipwreck timber in nature (Jong 1979a, b; Jespersen 1985) using two approaches: (1) moving the object from its original site and reburying it at a new site (reburial) and (2) physically protecting the object at its find site (*in situ* preservation). These methods were rarely used until 2002 when *in situ* preservation became an official issue after UNESCO published the proceedings from the *Convention on the Protection of Underwater Cultural Heritage* (UNESCO 2001). This document probably promoted the establishment and funding of several European research projects focusing on marine archaeology and the protection of shipwrecks and underwater archaeological sites, such as the Moss, the WreckProtect and the SASMAP projects. The WreckProtect project (2009–2011) focused on climate change and the potential risk of shipworm spread into the Baltic Sea, as well as on available methods for protection *in situ* (Björdal *et al.* 2012a, b). This was followed by the SASMAP project (2012–2015), which adopted a broader approach, including development of *in situ* protection methods (Gregory 2015). A Swedish project, 'Reburial and Analyses of Archaeological Remains' (RAAR) explored the idea of marine reburial and included a long-term study on the deterioration of several materials buried in sediment (Bergstrand *et al.* 2004).

The methods used for *in situ* protection of wooden remains in marine environments are based on the premise that a 'near anoxic' environment will prevent or strongly inhibit biological as well as physical degradation. A well-established method is to cover the wreck with sediment and add sandbags to keep the sediment in place. Other methods focus on artificial development of sediment layers on the wreck site using artificial sea grass or debris netting. All methods include covering items with specific types of geotextile to protect the wood and exclude marine borers. However, success depends to a large degree on the environment and the dynamics of the site.[2]

Several projects have reported difficulties protecting terrestrial archaeological sites containing wood. One of the biggest threats is drainage of the local area, which alters the site from a stable, waterlogged environment to an aggressive damp or dry environment. Understanding the hydrogeological dynamics of a site can often be difficult but is necessary in order to preserve the archaeological remains. At the Nydam site in Denmark (a sacrificial lake from the Iron Age), problems with reduced water levels and the growth of horsetail roots endangered the artefacts in the soil. Raising the water table would have secured the long-term preservation of the site but reasonable and economically feasible solutions were not found. Instead, long-term monitoring devices were installed to record the environmental changes. To avoid penetration of wooden objects by horsetail, the soil was covered with geotextile which proved successful (Gregory and Matthiesen 2012). In the UK, Sweet Track, a Stone Age wetland site in the Somerset Levels with a wooden causeway, was protected by artificial and localized rising of the water level (Brunning *et al.* 2000; Brunning 2012). The foundation including wooden construction of the Elizabethan theatre, the Rose in London, is another example of active management, monitoring and protection of a unique historic site (Corfield 2012). More advanced technical solutions were found for the UNESCO protected heritage site Bryggen, in Norway, where similar problems endangered the historical structures in the cultural layers beneath the present old town area. In this case, it was possible to raise the water level using rainwater low in oxygen (Rytter and Schonhowd 2015).

There are many sites with wooden archaeological materials that are without protection, and some are tourist attractions where the consequence is total decomposition (Dal *et al.* 2012). A few are fully excavated if the site is considered to be outstanding from an archaeological perspective, and even then a successful *in situ* preservation cannot be guaranteed as in the case of Flag Fen in the UK, which was found in 1999 but only recently excavated when it was finally concluded that *in situ* preservation was not possible (Kennedy 2016).

CONCLUSIONS

Today, wooden material exposed at terrestrial and aquatic sites are at the mercy of nature. Some, by good fortune, are well protected by sediments, whereas others are degrading fast. Preservation of archaeological sites using *in situ* or reburial approaches are still rare, despite the fact that knowledge on decay processes and methods for protection does exist. Future efforts for the preservation of wooden structures must focus on communication and discussions with stakeholders so that these issues can be further integrated into cultural heritage management. More multi-and interdisciplinary collaboration is required for greater understanding of the decay processes and refining the methods available for securing these unique archaeological objects for future generations.

NOTES

1. Readers with a special interest in this topic are encouraged to take part in the proceedings: http://www.icom-cc.org/42/working-groups/wet-organic-archaelogical-materials (accessed 5 July 2018).
2. More on the subject can be found in Björdal *et al.* (2012a, b) or in the guidelines available at http://wreckprotect.org/index.php?id=12679 (accessed 5 July 2018).

REFERENCES

Almkvist, G. and Persson, I. 2011. 'Distribution of iron and sulfur and their speciation in relation to degradation processes in wood from the Swedish warship *Vasa*', *New Journal of Chemistry* 35(7): 1491–502.

Babiński, L. 2011. 'Investigations on pre-treatment prior to freeze-drying of archaeological pine wood with abnormal shrinkage anisotropy', *Journal of Archaeological Science* 38(7): 1709–15.

Bergstrand, T., Björdal, C.G., Nilsson, T., Bohm, C., Gregory, D., Nyström, I., Peacock, E., Turner-Walker, G., Richards, V. and MacLeod, I. 2004. 'Reburial as a method of preserving archaeological remains: a presentation of the Marstrand project', in *Proceedings of the 9th ICOM Group on Wet Organic Archaeological Materials Conference*, 9–39. Bremerhaven: H M Hauschild.

Björdal, C.G. 2012a. 'Evaluation of microbial degradation of shipwrecks in the Baltic Sea', *International Biodeterioration and Biodegradation* 70: 126–40.

Björdal, C. 2012b. 'Microbial degradation of waterlogged archaeological wood', *Journal of Cultural Heritage* 13(3): 118–22.

Björdal, C.G. and Nilsson, T. 2002. 'Waterlogged archaeological wood: a substrate for white rot fungi during drainage of wetlands', *International Biodeterioration and Biodegradation* 50: 17–23.

Björdal, C.G. and Nilsson, T. 2008. 'Reburial of shipwrecks in marine sediments: a long term study on wood degradation', *Journal of Archaeological Science* 35: 862–72.

Björdal, C.G., Nilsson, T. and Daniel, G. 1999. 'Microbial decay of waterlogged archaeological wood found in Sweden', *International Biodeterioration and Biodegradation* 43(1–2): 63–73.

Björdal, C., Daniel, G. and Nilsson, T. 2000. 'Depth of burial: an important factor in controlling bacterial decay of waterlogged archaeological poles', *International Biodeterioration and Biodegradation* 45: 15–26.

Björdal, C., Gregory, D., Manders, M., Al-Hamdani, Z., Christin, A., Jonathan, N.H. and Dencker, J. 2012a. 'Strategies for protection of wooden underwater cultural heritage in the Baltic Sea against marine borers: the EU project WreckProtect', *Conservation and Management of Archaeological Sites* 14(1–4): 201–14.

Björdal, C.G., Gregory, D. and Trakadas, A. (eds) 2012b. *WreckProtect: Decay and Protection of Wooden Archaeological Shipwrecks*. London: Archaeopress.

Blanchette, R.A. 2000. 'A review of microbial deterioration found in archaeological wood from different environments', *International Biodeterioration and Biodegradation* 46: 189–204.

Blanchette, R.A. 2010. 'Microbial degradation of wood from aquatic and terrestrial environments', in *Cultural Heritage Microbiology: Fundamental Studies in Conservation Science*, R. Mitchell and C.J. McNamara (eds), 179–90. Washington DC: American Society of Microbiology Press.

Blanchette, R.A., Nilsson, T., Daniel, G. and Abad, A.R. 1990. 'Biological degradation of wood', in *Archaeological Wood*, R.J. Rowell and R.J. Barbour (eds), 141–74. Washington DC: American Chemical Society.

Borges, L.M.S. 2014. 'Biodegradation of wood exposed in the marine environment: evaluation of the hazard posed by marine wood-borers in fifteen European sites', *International Biodeterioration and Biodegradation* 96: 97–104.

Boutelje, J. and Bravery, A.F. 1968. 'Observations on the bacterial attack of piles supporting a Stockholm building', *Journal of the Institute of Wood Science* 20: 47–57.

Bowyer, J., Shmulsky, R. and Haygreen, J.G. 2007. *Forest Products and Wood Science: An Introduction.* Ames IA: Blackwell.

Braovac, S. and Kutzke, H. 2012. 'The presence of sulfuric acid in alum-conserved wood: origin and consequences', *Journal of Cultural Heritage* 13(3): S203–S208.

Brunning, R. 1996. 'An *in situ* preservation project in Somerset', in *Proceedings of the 6th ICOM Group on Wet Organic Archaeological Materials Conference.* Bremerhaven: ICOM.

Brunning, R. 2012. 'Partial solutions to partially understood problems: the experience of *in situ* monitoring and preservation in Somerset's peatlands', *Conservation and Management of Archaeological Sites* 14(1/4): 397–405.

Brunning, R., Hogan, D., Jones, J., Jones, M., Maltby, E., Robinson, M. and Straker, V. 2000. 'Saving the Sweet Track: the *in situ* preservation of a Neolithic wooden trackway, Somerset, UK', *Conservation and Management of Archaeological Sites* 4(1): 3–20.

Caple, C. 1994. 'Defining a reburial environment: research problems characterising waterlogged anoxic environments', in *Proceedings of the 5th ICOM Group on Wet Organic Archaeological Materials Conference*, 407–21. Bremerhaven: Deutsches Schiffahrtsmuseum.

Christensen, B.B. 1970. *The Conservation of Waterlogged Wood in the National Museum of Denmark.* Copenhagen: National Museum of Denmark.

Clausen, A.C. 1996. 'Bacterial associations with decaying wood: a review', *International Biodeterioration and Biodegradation* 37(1–2): 101–7.

Corfield, M. 2012. 'The Rose Theatre: twenty years of continuous monitoring, lessons, and legacy', *Conservation and Management of Archaeological Sites* 14(1/4): 384–96.

Cox, M., Earwood, C., Jones, E.B.G., Jones, J., Straker, V., Robinson, M., Tibbett, M. and West, S. 2001. 'An assessment of the impact of trees upon archaeology within a relict wetland', *Journal of Archaeological Science* 28(10): 1069–84.

Cragg, S.M., Pitman, A.J. and Henderson, S.M. 1999. 'Developments in the understanding of the biology of marine wood-boring crustaceans and in methods of controlling them', *International Biodeterioration and Biodegradation* 43: 197–205.

Dal, R.C., Fruet, S., Bellintani, P., Pisu, N., Macchioni, N., Pizzo, B. and Capretti, C. 2012. 'Preserving archaeological remains *in situ*: three case studies in Trentino, Italy', *Conservation and Management of Archaeological Sites* 14(1/4): 239–48.

Daniel, G. 1994. 'Use of electron microscopy for aiding our understanding of wood biodegradation', *FEMS Microbiology Reviews* 13: 199–233.

Daniel, G. and Nilsson, T. 1985. *Ultrastructural and T.E.M.-EDAX Studies on the Degradation of CCA Treated Radiata Pine by Tunnelling Bacteria.* The International Research Group on Wood Preservation Document No: IRG/WP/1264.

Daniel, G.F. and Nilsson, T. 1986. *Ultrastructural Observations on Wood Degrading Erosion Bacteria.* The International Research Group on Wood Preservation Documents No: IRG/WP/1283.

Daniel, G.F., Nilsson, T. and Singh, A.P. 1987. 'Degradation of lignocellulosics by unique tunnel-forming bacteria', *Canadian Journal of Microbiology* 33: 943–8.

Eaton, R.A. and Hale, M.D.C. 1993. *Wood: Decay, Pests and Protection.* London, Chapman & Hall.

Einarsson, L. 1990. '*Kronan*: underwater archaeological investigations of a 17th-century man-of-war. The nature, aims and development of a maritime cultural project', *International Journal of Nautical Archaeology* 19(4): 279–97.

Eriksen, A.M., Gregory, G. and Shashoua, Y. 2015. 'Selective attack of waterlogged archaeological wood by the shipworm, *Teredo navalis* and its implications for *in-situ* preservation', *Journal of Archaeological Science* 55: 9–15.

Eriksson, N. 2015. 'Skeppsarkeologisk analys. Gribshunden (1495) Skeppsvrak vid Stora Ekön, Ronneby, Blekinge', in *Marinarkeologisk undersökning 2013–2015*, J. Rönnby, 13–32. Karlskrona: Blekinge Museum.

Eriksson, N. and Rönnby, J. 2012. 'The ghost ship: an intact fluyt from c.1650 in the middle of the Baltic Sea', *International Journal of Nautical Archaeology* 41(2): 350–61.

Fetter, C.W. 1994. *Applied Hydrogeology.* New Jersey: Prentice Hall.

Fors, Y. 2008. *Sulfur-Related Conservation Concerns for Marine Archaeological Wood: The Origin, Speciation and Distribution of Accumulated Sulfur with Some Remedies for the Vasa. Structural Chemistry*, PhD dissertation, Stockholm University.

Greaves, H. 1969. 'Micromorphology of the bacterial attack of wood', *Wood Science and Technology* 3: 150–66.

Gregory, D.J. 2015. 'Development of tools and techniques to survey, assess, stabilise, monitor and preserve underwater archaeological sites: Sasmap'. Available at: http://journals.sagepub.com/doi/abs/10.1260/2047-4970.1.0.367?journalCode=hdea (accessed 7 June 2018).

Gregory, D. and Matthiesen, H. 2012. 'Nydam Mose: *in situ* preservation at work', *Conservation and Management of Archaeological Sites* 14(1/4): 479–86.

Gregory, D., Matthiesen, H. and Björdal, C. 2002. '*In situ* preservation of artefacts in Nydam Mose: studies into environmental monitoring and the deterioration of wooden artefacts', in *Proceedings of the 8th ICOM Group on Wet Organic Archaeological Materials Conference*, 213–23. Bremerhaven: Deutsches Schiffahrtsmuseum.

Helms, C.A., Martiny, C.A., Hofman-Bang, J., Ahring, K.B. and Kilstrup, M. 2004. 'Identification of bacterial cultures from archaeological wood using molecular biological techniques', *International Biodeterioration and Biodegradation* 53: 79–88.

Hoadley, B.R. 1990. *Identifying Wood: Accurate Results with Simple Tools*. Newtown CT: Taunton Press.

Hoffmann, P. 2013. *Conservation of Archaeological Ships and Boats: Personal Experiences*. London: Archetype Publications in association with Deutsches Shiffahrtsmuseum.

Hoffmann, P. and Fortuin, G. 1991. 'An evaluation study on the freeze-drying of waterlogged wood', in *Proceedings of the 4th ICOM Group on Wet Organic Archaeological Materials Conference*, 331–58. Bremerhaven: ICOM.

Håfors, B. 1990. 'The role of the *Wasa* in the development of the polyethylene glycol preservation method', in *Archaeological Wood*, R.M. Rowell and R.J. Barbour (eds), 195–216. Washington DC: American Chemical Society.

Håfors, B. 2010. *Conservation of the Wood of the Swedish Warship Vasa of A.D. 1628*. Evaluation of Polyethylene Glycol Conservation Programmes. Gothenburg: University of Gothenburg.

Jespersen, K. 1985. 'Extended storage of waterlogged wood in nature', in *Proceedings of the 2nd ICOM Waterlogged Wood Working Group Conference*. Grenoble: Centre Dètude et de traitement des bois gorges d'eau.

Jones, E.B.G., Kuhne, H., Trussel, P.C. and Turner, R.D. 1972. 'Results of an international cooperative research programme on the biodeterioration of timber submerged in the sea', *Material und Organismen* 7: 93–118.

Jones, M.A. and Rule, M.H. 1990. 'Preserving the wreck of the *Mary Rose*', in *Proceedings of the 4th ICOM Group on Wet Organic Archaeological Materials Conference*. Bremerhaven: ICOM.

Jong, J.D. 1979a. 'Protection and conservation of shipwrecks', in *Medieval Ships and Harbours*, M. Grail (ed.), 247–60. Oxford: BAR.

Jong, J.D. 1979b. 'The deterioration of waterlogged wood and its protection in the soil', paper presented at *Conservation of Waterlogged Wood: International Symposium on the Conservation of Large Objects of Waterlogged Wood, Amsterdam, UNESCO*.

Jordan, A.B. 2001. 'Site characteristics impacting the survival of historic waterlogged wood: a review', *International Biodeterioration and Biodegradation* 47(1): 47–54.

Kennedy, M. 2016. '"A bronze age Pompeii": archaeologists hail discovery of Peterborough site', *The Guardian* 12 January 2016.

Kim, Y.S. and Singh, A.P. 2000. 'Micromorphological characteristics of wood biodegradation in wet environments: a review', *IAWA Journal* 21(2): 135–55.

Kim, Y.S., Singh, A.P. and Nilsson, T. 1996. 'Bacteria as important degraders in waterlogged archaeological woods', *Holzforschung* 50: 389–92.

Kohlmeyer, J. and Kohlmeyer, E. 1979. *Marine Mycology: The Higher Fungi*. London: Academic Press.

Landy, E.T., Michell, J.I., Hotchkiss, S. and Eaton, R.A. 2008. 'Bacterial diversity associated with archaeological waterlogged wood: Ribosomal RNA clone libraries and denaturing gradient gel electrophoresis (DGGE)', *International Biodeterioration and Biodegradation* 61: 106–16.

Matthiesen, H., Dunlop, R., Jensen, J.A., Beer, H.D. and Christensson, A. 2008. 'Monitoring of preservation conditions and evaluation of decay rates of urban deposits: results from the first five years of monitoring at Bryggen in Bergen', in *Proceedings from the Conference: Preserving Archaeological Remains in Situ 3*, H. Kars and

R.M. van Heeringen (eds),163–74. Amsterdam: Institute for Geo and Bioarchaeology, Vrije Universiteit.

Mouzouras, R., Jones, E.B.G., Venkatasamy, R. and Moss, S.T. 1986. 'Decay of wood by micro-organisms in marine environments', in *British Wood Preserving Association Annual Convention 1986*, 27–44 . London: British Wood Preserving Association.

Mouzouras, R., Jones, A.M., Jones, E.B.G. and Rule, M.H. 1990. 'Non-destructive evaluation of hull and stored timbers from the Tudor ship *Mary Rose*', *Studies in Conservation* 34(4): 173–88.

Nair, N.B. and Saraswarthy, M. 1971. 'The biology of wood-boring teredinid mollusks', *Advances in Marine Biology* 9: 335–509.

Nilsson, T. and Daniel G. 1986. 'Lignolytic activity of wood-degrading bacteria', in *Biotechnology in the Pulp and Paper Industry: Third International Conference*, 54–6. Stockholm: STFI (Swedish Pulp and Paper Research Institute).

Nilsson, T. and Holt, D. 1983. 'Short note: bacterial attack occurring in the S2 layer of wood fibers', *Holzforschung* 37: 107–8.

Nilsson, T., Singh, A. and Daniel, G. 1992. 'Ultrastructure of the attack of *Eusideroxylon zwageri* wood by tunneling bacteria', *Holzforschung* 46(5): 361–7.

Pedersen, N.B., Björdal, C.G., Jensen, P. and Felby, C. 2013. *Bacterial Degradation of Archaeological Wood in Anoxic Waterlogged Environments*. Cambridge: RSC Publishing.

Pedersen, N.B., Gierlinger, N. and Thygesen, L.G. 2015. 'Bacterial and abiotic decay in waterlogged archaeological *Picea abies* (L.) Karst studied by confocal Raman imaging and ATR-FTIR spectroscopy', *Holzforschung* 69(1): 103–12.

Pointing, S.B. and Hyde, K.D. 2000. 'Lignocellulose degrading marine fungi', *Biofouling* 15(1–3): 221–9.

Powell, K.L., Pedley, S., Daniel, G. and Corfield, M. 2001. 'Ultrastructural observations of microbial succession and decay of wood buried at a Bronze Age archaeological site', *International Biodeterioration and Biodegradation* 47: 165–73.

Rosenquist, A.M. 1959. 'The stabilizing of wood found in the Viking ship of Oseberg, Part I', *Studies in Conservation* 1: 13–22.

Rowell, R.M. 2012. *Handbook of Wood Chemistry and Wood Composites*, 2nd edn. London: CRC Press.

Rytter, J. and Schonhowd, I. 2015. *Monitoring, Mitigation, Management: The Groundwater Project. Safeguarding the World Heritage Site of Bryggen in Bergen*. Oslo: Riksantikvaren.

Schnell, U. and Jensen, P. 2007. 'Determination of maximum freeze drying temperature for PEG-impregnated archaeological wood', *Studies in Conservation* 52(1): 50–58.

Schwarze, F.W.M.R. 2007. 'Wood decay under the microscope', *Fungal Biology Reviews* 21: 133–70.

Şen, S., Sivrikaya, H. and Yalçin, M. 2009. 'Natural durability of heartwoods from European and tropical Africa trees exposed to marine conditions', *African Journal of Biotechnology* 8(18): 4425–32.

Singh, A.P. and Butcher, J.A. 1991. 'Bacterial degradation of wood cell walls: a review of degradation patterns', *Journal of the Institute of Wood Science* 12(3): 143–57.

Singh, A.P., Nilsson, T. and Daniel, G.F. 1990. 'Bacterial attack of *Pinus sylvestris* wood under near anaerobic conditions', *Journal of the Institute of Wood Science* 11: 237–49.

UNESCO 2001. *UNESCO Convention on the Protection of Underwater Cultural Heritage.* Available from: http://www.unesco.org/culture/underwater/infokit_en/.

Zabel, R.A. and Morrell, J.J. 1992. *Wood Microbiology: Decay and its Prevention*. San Diego: Academic Press.

AUTHOR'S ADDRESS

Charlotte Björdal, Department of Marine Sciences, University of Gothenburg, Carl Skottbergs gata 22 B, Box 461, 405 30 Göteborg, Sweden (charlotte.bjordal@marine.gu.se)

NOVEL APPROACHES TO PRESERVATION

BIOCLEANING OF CULTURAL HERITAGE TREASURES

Pilar Bosch-Roig and Giancarlo Ranalli

INTRODUCTION

In the context of biotechnologies applied to the fields of conservation and restoration, techniques for the biological cleaning (biocleaning) of cultural heritage items continue to be improved and perfected. This contribution discusses scenarios in which biotechnologies can act in support of, or as an alternative to, traditional methods of restoration, when such methods prove to be ineffective or dangerous for the operators or the environment. The perception of potential microbial risk, such as biodeteriorating agents on heritage items, is gradually changing thanks to the promising results achieved using innovative, advanced methodologies, both in the laboratory and from on-site applications.

EwaGlos: European Illustrated Glossary of Conservation Terms for Wall Paintings and Architectural Surfaces defines cleaning as the 'action performed to remove dirt deposits, foreign matter and products of alteration present on artwork surfaces that can be a source of decay, deterioration or aesthetic disturbance' (EwaGlos 2016: 304). This working group of restorers and researchers clarify that the main cleaning techniques are mechanical, physical-chemical and biological. The latter includes the use of living organisms, and/or their enzymes, as cleaning agents by controlled application of their metabolic processes.

Microbial biotechnologies used in conservation and restoration science and technology interact with art and cultural heritage in many ways. The arts are also influenced, and/or inspired, by new materials developed from scientific investigation. Scientific evidence of artworks allows us to understand the progress of human civilizations (Frazzetto 2004). In this context, the biological sciences assume an even more important role in protecting and conserving cultural heritage for future generations (Rinaldi 2006). This field represents a considerable challenge when considering the magnitude and diversity of objects involved and the enormous variety of materials used (including stone, metal, ceramics, paper, synthetic substances and organic matter). Many disciplines are involved including researchers, scientists and politicians as well as conservators, art historians and restorers.

The use of living organisms, or derivatives thereof, to make or modify products or processes for specific uses, falls under the general heading of biotechnology. From a microbiological viewpoint, the main concepts of biotechnologies in relation to cultural heritage biocleaning applications involve establishing protocols which provide information on the microbial communities that colonize a material. This includes their biodeterioration potential along with the capacity to control and prevent excess colonization. These concepts

Table 1 Summary of the main biocleaning literature.

Main decay agents	Type of materials	Biocleaning bacteria
STONE BUILDINGS AND SCULPTURES		
Black crusts	Marble (Candoglia stone) (I)	*D. vulgaris* subsp. *vulgaris* (An)
	Marble (Candoglia stone) (I)	*D. vulgaris* subsp. *vulgaris* (An)
	Marble sculpture, Milan (I)	*D. vulgaris* subsp. *vulgaris* (An)
	Limestone sculpture, Trento (I)	*D. vulgaris* subsp. *vulgaris* (An)
	Colored lithotypes, Florence (I)	*D. vulgaris* subsp. *vulgaris* (An)
Grey deposits and black crust	Marble column and statue, Cemetery of Milan (I)	*D. vulgaris* subsp. *vulgaris* (An)
Nitrates	Brickworks and calcareous stones (I) (marble and Vicenza stones)	*P. stutzeri* (Ae)
	Sandstone walls, Matera (I)	*P. pseudoalcaligenes* (An)
Nitrates and Sulfates	Sandstone walls, Matera (I)	*P. pseudoalcaligenes* (Ae) *D. vulgaris* (An)
Sulfates	Marble (Georgia) stone and statue (US)	*D. desulfuricans* (An)
	Marble and sandstone (G)	*D. vulgaris* (An)
	Marble (I)	*D. desulfuricans,* (An) *D. vulgaris* (An)
WALL PAINTINGS		
Animal glue	Camposanto Monumentale di Pisa frescoes, Pisa (I)	*P. stutzeri A29* (Ae)
Calcium sulfate and carbonate	Casina Farnese wall paintings (Palatine Hill, Rome) (I)	*Cellulosimicrobium cellulans* (Ae)
Inorganic compounds and proteins	Casina Farnese wall paintings (Palatine Hill, Rome) (I)	*Stenotrophomonas maltophilia* (Ae) *Pseudomonas koreensis* (Ae)
Animal glue and salt efflorescence	Frescoes, Valencia (SP)	*P. stutzeri* DSMZ5190 (Ae)
	Frescoes, Valencia (SP)	*P. stutzeri* DSMZ5190 (Ae)
MANUSCRIPTS		
Animal glue	Ancient paper (I)	*Ochrobac- trum* sp. TNS15
GRAFFITI		
Graffiti paintings	Glass slides (I and US)	*Desulfovibrio desulfuricans* ATCC 13541 and others
ARCHAEOLOGICAL ARTEFACTS		
Sulfides	Archaeological waterlogged wood (S and FR)	*Thiobacillus denitrificans*
Iron corrosion layer	Iron archaeological artefacts (S)	*Desulfitobacterium hafniense* (An)
Efflorescence and animal glue	Archaeological frescoes, ceramic and bones (SP)	*P. stutzeri* DSMZ 5190
OTHERS		
Black organic layer (mainly lichens)	Concrete (B)	*Thiobacillus* sp. (Ae)

Key: (I): Italy; (SP): Spain, (G): Germany; (B): Belgium; (US): United States of America; (S): Switzerland; (FR): France; (Ae): Aerobic metabolism; (An): Anaerobic metabolism

Matrices used as delivery system	**References**
STONE BUILDINGS AND SCULPTURES	
Sepiolite	Cappitelli *et al.* 2006
Hydrobiogel.97	Cappitelli *et al.* 2006; Ranalli *et al.* 2000
Carbogel	Cappitelli *et al.* 2005
Carbogel	Polo *et al.* 2010
Carbogel	Gioventù *et al.* 2011
Arbocel	Troiano *et al.* 2013
Sepiolite	Ranalli *et al.* 1996
Mortar and alginate beads	May *et al.* 2008
Carbogel	Alfano *et al.* 2011
Immersion	Gauri *et al.* 1989, 1992
Immersion	Heselmeyer *et al.* 1991
Sepiolite	Ranalli *et al.* 1996, 1997
WALL PAINTINGS	
Cotton Wool	Ranalli *et al.* 2005; Antonioli *et al.* 2005 Lustrato *et al.* 2012
Laponite	Mazzoni *et al.* 2014
Laponite	Mazzoni *et al.* 2014
Cotton wool	Bosch-Roig *et al.* 2010
Agar	Bosch-Roig *et al.* 2012, 2013a, b
MANUSCRIPTS	
Agar	Barbabietola *et al.* 2012, 2016
GRAFFITI	
Immersion	Giacomucci *et al.* 2012. Sanmartín *et al.* 2014, 2015
ARCHAEOLOGICAL ARTEFACTS	
Immersion	Albelda-Berenguer *et al.* 2017
Immersion	Comensoli *et al.* 2017
Agar	Martín Ortega 2015
OTHERS	
Immersion	De Graef *et al.* 2003, 2005; De Belie *et al.* 2005

offer new procedures for biocleaning and bioconsolidation application (Palla 2006).

Bioremediation is another term that refers to a range of processes using microorganisms to return contaminated environments to their original condition (Ramirez *et al.* 2005). The biotreatment of contaminated soils or chemical waste by microorganisms is well developed. The application of bioremediation processes to improve the condition of deteriorated artwork is known as biorestoration and significant developments have been made in this area (Webster and May 2006).

The biocleaning of frescoes (Ranalli *et al.* 2005), the removal of black crusts and salts from stone (Heselmeyer *et al.* 1991; Ranalli *et al.* 1997; Cappitelli *et al.* 2007), and the bioconsolidation of carbonatic materials (Tiano *et al.* 1999; Castanier *et al.* 2000; Rodriguez-Navarro *et al.* 2003; Barabesi *et al.* 2006; Junier and Joseph 2017) represent examples of how microbial technology may contribute positively. One of the most interesting aspects of these new methods is that the microbes involved in the biorestoration processes can be isolated in nature and/or selected from the primitive microbial populations inhabiting historical artwork (bioaugmentation). Therefore, a potentially negative factor may in fact prove to be a positive approach to conservation.

One major advantage of biological cleaning over chemical treatments is safety. A number of toxic organic solvents are still used in some traditional methods of cleaning artwork. In numerous examples of biocleaning procedures cited, the main processes are water-based, which is suitable for living microorganisms or for their associated enzymatic activity. Campani *et al.* (2007) reported that water is necessary, sometimes irreplaceable, in performing a number of functions in artwork conservation. However, in the presence of porous materials, or of superficial alterations (such as holes, craquelures or lacunae), the use of water can represent a risk due to increased water diffusion into supports and the increased susceptibility to chemical-physical changes or variations in size (Bellucci and Cremonesi 1994). Therefore, controlled addition of water to materials and its regulated use is necessary in order to reduce diffusion and absorption by surface materials.

Bearing in mind some important properties of water (such as high surface tension and low wetting power), it is possible to obtain a range of water viscosity by adding mineral or organic substances. Among these, simple thickening agents, such as sepiolite or cellulose chemical pulp, can be chosen or gelling agents, such as cellulose ethers, added (Cremonesi 1999, 2004). These products are inexpensive, simple to prepare, long lasting and most commonly used, although their adhesive qualities are limited at concentrations below 10–15%. Improved delivery systems, such as polyacrylic acid derivatives have been used since 1980. At concentrations of 1–1.5% (w/v), they have a very high viscosity – at least 40–50 times greater than cellulose ethers – allowing for easy removal (Wolbers 2000: 198). Agar gels have also recently been used for biocleaning. These gels permit delivery of a controlled supply of water and good contact and adhesion with the surface to be cleaned, regardless of the inclination of the artwork, so no residue is then left on the cleaned surface (Bosch-Roig *et al.* 2013b, 2017).

Studies carried out on the biocleaning of artwork over the last few decades report on numerous delivery systems. A summary of the primary biocleaning literature is represented in Table 1 including the principal agents of decay, the materials cleaned, the bacteria used, the type of matrices and references.

In a critical review by Bosch-Roig and Ranalli (2014), the main characteristics and properties of the delivery systems adopted are described, compared and supported by evidence in order to aid conservator-restorers and researchers to choose the most appropriate system according to safety requirements for specific biocleaning applications.

An important question relates to the safety of the application of microorganisms on artwork (Bosch-Roig and Ranalli 2014). In all publications, microbiological control of the cleaned surface can be found, yet microbiological controls implemented after biocleaning processes are also very important in order to ensure that no microorganisms are left on the cleaned

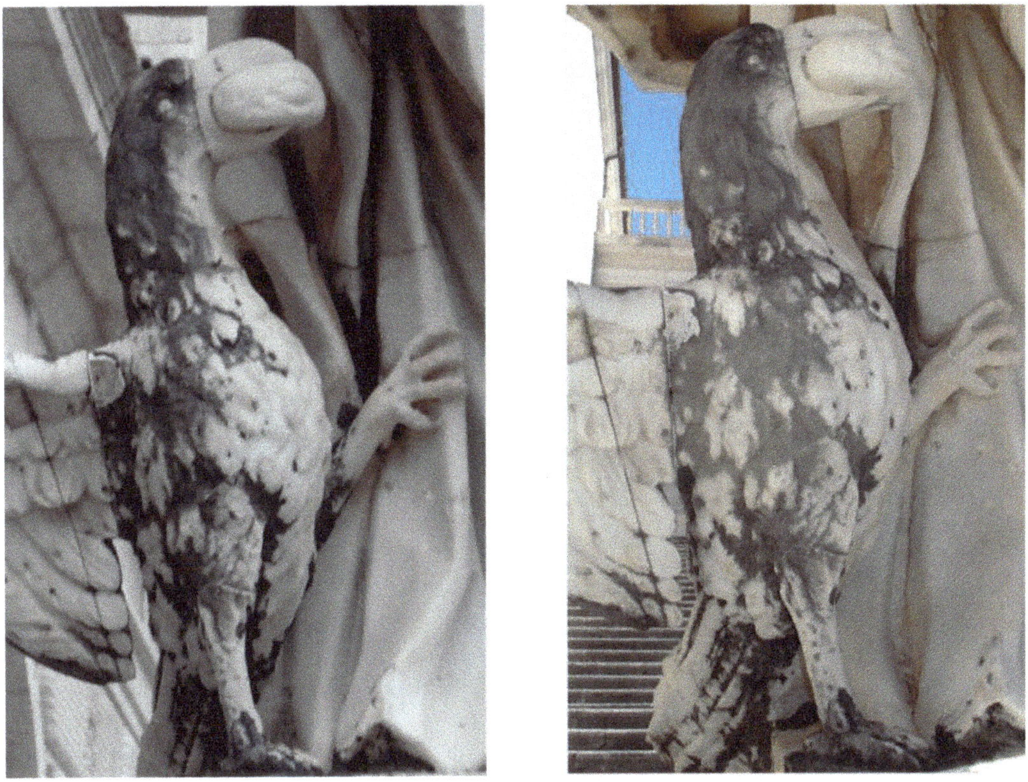

Figure 1 The effect of a one-step bioremoval (24 hours) of black crust on the surface of a crow figure (images taken from a different angle), part of the San Benedetto sculpture at Montecassino Abbey in Italy (Pontone 2014).

surface after treatment. A variety of techniques to evaluate microbial populations can be used, such as ATP surface analysis and contact plate microbial counts.

BIOCLEANING OF STONE BUILDINGS AND SCULPTURES: CASE STUDIES

The first examples of viable bacterial cell applications used in biocleaning procedures were carried out on stone for the bioremoval of black crusts and salts. In 1970, Moncrieff and Hempel described the use of a biological pack and the role of microorganisms in a poultice. The first successful application of anaerobic sulfate-reducing bacteria (Atlas *et al.* 1988; Atlas 1995; Gauri *et al.* 1989, 1992) used *Desulfovibrio desulfuricans* for the removal of sulfates from black crusts. Heselmeyer *et al.* (1991) applied *D. vulgaris* to gypsum crusts, resulting in the conversion of gypsum to calcite, in addition to removing the black deposits. Furthermore, Ranalli *et al.* (1996, 1997) demonstrated the possibility of using different strains of sulfate-reducing bacteria, such *as D. vulgaris* and *D. desulfuricans*, and denitrifying bacteria, *Pseudomonas stutzeri* and *P. pseudoalcaligenes* (May *et al.* 2008), in order to optimize a strategy for the removal of sulfates and nitrates, respectively, from artwork. In the experiment carried out on marble samples, about 80% of the nitrates and sulfates were removed as a result of bacterial activity within 36 hours of application.

The biotechnological procedure for black crust removal was first proposed using colonized sepiolite (Ranalli *et al.* 1997). This was improved, at Milan Cathedral, by the use of washed sepiolite which, because of its low ion content, permitted bacterial adhesion to the micropores of the matrix thereby reducing the risk of colour changes on the artwork's surface (Cappitelli *et al.* 2006). Conventional, chemical and mechanical techniques for the removal of black salt crusts from stone all have a number of disadvantages in that colour changes, salt transposition within the stone structure and even the excessive removal of the original materials may occur.

In a study performed by Cappitelli *et al.* (2007) on altered fragments of Angera stone at Milan Cathedral, the sepiolite was replaced with Carbogel and Carbopol which enabled greater sulfate removal (98%) and simplified the final removal of the activated, bacterial cell-enriched biogel from the surface of the artwork when the treatment was concluded (Cappitelli *et al.* 2006; Lustrato *et al.* 2012). In a comparative test, the relative merits of chemical and biotechnological procedures were examined: the chemical examination consisted of cleaning with a mixture of 30 g ethylenediaminetetraacetic acid (EDTA), 50 g ammonium carbonate and 10 mL of a surfactant, while the biotechnological method used a suspension of viable *D. vulgaris* cells entrapped in Carbogel. Both treatments converted gypsum to calcite, allowing consolidation of the stone. The chemical method left small, diffuse, black-grey residues of crust and the biological treatment resulted in a more homogeneous removal of the surface deposits, while preserving the noble patina under the black crust (Troiano *et al.* 2013). Figure 1 shows the effect of the bioremoval of black crusts on a sculpture.

Another case study of the successful application of biocleaning techniques on stone buildings was performed at Matera Cathedral. The 12th-century Romanesque-Apulian style church is located in the ancient part of Matera, called I Sassi, an agglomeration of almost completely abandoned dwellings hewn from the soft calcareous tuff rock of the Gravina Torrent valley. Attention focused on the outer wall of Matera Cathedral showed decay caused principally by water seepage and salt efflorescence (May *et al.* 2008; Alfano *et al.* 2011). The main conservation problems are due to the abundant soluble nitrates and sulfates in the groundwater that are causing reduction in stone cohesion, loss of material and darkening of the lower 7 m of the outer walls due to upward salt migration. Until the 18th century, this area was used as a cemetery. The nitrates originate from the oxidation of various nitrogen-based compounds (proteins, amino acids and urea) released from the bodies that were buried there. Similarly, the sulfates are mainly derived from sulfur-containing amino acids, rather than from environmental deposits. Moreover, egg residues are present from historical treatments that sought to consolidate the tuff surface. Bioremoval of nitrate and sulfate salts from the tuff stone surfaces was performed using nitrate- and sulfate-reducing bacteria. The treatment was based on the direct application of the strains *P. pseudoalcaligenes* (nitrate reducing) and *D. vulgaris* ATCC 29579 (sulfate reducing) onto the altered stone surfaces (Cappitelli *et al.* 2006). In order to permit uniform application of bacteria, maintain good contact between the cells and the surface area to be treated, and easily remove the cells after treatment, Carbogel was adopted as the delivery system. Bacterial cells were entrapped in the organic gel and each was applied, either individually or together. This resulted in the efficient, homogeneous removal of surface deposits and salt alterations. In fact, at 24 hours *P. pseudoalcaligenes* and *D. vulgaris* had removed 55% of the nitrate and 85% of the sulfate deposits, respectively.

Polo *et al.* (2010) carried out an applied research study on altered stone sculptures using a microbiological control and a biotechnological method for the removal of undesirable chemical substances. The Demetra and Cronos sculptures represent two of the 12 stone statues decorating the courtyard of Castello del Buonconsiglio in Trento, Italy. Initial inspection of the statues revealed putative black crusts and microbial contamination causing discoloration. In order to remove the black crusts, a remediation treatment of sulfate-reducing bacteria was applied in a polymer, which removed the chemical alteration while preserving the original stone and noble patina. The results of this case study, combining traditional and innovative methods,

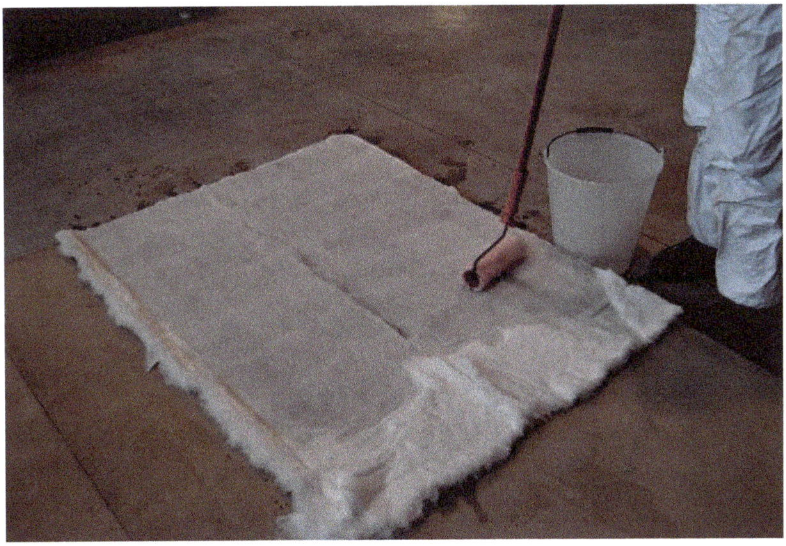

Figure 2 Bioapplication of a cotton layer, used as a delivery system, enriched with a bacterial suspension of *Pseudomonas stutzeri* A29 on the surface of *The Conversion of Saint Ephysius and Battle* by Spinello Aretino, a 14th-century *ex situ* fresco at the Monumental Cemetery in Pisa (Italy) that was altered by animal glue.

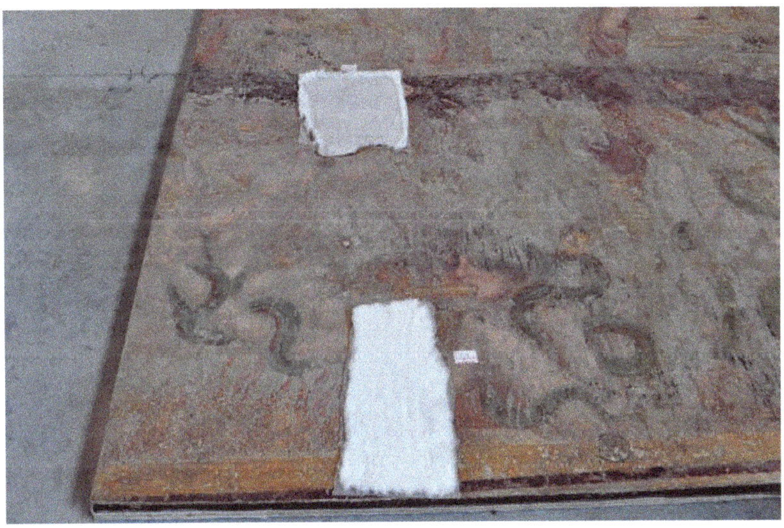

Figure 3 Detail of the *in situ* biocleaning phase on a section of *The Conversion of Saint Ephysius and Battle*. Biocleaning was performed using packs of cotton layer as a delivery system with a viable suspension of *Pseudomonas stutzeri* bacterial cells.

demonstrated the benefits of an integrated biotechnological approach (Polo *et al.* 2010).

On the Pietà Rondanini, a marble sculpture by Michelangelo Buonarroti, sulfate crusts were efficiently biocleaned using *D. vulgaris* subsp. *vulgaris* ATCC 29579 and Carbogel (Cappitelli *et al.* 2005). Other favourable results obtained involved the coloured lithotypes on the outer walls of the Duomo in Florence (Gioventù *et al.* 2011). Black crusts on the green serpentine, red marlstone and Carrara white marble were biocleaned. This work compared three different cleaning treatments: chemical poultice, laser and biocleaning. Results showed biocleaning using *D. vulgaris* subsp. *vulgaris* ATCC 29579 with a Carbogel carrier to be the best cleaning method.

BIOCLEANING OF WALL PAINTINGS

The first application of a biocleaning technique on frescoes was made in the Monumental Cemetery in Pisa, Italy (Ranalli *et al.* 2005). In these pioneering studies, a protocol that envisaged the synergistic combination of hydrolytic enzymatic action with microbial metabolites was described for the restoration of paintings. The case study featured the restoration of the fresco, *The Conversion of Saint Ephysius and Battle* (*Conversione di S. Efisio e Battaglia*) by Spinello Aretino. As a result of the dramatic consequences of bombing during World War II, this fresco had been removed from the walls using the 'tear-off' technique: covering the surface with a strong cloth bound with generous layers of animal glue (Antonioli *et al.* 2005; Ranalli *et al.* 2005). Twenty years later, restorers attempted to remove the cloth and found that the glue had polymerized due to the addition of formaldehyde to the surface, thus prohibiting effective use of a number of surfactants and solubilizing agents. Preliminary tests with individual or mixtures of proteolytic and lipolytic enzymes showed no positive effects – a bioremediation method was the only way to restore the fresco. This was done by direct application of whole bacterial cells of *P. stutzeri* A29, saturated in a cotton layer, onto the artwork surface (Figs 2 and 3).

The biological pack removed 80–100% of the glue, leaving a slight residue on a limited surface area, which was then eliminated with purified collagenase and protease. Complete biocleaning of the entire fresco was achieved by three manual, soft-brush applications of 10 minutes each. Similar results were reported in a study using a single fast biocleaning process (three hours) performed on a historic fresco *Stories of the Holy Fathers* (*Storie dei Santi Padri*), which had been altered by traces of animal glue (Lustrato *et al.* 2012) (Fig. 4).

In a recent biocleaning treatment performed at the Monumental Cemetery in Pisa, the success of a 'hi-tech' complex restoration, including a two-step biocleaning process, was reported. In three hours, *P. stutzeri* A29 strain removed traces of both the animal glue from the front, and residues of calcium caseinate from the back, of the noble fresco *Triumph of Death* (*Trionfo della Morte*) by Buoanamico Buffalmacco, after detachment from its asbestos-cement (Eternit) support (Ranalli *et al.* 2018).

Another successful biocleaning application on wall paintings was carried out at the church of St Johns in Valencia, Spain. The frescoes suffered from nitrate salt efflorescence and animal glue residues that required cleaning. Animal glue residues had appeared on the surface of the fresco as a consequence of inappropriate restoration attempts: detachment of the paint layer from the vault had been remediated by the application of animal glue, which was subsequently not removed correctly. Over time, these glue residues physically and aesthetically damaged the paint of the fresco and needed to be removed. Traditional cleaning techniques were unable to adequately remove the insoluble animal glue therefore a biocleaning technique was selected (Bosch-Roig *et al.* 2012, 2013a). Salt efflorescence was also found on some fragments from the fresco. Efflorescence can cause significant deterioration due to the formation of salt crystals on the surface, causing microfractures in the paint layer and the consequent loss of material from the artwork. To remedy this, an innovative biocleaning technique was used: *P. stutzeri* DSMZ 5190 delivered in an agar gel cleaned the nitrates from the frescoes (Bosch-Roig *et al.* 2013b). A similar situation occurred on the church of St Nicholas frescoes in Valencia: organic matter residues were damaging some fresco fragments that had detached and were subsequently reattached. Biocleaning with *P. stutzeri* DSMZ 5190 and agar gel removed the residues during a four-hour treatment (Sáiz Manleón *et al.* 2017).

One final example of successful biocleaning of wall paintings is found in Rome. Biocleaning of gypsum, weddellite, calcium carbonate, apatite and nitrate deposits as well

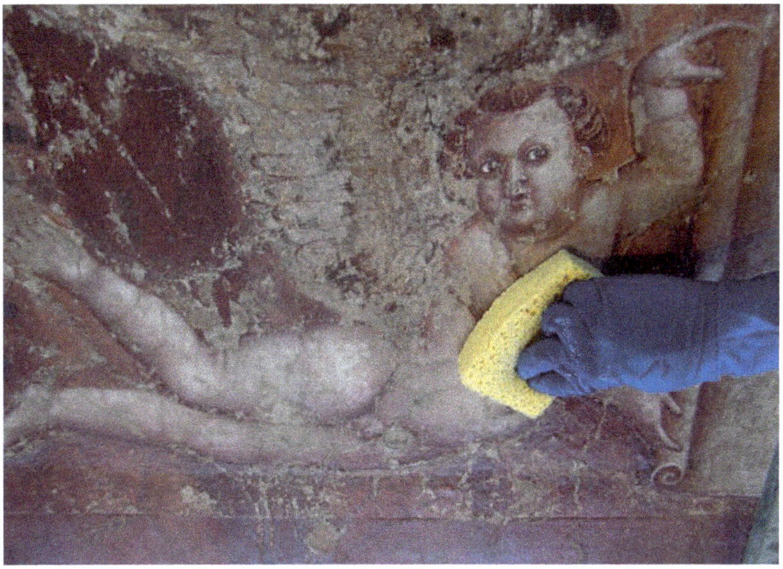

Figure 4 Detail of the final soft cleaning of the historic fresco *Stories of the Holy Fathers* with distilled water and sterile sponges after biocleaning treatment for the removal of surface glue residues.

as proteinaceous matter from the wall paintings of the lower loggia of the Casina Farnese was performed (Mazzoni *et al.* 2014). Three microorganisms were selected to clean the mixed deposits: *Cellulosimicrobium cellulans* TBF11E for the calcium sulfate (inorganic black crust), *Stenotrophomonas maltophilia* UI3E for the protein deposits (brown layers), and *Pseudomonas koreensis* UT30E was used to remove mixed inorganic and protein deposits. The cleaning agents were delivered using micro-packs made of laponite and one of the selected bacterial strains was applied individually or in succession.

BIOCLEANING OF MANUSCRIPTS

Barbabietola *et al.* (2016) reported on the biocleaning of aging animal glues from historical artwork on paper. This type of animal glue residue, often used in the manufacture and restoration of paper, can create tension, alterations, discoloration and cockling over time, thereby degrading paper-based artefacts. The bacterium selected was *Ochrobacterium* sp. TNS15E, which is non-pathogenic, non-spore-forming, and non-cellulolytic, in an agar gel delivery system. Successful removal of glue deposits, following a four-hour biocleaning treatment, without damaging the original cellulose fibres or leaving any undesirable residues, was reported.

This biocleaning method is easy to apply, inexpensive and safe for conservation staff, the environment and the artwork art itself. It is an interesting alternative to existing mechanical and chemical methods, most of which are aggressive to works of art and toxic to humans, and to enzyme cleaning methods that are more expensive, need expert operators and optimal conditions for application (Barbabietola *et al.* 2016).

BIOCLEANING OF OTHER ARCHAEOLOGICAL ARTWORK

Some investigations have focused on the microbiological cleaning of archaeological artefacts derived from other varied materials. The following preliminary studies show biotechnological procedures to be viable alternatives to current archaeological treatment techniques. One such study demonstrated that *P. stutzeri* was able to efficiently clean salt efflorescence and animal glue from small fragments of archaeological frescoes, ceramic, and bone (Martín Ortega 2015). Recently, Albelda-Berenguer *et al.* (2017) reported on a study that focused on microbial preventive treatments on simulated archaeological waterlogged wood. Archaeological waterlogged wood artwork accumulates iron sulfides during burial. Unearthing of these pieces of wood can result in salt precipitation and acidification, causing structural damage to the artefacts. The selection and application of *Thiobacillus denitrificans* bacteria was shown to halt this process through the oxidation of the sulfur compounds present in the wood

Another recent study has shown the potential of *Desulfitobacterium hafniense* bacteria to stabilize archaeological iron. Archaeological excavations commonly reveal iron artworks covered with a layer of corrosion. A method is therefore required to stabilize corrosion in recovered iron objects in order to prevent further deterioration. Most of the existing methods of stabilization generate hazardous waste products. Biocleaning, however, appears to be a very promising alternative method to stabilize corrosion layers on iron artwork (Comensoli *et al.* 2017).

BIOCLEANING OF GRAFFITI

Graffiti is defined as 'pictures or words painted or drawn on a wall, building, etc.'. Many types of graffiti exist ('Tags', 'Bombing', 'Pieces', 'Characters', 'Stickers', etc.) and it can be observed on any type of surface (walls, trains, urban furniture, garages) (Canales and Cueto Lominchar 2007; Hidetsugu and Yasumasa 2015). The graffiti movement started in the New York subway and rapidly expanded to the rest of the world (Castleman 1987). Graffiti represents a growing problem affecting ~3,500,000 protected monuments in Europe alone. Graffiti removal is an important public and private problem, which can be extremely costly. Between 1998 and 2000, graffiti removal cost the Paris Metro 26–50 million Euros per year and the national train company (SNCF) 6–15 million Euros (Fieni 2015). In 2013, San Francisco city authorities spent $20 million and Morelia (Mexico) spends hundreds of thousands of pesos per year to clean graffiti from the historic city centre (Sanmartín *et al.* 2014).

Graffiti removal is very difficult due to the variable composition of the materials used and their interaction with a substrate. Pigments, binders, solvents, additives and other minor components make up graffiti paint and these components can vary greatly: pigments may be inorganic or organic and binders can be natural (from plants or animals), synthetic (mostly resins such as alkyds, acrylics and polyvinyl acetates) or semi-synthetic polymers (e.g. nitrocellulose). Commonly used solvents are hydrocarbon solvents, oxygenated solvents and water. Many different additives are included in small quantities such as plasticizers, dispersants and pH buffers.

A variety of different methods exists for graffiti removal: chemical, physical and, biological. Chemical cleaning involves the use of solvents and strippers. Physical methods include scalpels; abrasive dust; blasting with water, sand, soda and dry ice; laser, ultrasonic/megasonic agitation; plasma sprays, arcs or thermal sprays. These cleaning methods all have limitations – they can cause undesirable side effects on the substrate, resulting in deterioration, and may be toxic both for the environment and the cleaners. Furthermore, some substrates may be incompatible with some cleaning methods. Research investigating the development of biological methods to overcome these limitations is ongoing. Biological cleaning actions claim to be very specific, easy to apply and non-toxic for the environment and the cleaners (Sanmartín *et al.* 2014).

Two studies have shown the potential for biocleaning of graffiti. Giacomucci *et al.* (2012) demonstrated the degradation of nitrocellulose-based, red, graffiti spray paint binder on glass slides by *D. desulfuricans* ATCC 1341, while Sanmartín *et al.* (2015) reported on the successful isolation of environmental microorganisms and their capacity to degrade black graffiti spray paint containing resin binders. Nine bacterial strains from the *Arthrobacter, Bacillus, Gordonia, Microbacterium, Pantoea and Pseudomonas* genera and one fungal strain from the *Alternaria* genus were tested.

This biological method is an interesting new approach to cleaning. The cost is similar to traditional methods and it has proved superior in terms of substrate integrity and respect for environmental and human safety. However, this method has not yet been fully developed and still has some limitations: complete cleaning of the graffiti is not achieved and the glass slides used as a research model do not represent the commonly used surfaces for graffiti. In laboratory testing, immersion of the item in nutrient medium is required, making application on real-world graffiti very difficult and lengthy treatments are required. Further research is needed to examine specific applicability for each substrate–cleaner interaction in order to develop a suitable biological method for effective graffiti removal.

BIOCLEANING OF OTHER ARTWORK

Weathered concrete (Portland cement or blast-furnace slag cement) covered by a black organic layer formed mainly of lichens was successfully biocleaned by submerging concrete specimens in, or sprinkling with, a nutrient solution containing *Thiobacillus* bacteria (De Belie *et al*, 2005; De Graef *et al.* 2003, 2005). The authors showed that selected sulfur-oxidizing bacteria used concrete as a substratum, producing sulfuric acid, which provided the cleaning effect. Although the formation of a white-coloured gypsum layer on some concrete specimens was reported as an unwanted side effect, this innovative biological treatment has good potential.

A number of other studies are in progress, including those looking at the potential application of advanced biocleaning on specific types of works of art. Among these is the biocleaning of ancient canvas and standards that have been altered by a mixture of dusts and organic substances, some of which were added in previous conservation attempts.

CONCLUSIONS

European policy has sought to identify the best, most sustainable means of conserving artwork including a number of key conservation principles relating to minimum intervention.[1] These include the reversibility and repeatability of products and/or procedures adopted together with supporting the final decision on acceptable and unacceptable conservation interventions. However, many conservation principles have evolved over the last 50 years as a consequence of both the internal development of professional conservation and changes in human perception of their safety and role in the environment. Many conservation principles have been reinterpreted. In the past, conservators attempted to achieve absolute control of the environment. Today, we accept the principle of minimum intervention, the notion of the life cycle of materials, including heritage materials, and the concept of acceptable levels of damage.

The biocleaning processes used in restoration applications and described in the literature show encouraging results and highlight the potential advantages of using biological approaches rather than traditional chemical-physical methods. These are considered to be advanced techniques with respect to those used in the past given their low environmental impact, the absence of toxic effects, the high level of selectivity for the altered materials, economic considerations and the actual preservation of artworks.

The remarkable results achieved in large-scale investigations (e.g. on frescoes in Pisa, Trento and Valencia), using the advanced biotechniques proposed, clearly show that the goal of offering hope for the 'resurrection' of degraded works of art is clearly being achieved.

ACKNOWLEDGEMENTS

The authors wish to thank the restorers involved in the above-mentioned case studies. Without their confidence, help and dedication, none of this work could have been successfully applied on actual artwork.

NOTES

1. European Parliament, 2001. Working Paper for the STOA Unit. Technological Requirements for Solutions in the Conservation and Protection of Historic Monuments and Archaeological Remains. European Parliament (EP/IV/A/STOA/2000/13/04), Luxembourg. Available from: http://www.europarl.eu.int/stoa/publi/pdf/00-13-04_en.pdf. The Venice Charter 1964. International Charter for the Conservation and Restoration of Monuments and Sites. 1964, Paris: http://www.international.icomos.org/charters/venice_e.pdf.

REFERENCES

Albelda-Berenguer, A.M., Monachon, M.M., Jacquet, C., Junier, P., Remazeilles, C., Schofield, E.J. and Joseph, E. 2017. 'Microbiological extraction of sulphur compounds from simulated waterlogged wood', poster contribution in *TECHNART 2017: Non-Destructive and Microanalytical Techniques in Art and Cultural Heritage Congress. Bilbao, 2–7 May 2017.*

Alfano, G., Lustrato, G., Belli, C., Zanardini, E., Cappitelli, F., Mello, E., Sorlini, C. and Ranalli, G. 2011. 'The bioremoval of nitrate and sulfate alterations on artistic stonework: the case study of Matera Cathedral after six years from the treatment', *International Biodeterioration and Biodegradadation* 65(7): 1004–11.

Antonioli, P., Zapparoli, G., Abbruscato, P., Sorlini, C., Ranalli, G. and Righetti, P.G. 2005. 'Art-loving bugs: the resurrection of Spinello Aretino from Pisa's cemetery', *Proteomics* 5: 2453–9.

Atlas, R.M. 1995. 'Bioremediation', *Chemical Engeenering News* 73: 32–42.

Atlas, R.M., Chowdhury, A.N. and Gauri, K.L. 1988. 'Microbial calcification of gypsum-rock and sulfated marble', *Studies in Conservation* 33(3): 149–53.

Barabesi, C., Mastromei, G. and Perito, B. 2006. 'La carbonatogenesi batterica nella conservazione delle opere d'arte lapidee', in *Biologia e archeobiologia nei beni culturali, conoscenze, problematiche e casi di studio*, 236–47. Como: AIAR e Musei Civici.

Barbabietola, N. 2012. *Development of Microbial Technology in Conservation and Restoration*, PhD dissertation, Florence University.

Barbabietola, N., Tasso, F., Grimald, M., Alisi, C., Chiavarini, S., Marconi, P., Perito, B. and Sprocati, A.R. 2012. 'Microbe-based technology for a novel approach to conservation and restoration', *EAI Speciale II. Knowledge, Diagnostics and Preservation of Cultural Heritage*: 69–76.

Barbabietola, N., Tasso, F., Alisi, C., Marconi, P., Perito, B., Pasquariello, G. and Sprocati, A.R. 2016. 'A safe microbe-based procedure for a gentle removal of aged animal glues from ancient paper', *International Biodeterioration and Biodegradation* 109: 53–60.

Bellucci, R. and Cremonesi, P. 1994. 'L'uso degli enzimi nella conservazione e nel restauro dei dipinti', *Arte e Tecnica del Restauro* 7(21): 45–64.

Bosch-Roig, P. and Ranalli, G. 2014. 'The safety of biocleaning technologies for cultural heritage', *Frontiers in Microbiology* 5: 155, doi:10.3389/fmicb.2014.00155.

Bosch-Roig, P., Regidor-Ros, J.L., Soriano-Sancho, P., Doménech-Carbó, M.T. and Montes-Estelles, R.M. 2010. 'Ensayos de biolimpieza con bacterias en pinturas murales', *Arché* 4–5: 115–24.

Bosch-Roig, P., Montes-Estellés, R.M., Regidor-Ros, J.L., Rog Picazo, P. and Ranalli, G. 2012. 'New frontiers in the microbial bio-cleaning of artworks', *The Picture Restorer* 41: 37–41.

Bosch-Roig, P., Regidor-Ros, J.L., Soriano-Sancho, P. and Montes-Estellés, R.M. 2013a. 'Biocleaning of animal glue on wall paintings by *Pseudomonas stutzeri*', *Chimica Oggi* 31(1): 50–53.

Bosch-Roig, P., Regidor-Ros, J.L. and Montes-Estellés, R.M. 2013b. 'Biocleaning of nitrate alterations on wall paintings by *Pseudomonas stutzeri*', *International Biodeterioration and Biodegradation* 84: 266–74.

Bosch-Roig, P., Lustrato, G., Zanardini, E. and Ranalli, G. 2015. 'Biocleaning of cultural heritage stone surfaces and frescoes: which delivery system can be the most appropriate?', *Annals of Microbiology* 65: 1227–41.

Bosch-Roig, P., Regidor-Ros, J.L., Soriano-Sancho, M.P., Montes-Estelles, R.M. and Roig-Picazo, P. 2017. 'Biocleaning of wall paintings on uneven surfaces with warm agar gels', in *Gels in the Conservation of Art*, L.V. Angelova, B. Ormsby, J.H. Townsend and R. Wolbers (eds), 119–22. London: Archetype Publications.

Campani, E., Casoli, A., Cremonesi, P., Saccani, I. and Signorini, E. 2007. *L'uso di agarosio e agar per pa preparazione di gel rigidi*, D. Kunzelman (tr.), *Quaderni del Cesmar 7*, no. 4. Padua: Il Prato.

Canales, J. and Cueto Lominchar, J.L. 2007. *De cara a la paret! Imatges de Daniel Magraner sobre el graffiti (i altres textures) de la ciutat de València*. Valencia: Universitat Politècnica de València.

Cappitelli, F., Zanardini, E., Toniolo, L., Abbruscato, P., Ranalli, G. and Sorlini, C. 2005. 'Bioconservation of the marble base of the Pietà Rondanini by Michelangelo Buonarroti', *Geophysical Researches Abstract* 7: 06675–6.

Cappitelli, F., Zanardini, E., Ranalli, G., Mello, E., Daffonchio, D. and Sorlini, C. 2006. 'Improved methodology for bioremoval of black crusts on historical stone artworks by use of sulfate-reducing bacteria', *Applied Environmental Microbiology* 72: 3733–7.

Cappitelli, F., Toniolo, L., Sansonetti, A., Gulotta, D., Ranalli, G., Zanardini, E. and Sorlini, C. 2007. 'Advantages of using microbial technology over traditional chemical technology in the removal of black crusts from stone surfaces of historical monuments', *Applied Environmental Microbiology* 73: 5671–5.

Castanier, S., Le Metayer-Levrel, G., Orial, G., Loubiere, J.F. and Perthuisot, J.P. 2000. 'Carbonatogenesis and applications to preservation and restoration of historic property', in *Of Microbes and Art: The Role of Microbial Communities in the Degradation and Protection of Cultural Heritage*, O. Ciferri, P. Tiano and G. Mastromei (eds), 203–18. New York: Kluwer Academic/Plenum Publishers.

Castleman, C. 1987. *Los Graffiti*. Madrid: Hermann Blume.

Comensoli, L., Maillard, J., Albini, M., Sandoz, F., Junier, P. and Joseph, E. 2017. 'Use of bacteria to stabilize archaeological iron', *Applied Environmental Microbiology* 83(9), doi:10.1128/AEM.03478-16.

Cremonesi, P. 1999. 'L'uso degli enzimi nella pulitura di opere policrome', in *I Talenti 4: metodologie, tecniche e formazione nel mondo del restauro*, 96. Padua: Il Prato.

Cremonesi, P. 2004. 'L'uso dei solventi organici nella pulitura di opere policrome', in *I Talenti 7: metodologie, tecniche e formazione nel mondo del restauro*, 166. Padua: Il Prato.

De Belie, N., De Graef, B., De Muynck, W., Dick, J., De Windt, W. and Verstraete, W. 2005. 'Biocatalytic processes on concrete: bacterial cleaning and repair', in *10DBMC International Conférence on Durability of Building Materials and Components, 17–20 April 2005, Lyon*. Available at: https://www.irbnet.de/daten/iconda/06059019714.pdf (accessed 16 July 2018).

De Graef, B., De Windt, W., De Belie, N. and Verstraete, W. 2003. 'Bacterial cleaning of concrete surfaces', in *International RILEM Conference on Microbial Impact on Building Materials RILEM PRO 34, RILEM TC 183-MIB, Lisbon, 8–9 September 2003*, 27–35. Available at: https://www.rilem.net/gene/main.php?base=500218&id_publication=39 (accessed 16 July 2018).

De Graef, B., De Windt, W., Dick, J., Verstraete, W. and De Belie, N. 2005. 'Cleaning of concrete fouled by lichens with the aid of *Thiobacilli*', *Materials and Structures* 38: 875–82.

EwaGlos 2016. *EwaGlos: European Illustrated Glossary of Conservation Terms for Wall Paintings and Architectural Surfaces. English definitions with translations into Bulgarian, Croatian, French, German, Hungarian, Italian, Polish, Romanian, Spanish and Turkish*, vol. 17, A. Weyer, P. Roig Picazo, D. Pop, J. Carras, A. Özköse, J-M. Vallet and I. Srša (eds). Petersberg: Michael Imhof Verlag, doi:10.5165/hawk-hhg/300.

Fieni, D. 2015. *Graffiti and Street Art in Paris. Routledge Handbook of Graffiti and Street Art*. Oxford: Routledge International Handbooks.

Frazzetto, G. 1994. 'Different and yet alike', *EMBO Reports* 5(3): 233–5.

Gauri, K.L, Chowdhury, A.N, Kulshreshtha, N.P. and Punuru, A.R. 1989. 'The sulfation of marble and the treatment of gypsum crusts', *Studies in Conservation* 34: 201–6.

Gauri, K.L, Parks, L., Jaynes, J. and Atlas, R. 1992. 'Removal of sulfated-crusts from marble using sulphate-reducing bacteria', in *Proceedings of the International Conference on Stone Cleaning and the Nature, Soiling and Decay Mechanisms of Stone*, R.G.M. Webster (ed.), 160–65. Edinburgh: Donhead.

Giacomucci, L., Toja, F., Sanmartín, P., Toniolo, L., Prieto, B., Villa, F. and Cappitelli, F. 2012. 'Degradation of nitrocellulose-based paint by *Desulfovibrio desulfuricans* ATCC 13541', *Biodegradation* 23: 705–16.

Gioventù, E., Lorenzi, P.F., Villa, F., Sorlini, C., Rizzi, M., Cagnini, A., Griffo, A. and Cappitelli, F. 2011. 'Comparing the bioremoval of black crusts on colored artistic lithotypes of the cathedral of Florence with chemical and laser treatment', *International Biodeterioration and Biodegradation* 65: 832–9.

Heselmeyer, K., Fischer, U., Krumbein, W.E. and Warsheid, T. 1991. 'Application of *Desulfovibrio vulgaris* for the bioconversion of rock gypsum crusts into calcite', *Bioforum* 1: 89.

Hidetsugu, Y. and Yasumasa, S. 2015. *Graffiti/ Street Art in Tokyo. Routledge Handbook of Graffiti and Street Art.* Oxford: Routledge International Handbooks.

Junier, P. and Joseph, E. 2017. 'Microbial biotechnology approaches to mitigating the deterioration of construction and heritage materials', in *Microbial Biotechnology* 10(5): 1145–8.

Lustrato, G., Alfano, G., Andreotti, A., Colombini, M.P. and Ranalli, G. 2012. 'Fast biocleaning of mediaeval frescoes using viable bacterial cells', *International Biodeterioration and Biodegradation* 69: 51–61.

Martín Ortega, S. 2015. *Neteges raonades: experimentació amb nou tractaments de neteja aplicats sobre 6 peces de material arqueologic divers*, Treball final de grau ESCRBCC, Barcelona.

May, E., Webster, A.M., Inkpen, R., Zamarreño, D., Kuever, J., Rudolph, C., Warscheid, T., Sorlini, C., Cappitelli, F., Zanardini, E., Ranalli, G., Krage, L., Vgenopoulos, A., Katsinis, D., Mello, E. and Malagodi, M. 2008. 'The BIOBRUSH project for bioremediation of heritage stone', in *Heritage Microbiology and Science: Microbes, Monuments and Maritime Materials*, E. May, M. Jones and J. Mitchell (eds), 76–93. Cambridge: RSC Publishing.

Mazzoni, M., Alisi, C., Tasso, F., Cecchini, A., Marconi, P. and Sprocati, A.R. 2014. 'Laponite micro-packs for the selective cleaning of multiple coherent deposits on wall paintings: the case study of Casina Farnese on the Palatine Hill (Rome-Italy)', *International Biodeterioration and Biodegradation* 94: 1–11.

Moncrieff, A. and Hempel, K. 1970. 'Work on deterioration of sculptures: stone', in *Proceedings of the Conference on Stone and Wooden Objects, New York*, 103–14. London: IIC.

Palla, F. 2006. 'Le biotecnologie molecolari per la caratterizzazione e la valutazione del ruolo dei microrganismi nei processi di degrado di manufatti di interesse storico artistico', *Quaderni di Scienza della conservazione*, 183–94.

Polo, A., Cappitelli, F., Brusetti, L., Principi, P., Villa, F., Giacomucci, L., Ranalli, G. Sorlini, C. 2010. 'Feasibility of removing surface deposits on stone using biological and chemical remediation methods', *Microbial Ecology* 60:1–14.

Pontone, M. 2014. *Biorestauro: Applicazione di batteri alle opere d'arte. Il caso dell'Abbazia di Montecassino*, thesis, University of Molise.

Ramirez, J.L., Santana, M.A., Galindo-Castro, I. and Gonzalez, A. 2005. 'The role of biotechnology in art preservation', *Trends in Biotechnology* 23(32): 564–88.

Ranalli, G., Chiavarini, M., Guidetti, V., Marsala, F., Matteini, M., Zanardini, E. Sorlini, C. 1996. 'The use of microorganisms for the removal of nitrates and organic substances on artistic stoneworks', in *8th International Congress of Deterioration and Conservation of Stone*, J. Riederer (ed.), 1415–20. Berlin: Möller.

Ranalli, G., Chiavarini, M., Guidetti, V., Marsala, F., Matteini, M., Zanardini, E. Sorlini, C. 1997. 'The use of microorganisms for the removal of sulfates on artistic stoneworks', *International Biodeterioration and Biodegradation* 40(2–4): 255–61.

Ranalli, G., Matteini, M., Tosini, I., Zanardini, E. Sorlini, C. 2000. 'Bioremediation of cultural heritage: removal of sulphates, nitrates and organic substances', in *Of Microbes and Art: The Role of Microbial Communities in the Degradation and Protection of Cultural Heritage*, O. Ciferri, P. Tiano, G. Mastromei (eds), 231–45. New York: Kluwer Academic/Plenum Publishers.

Ranalli, G., Alfano, G., Belli, C., Lustrato, G., Colombini, M.P., Bonaduce, I., Zanardini, E., Abbruscato, P., Cappitelli, F. Sorlini, C. 2005. 'Biotechnology applied to cultural heritage: biorestoration of frescoes using viable bacterial cells and enzymes', *Journal Applied Microbiolgy* 98: 73–83.

Ranalli, G., Zanardini, E., Andreotti, A., Colombini, M.P., Corti, C., Bosch-Roig, P., De Nuntiis, P., Lustrato, P., Mandrioli, P., Rampazzi, L., Giantomassi, C., Zari, D. 2018. 'Hi-tech restoration by two-steps biocleaning process of *Triumph of Death* fresco at the Camposanto Monumental Cemetery (Pisa, Italy)', *Journal of Applied Microbiology*, May 13, doi:10.1111/ jam.13913.

Rinaldi, A. 2006. 'Saving a fragile legacy', *Embo Reports* 7(11): 1075–9.

Rodriguez-Navarro, C., Rodriguez-Gallego, M., Ben Chekroun, K. and Gonzalez-Muñoz, M.T. 2003. 'Conservation of ornamental stone by *Myxococcus xanthus*-induced carbonate biomineralization', *Applied Environmental Microbiology* 4: 2182–93.

Saiz Manleón, M.B., Bosch-Roig, P. and Montes Estellés, M.R. 2017. 'Tecnologías especiales de limpieza; láser y biolimpieza', in *Intervención Arquitectónica y Pictórico-Ornamental en la Iglesia Parroquial de San Nicolás Obispo y San Pedro Mártir de Valencia*, 139–45. Valencia: Parroquia de San Nicolás.

Sanmartín, P., Cappitelli, F. and Mitchell, R. 2014. 'Current methods of graffiti removal: a review', *Construction and Building Materials* 71: 363–74.

Sanmartín, P., De Araujo, A., Vasanthakumar, A. and Mitchell, R. 2015. 'Feasibility study involving the search for natural strains of microorganisms capable of degrading graffiti from heritage materials', *International Biodeterioration and Biodegradation* 103: 186–90.

Tiano, P., Biagiotti, L., and Mastromei, G. 1999. 'Bacterial bio-mediated calcite precipitation for monumental stones conservation: methods of evaluation', *Journal of Microbiological Methods* 36: 139–45. doi: 10.1016/S0167-7012(99)00019-6

Troiano, F., Gulotta, D., Balloi, A., Polo, A., Toniolo, L., Lombardi, E., Daffonchio, D., Sorlini, C. and Cappitelli, F. 2013. 'Successful combination of chemical and biological treatments for the cleaning of stone artworks', *International Biodeterioration and Biodegradation* 85: 294–304.

Webster, A. and May, E. 2006. 'Bioremediation of weathered-building stone surface', *Trends in Biotechnology* 24: 255–60.

Wolbers, R. 2000. *Cleaning Painted Surfaces: Aqueous Methods*. London: Archetype Publications.

AUTHORS' ADDRESSES

- Pilar Bosch-Roig, University Politècnica de València, Spain (pboschroig@gmail.com)
- Giancarlo Ranalli, Department of Bioscience and Territory, University of Molise, Pesche (IS), Italy (ranalli@unimol.it)

THE APPLICATION OF MODERN TECHNOLOGY IN THE BIODETERIORATION AND PRESERVATION OF BUILT CULTURAL HERITAGE

Christine C. Gaylarde and B. Otto Ortega Morales

INTRODUCTION

This contribution seeks to cover the ways in which microorganisms can be involved not only in the deterioration but also the conservation of stone cultural heritage materials. The topics covered include modern methods for identifying microorganisms and their activities in biofilms; biological treatment of stone ('bioconservation'); and modern methods for treating and preventing colonization and recolonization.

MODERN METHODS FOR IDENTIFYING MICROORGANISMS AND THEIR ACTIVITIES IN BIOFILMS

Identification of the type and activity of microorganisms involved in biodeterioration is necessary in order to understand and control their effects on cultural assets. Recent investigations have focused on the development of qualitative and quantitative molecular tools and latest developments in this area have been reviewed by Sanmartín *et al.* (2016).

Most modern identification techniques involve nucleic acid sequencing. Ettenauer *et al.* (2014) developed a real-time PCR (qPCR) method to detect and quantify fungi present in materials used in the construction of historic buildings. Samples of thermal insulation materials were inoculated with *Cladosporium cladosporioides*, *Aspergillus niger* and *Penicillium chrysogenum*, and parchment samples from the 13th to the 19th century were used as original materials. Villa *et al.* (2015) employed 16S rRNA and sulfur metabolism genes to detect metabolically active microorganisms and provide insight into the diversity of sulfur-oxidizing bacteria in limestone tombstones from two differently polluted sites in urban (more polluted) and rural (less polluted) areas in Massachusetts, USA. The authors used denaturing gradient gel electrophoresis (DGGE) fingerprints of 16S rDNA, 16S rRNA, *soxB*, *aspA* and *dsrA* transcripts to demonstrate that Cyanobacteria was the main metabolically active group at both sites. Higher levels of phototrophic sulfur bacteria, other sulfate-oxidizing bacteria and strains within the sulfate-reducing genus *Desulfovibrio* occurred at the more polluted site.

Next-generation sequencing (NGS) is utilized for the analysis of biofilms on historic and cultural buildings. The ability to use small samples and analyze entire communities without using culture-dependent techniques is extremely attractive and the literature generated from the use of this technology is expanding rapidly. Halophilic bacteria are increasingly found in culturally important buildings. Salt efflorescences are common on stone and other lithic materials, and the use of NGS technology has allowed ready detection of these 'salt-loving' organisms, which frequently belong to the domain Archaea (Piñar *et al.* 2009; Ogawa *et al.* 2017; Otlewska *et al.* 2017). As they are often pigmented, they play an important role in the discoloration of surfaces.

Pigmented bacteria, such as *Rubrobacter* spp., and fungi, such as the Dothideomycetes, are frequent colonizers of stone heritage items (Duan *et al.* 2017; Gaylarde *et al.* 2017a, b). Gaylarde *et al.* (2017a) reported the presence of microorganisms with important survival and deteriogenic properties, such as pigment production, halophily and thermophily, in biofilms on historic churches in Rio de Janeiro, Brazil. These technologies have drawbacks, however (Ogawa *et al.* 2017b), and improvements can certainly be expected in the future. One of the many problems is the lack of discrimination of fungal populations on the stone (Sohlberg *et al.* 2015; Gaylarde *et al.* 2017a).

The activity, rather than the mere identity, of microbial cells in biofilms on the surfaces of historic buildings is a subject of obvious interest. The analysis of potentially deteriogenic activities within the biofilm would greatly increase our capacity to design control measures against biodeterioration. As of yet, methods for the determination of such activities directly within the biofilm are insufficiently developed. However, an initial study was made on buildings of the Auschwitz II-Birkenau concentration and extermination camp by Gutarowska *et al.* (2015), employing metabolomics techniques to determine the metabolic products present in biofilms on brick and wooden buildings at this important site.

The metabolome includes all the small metabolic products that are produced within a cell. Sophisticated chemical analytical techniques used in its study include mass spectrometry (MS), nuclear magnetic resonance (NMR) spectroscopy, and Fourier transform infrared (FTIR) spectroscopy, coupled with computerized database searches (Longnecker *et al.* 2015). Metabolomic studies can be grouped into two categories: targeted, where quantitative data on a pre-defined set of compounds is obtained, and untargeted, i.e. the search for a broader range of metabolites, some of which may not be present in the known databases (for an overview of this subject see Patti *et al.* 2012). Since thousands of such compounds may be detected, the work can be limited by throughput problems (Dunn *et al.* 2012). Other major drawbacks are the lack of suitable databases and software for analysis. Costa *et al.* (2016) suggested a new computational platform for the analysis of metabolomics data that include data loading in various formats, preprocessing, univariate and multivariate data analyses, data selection and even machine learning. This is a rapidly expanding area in the field of biodeterioration of heritage items.

BIOCONSERVATION

The use of biological organisms in the treatment of degraded stone is relatively new. Biotechnology is an innovative area in the conservation of historic and artistic heritage and biotechnological approaches have only relatively recently begun to be used for the analysis of lithotrophic organisms suitable for use in conservation (Fernandes 2006). The technologies include biomineralization, biocleaning and biomimetic surfaces.

Consolidation of stone by biomineralization, in which microorganisms (most commonly bacteria) are impregnated into deteriorated stone and allowed to produce new lithic material, is a practice already used in both conservation-restoration and civil engineering. Igneous, sedimentary and metamorphic stone have different physical characteristics that will influence the success or failure of consolidation procedures. The majority of this work has been done with limestone and concrete. Bacteria used in the process include those from the genera *Streptomyces*, *Ralstonia*, *Bradirhyzobium*, *Methylobacterium*, *Burkholderia*, *Caulobacter*, *Bacillus*, *Myxococcus* and *Xanthobacter* (Braissant 2005).

Bacterial calcium carbonate precipitation has been proposed as an environmentally and health and safety friendly biological technique to restore and consolidate ornamental calcareous stones and other architectural heritage (Fernandes 2006; Jroundi *et al.* 2014; Alberto *et al.* 2015). Under suitable conditions, most heterotrophic bacteria are able to induce $CaCO_3$ precipitation in artificial media. The metabolic pathways that produce carbonate and bicarbonate ions and increase pH favor $CaCO_3$ precipitation in the presence of free calcium (Knorre and Krumbein 2000). More can be learned about these activities and their potential use in biotechnological applications elsewhere (Dhami *et al.* 2014; Zhu and Dittrich 2016).

Three metabolic pathways related to the nitrogen cycle can increase pH by releasing ammonia into the surroundings: (1) the deamination of amino acids; (2) the dissimilatory reduction of nitrate; and (3) the degradation of urea (Perito and Mastromei 2011). The degradation of urea produces NH_3 and CO_2 that dissolve in water, forming HCO_3^- and CO_3^{2-} ions. Macromolecules present on the surface of bacteria, in the cell walls or in extracellular polymeric substance (EPS) have anionic functional groups that also play an important role in $CaCO_3$ precipitation. Cell-attached or free EPS can trap calcium ions and influence the development of the $CaCO_3$ precipitates (Braissant *et al.* 2003). However, emerging evidence suggests that EPS could also inhibit $CaCO_3$ precipitation. The final outcome may, at least partially, be reliant on the fact that the calcium binding property is dependent on pH and the physical state of the biopolymers (Braissant *et al.* 2007).

EPS occurs in different physical forms: loosely bound, tightly bound, as slime and capsular. The use of EPS biopolymers to enhance carbonate precipitation and subsequent stone consolidation appears to be dependent on the extraction method used for EPS recovery; different states of EPS are obtained depending on the type of method employed. Even if the correct type of EPS is extracted and a suitable formulation prepared, it is possible that the EPS, while enhancing $CaCO_3$ precipitation, may obstruct the pore network within the stone, altering the flux of water and dissolved ions, thereby weakening the structure. The weakening of the mineral lattice through repeated wetting and drying cycles and concurrent expansion and contraction induces damage in the material. Thus, from a technological perspective, it seems that the use of EPS or the management of microbial processes to control the precipitation of carbonate within EPS matrices still requires future research (Decho 2010).

Calcite 'biocement' produced by bacteria is compatible with calcareous stones as it mimics their natural formation process. However, it has important limitations (Rodriguez-Navarro *et al.* 2003). Biotic stone crystals are more disorderly than the abiotic ones. Plugging stone pores with biological remains and uncontrolled biofilm growth after application of biocalcifying bacteria may prevent airflow through the stone, leading to increased decay. The use of heterotrophic bacteria that do not survive on the stone after treatment, such as *M. xanthus*, may overcome the latter limitation. Rodriguez-Navarro *et al.* (2003) used this bacterium in the consolidation of calcarenite, a highly porous stone frequently employed in Spanish historic buildings. In this study, stone slabs were steam sterilized before incubation in a growth medium, a condition that could not be applied in the real world. However, the newly formed calcite and vaterite crystals were more resistant to sonication stress and were strongly adhered to the pore surfaces without any blocking or plugging, suggesting that further experiments could be worthwhile.

The migration of the bacteria in the pores determines the distribution of the newly formed material in the stone. Because most stone degradation occurs near the surface, the bacteria should ideally stay close to that surface. On the other hand, if the bacteria do not travel deep enough into the stone, the treatment will only be superficial and this is also an important consideration. The use of this technique on stone buildings in general has been reviewed by De Belie (2016).

An alternative to the addition of calcifying microorganisms is a process called bioaugmentation, in which the indigenous microorganisms are encouraged to proliferate by adding suitable nutrients to the stone. This approach was adopted by Jroundi *et al.* (2014) on decayed gypsum plasters from the Alcázar in Guadalajara, Spain. Treatment with conventional consolidants (ethyl silicate) was compared with bioaugmentation using a bacterial growth medium. Seven successive impregnations with the conventional consolidants were necessary because of the delapidated state of the plaster. For the bioaugmentation process, the growth medium was sprayed to saturation twice each day for six successive days. Both types of treatment were successful in terms of strengthening the plasters, but the bioaugmentation treatment produced greater depth of penetration and consolidation. The strengthening effect was associated mainly with the bacterial production of metastable vaterite. The authors later successfully used this technique *in situ* on limestone cresting on the roof of the Royal Chapel in Granada, Spain (Jroundi *et al.* 2015). The challenges of such treatments are to encourage growth of only the calcinogenic bacteria and ensure that they remain dominant in the stone over time.

Non-biologically produced calcium nanoparticles can also be used. A commercially available solution of $Ca(OH)_2$ nanoparticles has been shown to be effective in consolidating powdering painted layers of Mesoamerican wall paintings in Ixcaquixtla and Calakmul, Mexico (Baglioni *et al.* 2014). However, there can be problems associated with the carrier system: alcohol is generally recommended, but may result in localization of the consolidant too near the surface of the stone, an undesired outcome when mass consolidation is required (Borsoi *et al.* 2016).

Until recently, it was not known how the bioaugmented stone would react to the biodeteriogenic processes in the environment. Shirakawa *et al.* (2015) demonstrated that *B. sphaericus*-biocalcified fiber cement was sufficiently resistant to subsequent colonization in the subtropical environment of São Paulo, Brazil, over 22 months. The authors showed that this technology could be used not only to repair, but also to reduce the fouling-associated darkening of fiber cement. In all cases, regulatory barriers may be an impediment to the routine use of such new technologies. Regulations can vary from country to country and obtaining consent from officials usually involves considerable time and effort. This will obviously change as the technology becomes more accepted.

Lopez-Moreno *et al.* (2014) isolated $CaCO_3$-precipitating bacteria from ignimbrite stone (a siliceous stone containing small quantities of Ca) from the roof of Morelia Cathedral in Mexico, and suggested their possible use in bioconsolidation of that stone. They pointed out that EPS-producing bacteria in biofilms may be either advantageous or disadvantageous for the integrity of the stone, and that the final result from using these bacteria in bioconsolidation will vary depending on the type of stone. This has important implications for microbiological applications in stone consolidation. More recently, Tang *et al.* (2016) proposed a biological (living eukaryotic cells) autocementation kit called Euk.cement, based on the ability of silica-binding peptides on the surface of the dimorphic yeast *Yarrowia lipolytica* to immobilize onto silica particles. The yeast is modified by the integration of genes for calcium precipitation. The silica-immobilized recombinant cells then produce carbonate that cements the particles together.

Biocleaning systems utilize microbial activites, either extracted from or still within the living cells. Biocleaning aims to remove altered stone forms such as sulfate and nitrate crusts, organic substances like animal glue, and foreign materials such as graffiti pigments. Delivery systems may include sepiolite, Hydrobiogel-97, cotton, wool, Carbogel, mortar and alginate beads, agar and arbocel (Bosch-Roig *et al.* 2015). The safety of this methodology was discussed in 2014 by Bosch-Roig and Ranalli, who stressed

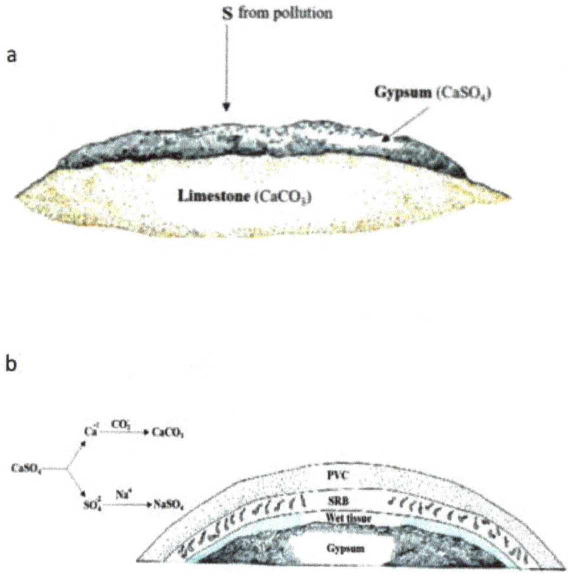

Figure 1 (a) Black gypsum formed on the surface of a limestone building by reaction with air pollutants. For the biocleaning treatment, the surface is covered with a moist tissue, followed by a carrier gel containing the aerotolerant sulfate-reducing bacterial cells (*Desulfovibrio vulgaris* subsp. vulgaris ATCC 29579) and a small amount of lactate carbon source. (b) The whole surface is then covered with a PVC film to help preserve humidity and reduce oxygen ingress. The overall reaction, changing black gypsum to white calcite and sodium sulfate, is shown on the left.

the importance of developing suitable strategies for inspection and monitoring of biocleaned buildings.

Martino *et al.* (2015) used sulfate-reducing bacteria to remove the gypsum black layer deposits from the surface of a limestone polychrome bas-relief in a gallery in Palermo, Sicily (Fig. 1). X-ray fluorescence (XRF) and FTIR investigations showed that the application of live *Desulfovibrio vulgaris* in gel form efficiently and selectively removed the sulfate deposits from large areas of the surface. Three applications of the bacteria in a Carbogel carrier were necessary to achieve the desired level of cleaning. Similar technology has been applied successfully on other important cultural monuments, including the sculpture *Allegoria della Morte* (Gioventù and Lorenzi 2013) and a tile from Milan Cathedral (Cappitelli *et al.* 2007).

Sanmartín *et al.* (2015) identified various bacterial genera and a fungal *Alternaria* sp. as candidates for removing graffiti from disfigured buildings by using painted glass slides as test substrates. Microorganisms were isolated from graffiti, an acrylic wall painting, a car in a scrapyard and the soil beneath it, and the interior of spray paint cans.

Other examples of successful biocleaning treatments include the use of *Thiobacillus* sp. to remove lichens (De Muynck *et al.* 2010), the application of enzyme extracts to whiten marble (Konkol *et al.* 2009), and the use of viable cells of *Pseudomonas stutzeri* for cleaning medieval frescoes (Lustrato *et al.* 2012). Troiano *et al.* (2013) successfully used a combination of chemical and microbiological treatments (application of *D. vulgaris* subsp. *vulgaris* ATCC 29579 after pre-treatment with the non-ionic detergent Tween 20) in the cleaning of stone. The considerable development required to produce a commercially viable biocleaning treatment is discussed by Balloi *et al.* (2017).

Biomimetic surfaces, which are protective films inspired by biologically produced

Table 1 A review of some methods used for removal or prevention of biofilms on stone.

Method	Prevention (P) or Treatment (T)	Main advantages	Main disadvantages	References
Physical				
Water jet	T	Simple, cheap	Potentially damaging, water retention is a problem Efficacy unproven Short-lived	Blasquez *et al.* 2000; Shirakawa *et al.* 2012
Irradiation:				
Visible light	T	Simple, cheap		Sanmartín *et al.* 2017;
UV	T	Simple, cheap	Unproven, ineffective against pigmented fungi	Borderie *et al.* 2014
Gamma	T	High penetration		Boniek *et al.* 2017
Laser	T	Versatile	Not always successful	Mascalchi *et al.* 2015; Senesi *et al.* 2016; Osticioli *et al.* 2015; Sanz *et al.* 2015, 2017
Chemical				
Liquid/gaseous biocides	T	Many options	Ecological/health hazards, require specialist help	Allsopp *et al.* 2004
Metal nanoparticles	P	Can be compatible with consolidants Questions on ecotoxological effects	May affect stone color	Gherardi *et al.* 2017; Gómez-Ortiz *et al.* 2013; van der Werf *et al.* 2015; Pinna *et al.* 2012; Ditaranto *et al.* 2011; La Russa *et al.* 2014; Carrillo-González *et al.* 2016; Bellissima *et al.* 2014; Munafò *et al.* 2015
Biological				
Natural biocides	T	More environmentally acceptable	May lack efficacy, unproven	Pinna *et al.* 2017; Sasso *et al.* 2016; Jeong *et al.* 2017, Rotolo *et al.* 2016
Biological agents	T/P	Non-polluting, long-lived	Efficacy unproven, unknown long-term ecological effects	Szewczyk *et al.* 2006; Silva *et al.* 2015

substances or surfaces, are interesting new approaches to stone conservation. The lithic materials may have uniform size, novel crystal morphology, specific crystallographic orientation, reduced weight, mechanical robustness and unique functions, as well as compatibility with the original stone. Liu *et al.* (2006) identified an inscription-protecting film of crude CaC_2O_4 on the Lingyin Tower in Hangzhou, China, and produced an artificial conservation film of the same material to mimic this property. Another novel biomimetic nanocomposite, comprised of amorphous silica and calcium oxalate, a protective patina often found on monuments, has been efficiently synthesized at low cost by modifying tetraethoxysilane by incorporating calcium oxalate monohydrate nanoparticles into the silica matrix (Maravelaki *et al.* 2015). The new material significantly improved porous limestone performance without changing the microstructure. Water capillary absorption coefficient, the total amount of capillary absorbed water, and water vapor permeability were decreased, while indirect tensile strength was increased.

Future developments for biomimics could include protective, but non-lithic, materials. For example, the skin of the gecko has anti-wetting and self-cleaning properties (Watson *et al.* 2015). Such characteristics could usefully be mimicked for antimicrobial coatings on historic buildings. Indeed, the superhydrophobicity of the lotus leaf surface has already led to the development of water-repellent-mimicking materials used commercially in other areas. An excellent discussion of these and other self-cleaning surfaces can be found in He (2017).

MODERN METHODS FOR TREATING AND PREVENTING COLONIZATION AND RECOLONIZATION

A variety of methods employed over the years for removing or preventing biological growths on stone is summarized in Table 1. It would appear obvious that modulation of light can affect the growth and activity of phototrophs, organisms that are immensely important in biofilms on historic buildings, and a considerable amount of work has been carried out to investigate this interaction.

Moderate light intensity can inactivate Cyanobacteria by causing photoinhibition, photobleaching and photodamage to the cells through reactive oxygen species formation. The effectiveness depends not only on light intensity but also on the particular Cyanobacteria species that are present in the biofilms. Red light is most effective against species rich in phycocyanin and allophycocyanin, such as those belonging to the genera *Leptolyngbya*, *Synechocystis* and *Scytonema*, while green light is more effective against species rich in phycoerythrin, such as *Oculatella subterranea* (Hsieh *et al.* 2014). These same authors state that blue light is the least effective, since 150 µmol photons $m^{-2} s^{-1}$ of blue light could not cause damage to biofilms produced from Cyanobacteria isolates treated with the photosensitizer D-Ala. This may be due, in part, to the fact that blue light has high energy but low penetration ability, whereas red light has low energy but high penetration ability.

Sanmartín *et al.* (2017), examining the possibility of using colored cellophane films to control growth and color of phototrophic biofilms in bioreactors, found that blue cellophane led to complete growth inhibition of an algal and cyanobacterial mixed culture derived from a natural biofilm on a granitic building in Spain. High light levels cause photoinhibition and radical formation, reducing phototroph growth, while dim light may promote photosynthesis (Hsieh *et al.* 2013). Therefore, the apparent paradox could be explained by the intensity, rather than the quality, of light. In the Sanmartín experiment, the quantity of blue light beneath the cellophane covers, expressed as photosynthetic photon flux density, was 12.60 µmol photons $m^{-2} s^{-1}$, suggesting that light intensity may play a role. In a word of warning, Urzì *et al.* (2014) pointed out that treatment with certain visible light wavelengths (blue light in the aforementioned case) may actually cause an increase in the growth of certain microorganisms that can be just as, or even more, damaging.

UV-C irradiation has been shown to have deleterious effects on the growth of mixed unicellular algae isolated from caves in the Dordogne, France (Borderie *et al.* 2011). This has been suggested for controlling algal blooms in water bodies (Tao *et al.* 2010); UV-C irradiation at 20–200 mJ cm^{-2} suppressed *M. aeruginosa* growth for 3–13 d in a dose-dependent manner and UV-C irradiation at 100 and 200 mJ cm^{-2} induced severe cell disintegration in more than 70% of the irradiated cells. In a show cave in Northeastern France, however, UV-C was used against terrestrial biofilms with alternating periods of 30 min on/15 min off over 12 h at night (effective dose 180 kJ m^{-2}). Although effective after one year, recolonization occurred after 16 months (Borderie *et al.* 2014).

Gamma irradiation has often been used for the disinfection of historic materials such as documents. This was tested for the first time as an application on stone cultural property by Boniek *et al.* (2017). The researchers treated soapstone blocks that had been exposed for 12 months to the humid environment of Minas Gerais, Brazil, with 6.2 Gy/h γ-radiation for 97 minutes at a 10-cm distance. Fungi isolated from the blocks before irradiation were of the genera *Acremonium, Alternaria, Aspergillus, Calcarisporium, Cladosporium, Curvularia, Epicoccum, Fusarium and Penicillium.* Of these, only *Cladosporium cladosporioides* survived and increased in number after irradiation. The authors emphasized that, although γ-radiation could be used to treat fungal colonization on stone surfaces, it would have serious limitations when the highly disfiguring, melanin-producing species are present.

The use of lasers has become common in stone conservation (Salimbeni *et al.* 2000; Bromblet *et al.* 2003). The ablation processes can utilize various wavelengths and pulse widths, giving versatility for a variety of conservation problems. Mascalchi *et al.* (2015 and references therein) used selective microwaves which proved effective and suggested that the treatment alone might be suitable to avoid future recolonization.

Senesi *et al.* (2016) used laser cleaning combined with laser-induced breakdown spectroscopy (LIBS) on the breakdown products of black crusts on a limestone castle in Italy in order to remove crusts only down to the appropriate depth. Recently, a Q-switched (QS) Nd:YAG laser operating at 532 nm successfully removed *Verrucaria nigrescens* growths from Carrara marble artefacts (Osticioli *et al.* 2015). Both fungal and algal cells were destroyed within about 100–200 μm of the surface. Other studies, however, have suggested that laser treatment is only partially successful in the removal of lichens from stone surfaces (Sanz *et al.* 2015, 2017).

Aesthetic disfiguration of stone is often related to the pigments produced by algae and Cyanobacteria (chlorophylls and carotenoids), fungi and Actinomycetes (melanins, carotenoids and others). Pigment inhibitors interfere with the production of physiologically important pigments, rendering organisms more vulnerable to environmental stress, but also reducing the disfiguring effects of microbial growth. Tricyclazole, arbutin and carpropamid are compounds that can disturb the formation of melanins or other dark pigments, while cerulenin is a potent inhibitor of carotenoid synthesis in bacteria. Although tricyclazole was shown to increase the sensitivity of the fungus *C. cladosporioides* to biocides, the effect in the field was erratic (Young *et al.* 2009). It appears that this approach has not yielded sufficiently promising results for further research.

Nanoparticles are appearing more frequently in the list of potential treatments for cultural heritage conservation. In an antimicrobial nanocoating, at least one dimension of the particle must be in the range of 10^{-9} m. One of the advantages of using nanoparticles is that stable suspensions can be achieved without the use of surfactants, which might leave unwanted residues on the stone surface (Baglioni *et al.* 2015). The suspensions can easily be brushed or sprayed onto the surface. They may also be applied as a gel, but this may have the disadvantage of leaving chemical residues that can be used by microorganisms for growth.

The antimicrobial activity of titanium dioxide (anatase) nanoparticles is well known and it is probably the most widely used nanocoating. The biocidal efficiency of TiO_2-based engineered nanoparticles, as well as their de-polluting and self-cleaning activities, has been reviewed by Munafò *et al.* (2015). Titanium dioxide (TiO_2) spiked with various metals was used by La Russa *et al.* (2014) for testing on cultural heritage materials. They incorporated silver, iron and strontium into TiO_2 gels and tested them *in vitro* against bacteria isolated from deteriorated surfaces: *P. stutzeri*, *Stenotrophomonas maltophilia*, *Micrococcus luteus* and *B. subtilis*. As expected, the addition of iron caused the greatest chromatic change, with silver and strontium inducing merely a grey discoloration, which was more notable at higher TiO_2 concentrations. Silver, with or without the other metals, showed the greatest antibacterial activity, and could be further investigated.

Silver nanoparticles were shown to be effective protectants of gypsum (Shirakawa *et al.* 2013) and they have also been studied by application onto Italian Serena sandstone surfaces (Bellissima *et al.* 2014). In the latter case, the nanoparticles were functionalized by condensing with a silane precursor (tetraethylorthosilicate), resulting in a 50–80% reduction in cell viability. Carrillo-González *et al.* (2016), using two different plant extracts for nanoparticle synthesis, determined that silver nanoparticles were effective antimicrobial agents for application on stucco, basalt and calcite conservation, but their efficacy was dose-dependent. The authors used isolates from the walls of buildings in the important archaeological site of Teotihuacan, Mexico. The isolates were first tested *in vitro* with subsequent evaluation of the nanoparticles' efficacy against the most sensitive and rapidly growing isolates, *P. carotovorum* and *A. alternata*, which were inoculated onto stucco, calcite and basalt blocks.

Li *et al.* (2017) demonstrated that silver ions could be more effective than silver nanoparticles, at least against bacteria grown in suspension in microtiter plates. Modified titanium and silver nanoparticles were compared by Aflori *et al.* (2013), who prepared the silsesquioxane-based hybrid nanocomposites with methacrylate units containing TiO_2 and/or silver nanoparticles from titanium isopropoxide and/or 3-(trimethoxysilyl) propyl methacrylate precursors. This preparation was done in the presence of silver nitrate and a primary amine surfactant dodecylamine to avoid fissuring of the hybrid gels while drying inside the stone. The nanocomposites containing TiO_2 showed higher activity against the tested *E. coli* and *C. albicans*.

Several authors have reported on the use of polymer matrices containing copper nanoparticles for the protection of stone heritage. Ditaranto *et al.* (2011) investigated the use of copper (Cu) nanoparticles dispersed in the consolidant and water-repellent Estel 1100 for the protection of several types of calcarious stone. They demonstrated that copper was released at the surface of the treated stones, a necessary step for the inhibition of microorganisms, and that, at low nanoparticle concentrations, there were no chromatic changes other than those that normally occur with Estel 1100. Another study reported that copper nanoparticles plus a water repellent were able to prevent recolonization of sandstone, marble and plaster by crustose and foliose lichens in the archaeological area of Fiesole, Italy, over a three-year period (Pinna *et al.* 2012).

Zinc oxide (ZnO) may be more appropriate for the conservation of light colored stones. Van der Werf *et al.* (2015) demonstrated that higher levels of ZnO nanoparticles could be used without affecting the color of the stone. Using a 12th-century church in Manfredonia, southern Italy, as the test site, they showed that a tenfold greater concentration of ZnO nanoparticles than Cu nanoparticles could be utilized without affecting the color of the stone substrate. As the nanoparticles are whitish and the carriers used, Silo 111 and Estel 1000, are colorless, this would be expected. In fact, the addition of the ZnO nanoparticles reduced the slight chromatic changes that are normally observed when using consolidants/water repellents alone.

Gómez-Ortíz *et al.* (2013) developed a treatment using $Ca(OH)_2$ mixed with ZnO and TiO_2 nanoparticles, all of which were synthesized in the laboratory. The ZnO-based systems showed the best antifungal activity against *P. oxalicum* and *A. niger* biofilms on glass slides and limestone coupons under simulated photoperiod conditions. Ti nanoparticles have also been shown to be compatible with both stone consolidants and organosiloxanes, thus allowing these traditional conservation tools to be combined with a photocatalytic cleaning component.

Gherardi *et al.* (2017) illustrated that the photoactivity range of TiO_2 could be extended to solar irradiation due to the presence of residual benzyl alcohol groups from its non-aqueous synthesis that allowed it to be combined with commercial organosiloxane consolidants. They emphasized the importance of field testing potential new treatments and designed a protocol for such trials, involving colorimetric and water absorption tests, optical microscopy, spectrophotometric measurements, and environmental scanning electron microscopy combined with energy dispersive X-ray analysis (ESEM-EDX). The two types of marble used in their tests (Candoglia and Crevoladossola) produced very different results, indicating the need to use the appropriate substrate in such testing.

Scott Mitchell and colleagues at the Spanish National Research Council in Spain perform ongoing investigations into surface treatments with liquids or gels constructed from polyoxometalates and quaternary ammonium cations. The superhydrophobic ionic liquids are composed of long-chain alkylammonium cations, of well-known antimicrobial activity, and polyoxometalate anions. These nanohybrid materials are being tested for activity on stone surfaces.

Synthetic biocides have been used widely by the cultural heritage community of conservators, restorers and biodeterioration scientists. Biocides vary in their ability to kill microorganisms, provide long-lasting effects and in their impact on stone materials. Their effectiveness in controlling microbial growth is related to the susceptibility of microorganisms and the surrounding environment. For instance, water-soluble compounds in biocidal formulations may be leached from treated surfaces by rain and run-off water. This could be particularly important in the case of buildings located in tropical zones with heavy rains (Ortega-Morales *et al.* 1999).

Those working in cultural heritage seek biocide formulations which are harmless for the users and have low environmental impact. This has prompted the search for natural molecules to develop novel, long-lasting, environmentally acceptable biocides. Emerging areas of research include the use of eco-friendly compounds derived from natural sources, such as plant metabolites (Jeong *et al.* 2017) and antimicrobial products of microbial origin. Isothyocyanates from plant sources have proved useful to prevent biofilm growth (Saravia and Gaylarde 1998) and other natural metabolites have also demonstrated their value in biofilm control in the laboratory. Sasso *et al.* (2016) found that a cell extract from *Trichoderma harzianum*, rich in a range of chemical compounds, was effective against the photosynthetic organisms *Chlorella*, *Stichococcus*, *Leptolyngbya* and *Pleurocapsa*, as confirmed by confocal laser scanning microscopy (CLSM) and chlorophyll a determinations. Antifouling compounds sourced from the sea (e.g. zosteric acid) and even lichen-associated metabolites also seem promising (Pinna *et al.* 2018 and references therein). However, it appears unlikely that such novel products will be manufactured in the near future (Pinna *et al.* 2018). However, novel treatments to eliminate and prevent biofilm colonization of stone substrates may be produced by using mixed formulations. These include the use of pigment and/or exopolysaccharide inhibitors and permeabilizers in order to increase the effectiveness of existing biocides, natural antimicrobial compounds or extracts (Young *et al.* 2009).

CONCLUSIONS

Microbial biodeterioration of built cultural heritage presents a significant problem for conservation scientists. The advent of novel technologies such as next-generation sequencing allows scientists and conservators to deal with minute amounts of material, enables the detection of specific microbial populations such as halophiles and has also contributed to a better understanding of the complexity of microbial communities. This technology has its drawbacks, however, and improvements are necessary. A proper assessment of the biodeterioration potential of microbial deteriogens is mandatory.

Most studies rely on culture-based methods that allow the detection and characterization of deteriorating microbial metabolites. This approach has limited applicability under real-life conditions. Metabolomic profiling on environmental samples in the study of biodeterioration of building materials using ultra-high performance liquid chromatography with high-resolution mass spectrometry (UPLC-HRMS) circumvents this problem and can certainly be expected to expand in the future.

An increasing number of approaches and types of materials for the conservation of stone cultural heritage have also been discussed. Some of these approaches and materials appear to be their infancy (e.g. biomimetic surfaces and natural antimicrobials), while others, such as nanocoatings, would seem to have reached maturity, particularly for consolidation and self-cleaning purposes. An increasing avenue of evidence shows that engineered nanoparticles could also be used to control microbial growth on stone; however, their ecotoxicological risks are still to be determined. Whether some of these approaches will find real-life application depends on a range of factors. The most important drive for ultimate use is their acceptance by the conservation community. Conservators may prefer approaches and materials that originate from biological sources (natural consolidants/protectants), are inspired by nature or supported by biological processes (biocarbonatogenesis), or have a low impact on the environment (natural antimicrobials vs synthetic biocides). Such new technologies need to be validated in the field.

The variety of approaches testifies to the growth of this sector. Biodeterioration scientists need to keep in mind that the main goal of conservators is to prolong the life of cultural heritage property using effective and, if possible, simple, inexpensive materials and methods. On the other hand, conservators must assess the scientific evidence generated by biodeterioration scientists and determine its value using cost-benefit analysis and the goals of their particular institutions. Multidisciplinary teams will certainly be involved in future developments.

ACKNOWLEDGEMENTS

This work was partially supported by CONACYT Ciência Básica 2015 grant 257449 'Influencia de tratamientos con nano y biomateriales en la colonización microbiana de roca monumental' to Benjamín Otto Ortega Morales.

REFERENCES

Aflori, M., Simionescu, B., Bordiani, I. and Olaru, M. 2013. 'Silesquioxane-based hybrid nanocomposite with methacrylate units containing titania and/or silver nanoparticles as antibacterial/antifungal coatings for monumental stones,' *Materials Science and Engineering B* 178: 1339–46, doi:10.1016/j.mseb.2013.04.004.

Alberto, L., Aguilera, P., Alberto, J., Zapata, N. and Ortega Morales, B.O. 2015. 'The bioprecipitation of calcium carbonate by native microbial population as a restoration method,' *Nexo* 28(1): 25–40.

Allsopp, D., Seal, K.J. and Gaylarde, C.C. 2004. *Introduction to Biodeterioration*, 2nd edn. Cambridge: Cambridge University Press.

Baglioni, P., Chelazzi, D., Giorgi, R., Carretti, E., Toccafondi, N. and Jaidar, Y. 2014. 'Commercial $Ca(OH)_2$ nanoparticles for the consolidation of immovable works of art,' *Applied Physics A* 114: 723, doi:10.1007/s00339-013-7942-6.

Baglioni, P., Chelazzi, D. and Giorgi, R. 2015. *Nanotechnologies in the Conservation of Cultural Heritage: A Compendium of Materials and Techniques*. Medford MA: Springer Science.

Balloi, A., Lombardi, E., Troiano, F., Polo, A., Caoitelli, F., Gulotta, D., Toniolo, L., Lucchini, A. and Daffonchio, D. 2017. 'Sulfate reducing bacteria as bio-cleaning agents: development of new methodologies and study cases,' *Conservation Science in Cultural Heritage* 15: 109–19, doi:10.6092/issn.1973-9494/7123.

Bellissima, F., Bonini, M., Giorgi, R., Baglioni, P., Barresi, P., Mastromei, G. and Perito, B. 2014. 'Antibacterial activity of silver nanoparticles grafted on stone surface', *Environmental Science and Pollution Research* 21: 13278–86, doi:10.1007/s11356-013-2215-7.

Blazquez, A.B., Lorenzo, J., Flores, M. and Gomez-Alarcon, G. 2000. 'Evaluation of the effect of some biocides against organisms isolated from historic monuments', *Aerobiologia* 16: 423–8.

Boniek, D., de Castro Mendes, I., dos Santos, A.F.B. and de Resende Stoianoff, M.A. 2017. 'Biocidal effect of gamma radiation on the ecology of filamentous fungal populations associated with stone deterioration', *Journal of Environmental Science and Engineering A* 6: 252–9, doi:10.17265/2162-5298/2017.05.005.

Borderie, F., Alaoui-Sehmer, L., Raouf, N., Bousta, F., Orial, G., Rieffel, D. and Alaoui-Sossé, B. 2011. 'UV-C irradiation as a tool to eradicate algae in caves', *International Biodeterioration and Biodegradation* 65: 579–84.

Borderie, F., Tête, N., Cailhol, D., Alaoui-Sehmer, L., Bousta, F., Rieffel, D., Aleya, L. and Alaoui-Sossé, B. 2014. 'Factors driving epilithic algal colonization in show caves and new insights into combating biofilm development with UV-C treatments', *Science of the Total Environment* 484: 43–52.

Borsoi, G., Lubelli, B., van Hees, R., Veiga, R. and Silva, A.S. 2016. 'Understanding the transport of nanolime consolidants within Maastricht limestone', *Journal of Cultural Heritage* 18: 242–9.

Bosch-Roig. P. and Ranalli, G. 2014. 'The safety of biocleaning technologies for cultural heritage', *Frontiers in Microbiology* 5: 155, doi:10.3389/fmicb.2014.00155.

Bosch-Roig, P., Lustrato, G., Zanardini, E. and Ranalli, G. 2015. 'Biocleaning of cultural heritage stone surfaces and frescoes: which delivery system can be the most appropriate?', *Annual Review of Microbiology* 65: 1227, doi:10.1007/s13213-014-0938-4.

Braissant, O. 2005. *Carbonatogénèse bactérienne liée au cycle biogéochimique oxalate-carbonate*, PhD dissertation, Université de Neuchâtel.

Braissant, O., Cailleau, G., Dupraz, C. and Verrecchia, E.P. 2003. 'Bacterially induced mineralization of calcium carbonate in terrestrial environments: the role of exopolysaccharides and amino acids', *Journal of Sedimentary Research* 73: 485–90.

Bromblet, P., Labouré, M. and Orial, G. 2003. 'Diversity of the cleaning procedures including laser for the restoration of carved portals in France over the last 10 years', *Journal of Cultural Heritage* 4: 17S–26S,

Cappitelli, F., Toniolo, L., Sansonetti, A., Gulotta, D., Ranalli, G., Zanardini, E. and Sorlini, C. 2007. 'Advantages of using microbial technology over traditional chemical technology in removal of black crusts from stone surfaces of historical monuments', *Applied and Environmental Microbiology* 73: 5671–5.

Carrillo-González, R., Martínez-Gómez, M.A., González-Chávez, M.D. and Mendoza Hernández, J.C. 2016. 'Inhibition of microorganisms involved in deterioration of an archaeological site by silver nanoparticles produced by a green synthesis method', *Science of the Total Environment* 16: 30320–5, doi:10.1016/j.scitotenv.2016.02.110.

Costa, C., Maraschin, M. and Rocha, M. 2016. 'An R package for the integrated analysis of metabolomics and spectral data', *Computer Methods and Programs in Biomedicine* 129: 117–24.

De Belie, N. 2016 'Application of bacteria in concrete: a critical review', *RILEM Technical Letters* 1: 56–61.

De Muynck, W., De Belie, N. and Verstraete, W. 2010. 'Microbial carbonate precipitation in construction materials: a review', *Ecological Engineering* 36: 118–36, doi: 10.1016/j.ecoleng.2009.02.006.

Decho, A.W. 2010. 'Overview of biopolymer-induced mineralization: what goes on in biofilms?', *Ecological Engineering* 36: 137–44.

Dhami, N.K., Reddy, M.S. and Mukherjee, A. 2014. 'Application of calcifying bacteria for remediation of stones and cultural heritages', *Frontiers in Microbiology* 5: 304, doi: 10.3389/fmicb.2014.00304.

Ditaranto, N., Loperfido, S., van der Werf, I., Mangone, A., Cioffi, N. and Sabbatini, L. 2011. 'Synthesis and analytical characterisation of copper-based nanocoatings for bioactive stone artworks treatment', *Analytical and Bioanalytical Chemistry* 399: 473–81, doi: 10.1007/s00216-010-4301-8.

Duan, Y., Wu, F., Wang, W., He, D., Gu, J.D., Feng, H., Chen, T., Liu, G. and An, L. 2017. 'The microbial community characteristics of ancient painted sculptures in Maijishan Grottoes, China', *PLOS* 5 July 2017, doi:10.1371/journal.pone.0179718.

Dunn, W.B., Erban, A., Weber, R.J.M., Creek, D.J., Brown, M., Breitling, R., Hankemeier, T., Goodacre, R., Neumann, S., Kopka, J. and

Viant, M.R. 2012. 'Mass appeal: metabolite identification in mass spectrometry-focused untargeted metabolomics', *Metabolomics* 9: S44–S66.

Ettenauer, J., Piñar, G., Tafer, H. and Sterflinger, K. 2014. 'Quantification of fungal abundance on cultural heritage using real time PCR targeting the β-actin gene', *Frontiers in Microbiology* 5: 262, doi: 26210.3389/fmicb.2014.00262.

Fernandes, P. 2006. 'Applied microbiology and biotechnology in the conservation of stone cultural heritage materials', *Applied Microbiology and Biotechnology* 73: 291–6.

Gaylarde, C., Baptista-Neto, J.A., Ogawa, A., Kowalski, M., Celikkol-Aydin, S. and Beech, I. 2017a. 'Epilithic and endolithic microorganisms and deterioration on stone church facades subject to urban pollution in a sub-tropical climate', *Biofouling* 33: 113–27, doi:10.1080/08927014.2016.1269893.

Gaylarde, C., Ogawa, A., Beech, I., Kowalski, M. and Baptista-Neto, J.A. 2017b. 'Analysis of dark crusts on the church of Nossa Senhora do Carmo in Rio de Janeiro, Brazil, using chemical, microscope and metabarcoding microbial identification techniques', *International Biodeterioration and Biodegradation* 117: 60–67.

Gherardi, F., Gulotta, D., Goidanich, S., Colombo, A. and Toniolo, L. 2017. 'On-site monitoring of the performance of innovative treatments for marble conservation in architectural heritage', *Heritage Science* 5: 4, doi:10.1186/s40494-017-0118-5.

Gioventù, E. and Lorenzi, P. 2013. 'Bio-removal of black crust from marble surface: comparison with traditional methodologies and application on a sculpture from Florence's English Cemetery', *Procedia Chemistry* 8: 123–9.

Gómez-Ortíz, N., De la Rosa-García, S., González-Gómez, W., Soria-Castro, M., Quintana, P., Oskam, G. and Ortega Morales, B. 2013. 'Antifungal coatings based on $Ca(OH)_2$ mixed with ZnO-TiO_2 nanomaterials for protection of limestone monuments', *ACS Applied Materials and Interfaces* 13: 1556–65, doi:10.1021/am302783h.

Gutarowska, B., Celikkol-Aydin, S., Bonifay V., Otlewska, A., Aydin, E., Oldham, A.L., Brauer, J.I., Duncan, K.E., Adamiak, J., Sunner, J.A. and Beech, I.B. 2015. 'Metabolomic and high-throughput sequencing analysis: modern approach for the assessment of biodeterioration of materials from historic buildings', *Frontiers in Microbiology* 6: 979, doi:10.3389/fmicb.2015.00979.

He, J. (ed.) 2017. *Self-Cleaning Coatings: Structure, Fabrication and Application*. Cambridge: Royal Society of Chemistry.

Hsieh, P., Pedersen, J.Z. and Albertano, P. 2013. 'Generation of reactive oxygen species upon red light exposure of Cyanobacteria from Roman hypogea', *International Biodeterioration and Biodegradation* 84: 258–65, doi:10.1016/j.ibiod.2012.11.007.

Hsieh, P., Pedersen, J.Z. and Bruno, L. 2014. 'Photoinhibition of Cyanobacteria and its application in cultural heritage conservation', *Photochemistry and Photobiology* 90: 533–43.

Jeong, M.H., Lee, Y.S., Cho, J.Y., Ahn, Y.S., Moon, J.H., Hyun, H.N., Cha, G.S. and Kim, K.Y. 2017. 'Isolation and characterization of metabolites from *Bacillus licheniformis* MH48 with antifungal activity against plant pathogens', *Microbiology and Pathology* 110: 645–53.

Jroundi, F., Gonzalez-Muñoz, M.T., Garcia-Bueno, A. and Rodriguez-Navarro, C. 2014. 'Consolidation of archaeological gypsum plaster by bacterial biomineralization of calcium carbonate', *Acta Biomaterialia* 10(9): 3844–54.

Jroundi, F., Gonzalez-Muñoz, M.T., Sterflinger, K. and Piñar, G. 2015. 'Molecular tools for monitoring the ecological sustainability of a stone bio-consolidation treatment at the Royal Chapel, Granada', *PLoS One* 10(7): e0132465, doi:10.1371/journal.pone.0132465.

Knorre, H. and Krumbein, W.E. 2000. 'Bacterial calcification', in *Microbial Sediments*, R.E. Riding and S.M. Awramik (eds), 25–31. Berlin: Springer-Verlag.

Konkol, N., Mcnamara, C., Sembrat, J., Rabinowitz, M. and Mitchell, R. 2009. 'Enzymatic decolorization of bacterial pigments from culturally significant marble', *Journal of Cultural Heritage* 10: 362–6.

La Russa, M.F., Macchia, A., Ruffolo, S.A., De Leo, F., Barberio, M., Barone, P., Crisci, G.M. and Urzì, C. 2014 'Testing the antibacterial activity of doped TiO_2 for preventing biodeterioration of cultural heritage building materials', *International Biodeterioration and Biodegradation* 96: 87–96, doi: 10.1016/j.ibiod.2014.10.002.

Li, W.R., Sun, T.L., Zhou, S.L., Ma, Y.K., Shi, Q.S., Xie, X.B. and Huang, X.M. 2017. 'A comparative analysis of antibacterial activity, dynamics, and effects of silver ions and silver nanoparticles against four bacterial strains', *International Biodeterioration and Biodegradation* 123: 304–10.

Liu, Q., Zhang, B., Shen, Z. and Lu, H. 2006. 'A crude protective film on historic stones and its artificial preparation through biomimetic synthesis', *Applied Surface Science* 253: 2625–32, doi:10.1016/j.apsusc.2006.05.032.

Longnecker, K., Futrelle, J., Coburn, E., Kido Soule, M.C. and Kujawinski, E.B. 2015. 'Environmental metabolomics: databases and tools for data analysis', *Marine Chemistry* 177: 366–73.

López-Moreno, A., Sepúlveda-Sánchez, J.D., Guzmán, E.M.M.A. and Le Borgne, S. 2014. 'Calcium carbonate precipitation by heterotrophic bacteria isolated from biofilms formed on deteriorated ignimbrite stones: influence of calcium on EPS production and biofilm formation by these isolates', *Biofouling* 30: 547–60, doi:10.1080/08927014.2014.888715.

Lustrato, G., Alfano, G., Andreotti, A., Colombini, M.P. and Ranalli, G. 2012. 'Fast biocleaning of mediaeval frescoes using viable bacterial cells', *International Biodeterioration and Biodegradation* 69: 51–61.

Maravelaki, N., Verganelaki, A., Kilikoglou, V. and Karatasios, I. 2015. 'Synthesis and characterization of a calcium oxalate-silica nanocomposite for stone conservation', in *Engineering Geology for Society and Territory*, vol. 8, G. Lollino, D. Giordan, C. Marunteanu, B. Christaras B, I. Yoshinori and C. Margottini (eds), 525–9. Cham: Springer.

Martino, M., Schiavone, S., Palla, F., Pellegrino, L., De Castro, E. and Balloi, A. 2015. 'Bioremoval of sulphate layer from a 15th century polychrome marble artifact', *Conservation Science in Cultural Heritage* 15: 235–52, doi:10.6092/issn.1973-9494/6177.

Mascalchi, M., Osticioli, I., Riminesi, C., Cuzman, O.A., Salvadori, B. and Siano, S. 2015. 'Preliminary investigation of combined laser and microwave treatment for stone biodeterioration', *Studies in Conservation* 60(suppl. 1): S19–S27, doi: 10.1179/0039363015Z.000000000203.

Munafò, P., Goffredo, G. and Quagliarini, E. 2015. 'TiO_2-based nanocoatings for preserving architectural stone surfaces: an overview', *Construction and Building Materials* 84: 201–18.

Ogawa, A., Celikkol-Aydin, S., Gaylarde, C., Baptista-Neto, J.A. and Beech, I. 2017. 'Microbial communities on painted wet and dry external surfaces of a historic fortress in Niterói, Brazil', *International Biodeterioration and Biodegradation* 123: 164–73.

Ortega Morales, B.O., Hernandez-Duque, G., Borges-Gomez, L. and Guezennec, J. 1999. 'Characterization of epilithic microbial communities associated with Mayan stone monuments in Yucatan, México', *Journal of Geomicrobiology* 16: 221–32.

Osticioli, I., Mascalchi, M., Pinna, D. and Siano, S. 2015. 'Removal of *Verrucaria nigrescens* from Carrara marble artefacts using Nd:YAG lasers: comparison among different pulse durations and wavelengths', *Applied Physics A* 118: 1517.

Otlewska, A., Adamiak, J., Stryszewska, T., Kanka, S. and Gutarowska, B. 2017. 'Factors determining the biodiversity of halophilic microorganisms on historic masonry buildings', *Microbes and Environments* 32: 164–73.

Patti, G.T., Yanes, O. and Siuzdak, G. 2012. 'Metabolomics: the apogee of the omics trilogy', *Nature Reviews Molecular Cell Biology* 13: 263–9.

Perito, B. and Mastromei, G. 2011. 'Molecular basis of bacterial calcium carbonate precipitation', *Progress in Molecular and Subcellular Biology* 52: 113–39.

Piñar, G., Ripka, K., Weber, J. and Sterflinger, K. 2009. 'The micro-biota of a sub-surface monument the medieval chapel of St. Virgil (Vienna, Austria)', *International Biodeterioration and Biodegradation* 63: 851–9.

Piñar, G., Kraková, L., Pangallo, D., Piombino-Mascali, D., Maixner, F., Zink, A. and Sterflinger, K. 2014. 'Halophilic bacteria are colonizing the exhibition areas of the Capuchin catacombs in Palermo, Italy', *Extremophiles* 18: 677–91, doi: 10.1007/s00792-014-0649-6.

Pinna, D., Salvadori, B. and Galeott, M. 2012. 'Monitoring the performance of innovative and traditional biocides mixed with consolidants and water-repellents for the prevention of biological growth on stone', *Science of the Total Environment* 15: 132–41, doi:10.1016/j.scitotenv.2012.02.012.

Pinna, D., Galeotti, M., Perito, B., Daly, G. and Salvadori, B. 2018. '*In situ* long-term monitoring of recolonization by fungi and lichens after innovative and traditional conservative treatments of archaeological stones in Fiesole (Italy)', *International Biodeterioration and Biodegradation* 132: 49–58.

Rodriguez-Navarro, C., Rodriguez-Gallego, M., Ben Chekroun, K. and Gonzalez-Muñoz, M.T. 2003. 'Conservation of ornamental stone by *Myxococcus xanthus*-induced carbonate biomineralization', *Applied and Environmental Microbiology* 69: 2182–93.

Rotolo, V., Barresi, G., Di Carlo, E., Giordano, A., Lombardo, G., Crimi, E., Costa, E., Bruno, M. and Palla, F. 2016. 'Plant extracts as green potential strategies to control the biodeterioration of cultural heritage', *International Journal of Conservation Science* Special Issue 2: 839–46.

Salimbeni, R., Pini, R., Siano, S. and Calcagno, G. 2000. 'Assessment of the state of conservation of stone artworks after laser cleaning: comparaison with conventional cleaning results on a two-decade follow up', *Journal of Cultural Heritage* 1: 385–91.

Sanmartín, P., DeAraujo, A., Vasanthakumar, A. and Mitchell, R. 2015. 'Feasibility study involving the search for natural strains of microorganisms capable of degrading graffiti from heritage materials', *International Biodeterioration and Biodegradation* 103: 186–90.

Sanmartín, P., De Araujo, A. and Vasanthakumar, A. 2016. 'Melding the old with the new: trends in methods used to identify, monitor, and control microorganism on cultural heritage materials', *Microbial Ecology* 7: 1–17, doi:10.1007/s00248-016-0770-4.

Sanmartín, P., Vázquez-Nion, D., Arines, J., Cabo-Domínguez, L. and Prieto, B. 2017. 'Controlling growth and colour of phototrophs by using simple and inexpensive coloured lighting: a preliminary study in the Light4Heritage project towards future strategies for outdoor illumination', *International Biodeterioration and Biodegradation* 122: 107–15.

Sanz, M., Oujja, M., Ascaso, C., de los Ríos, A., Pérez-Ortega, S., Souza-Egipsy, V., Wierzchos, J., Speranza, M., Cañamares, M.V. and Castillejo, M. 2015. 'Infrared and ultraviolet laser removal of crustose lichens on dolomite heritage stone', *Applied Surface Science* 346: 248–55.

Sanz, M., Oujja, M., Ascaso, C., de los Ríos, A., Pérez-Ortega, S., Souza-Egipsy, V., Wierzchos, J., Cañamares, M.V. and Castillejo, M. 2017. 'Influence of wavelength on the laser removal of lichens colonizing heritage stone', *Applied Surface Science* 399: 758–68.

Saravia, S.G.G. and Gaylarde, C.C. 1998. 'The antimicrobial activity of an aqueous extract of *Brassica nigra*', *International Biodeterioration and Biodegradation* 41: 145–8.

Sasso, S., Miller, A.Z., Rogerio-Candelera, M.A., Cubero, B., Couthino, M.I., Scrano, L. and Bufo, S.A. 2016. 'Potential of natural biocides for biocontrolling phototropic colonization on limestone', *International Biodeterioration and Biodegradation* 107 : 102–10.

Senesi, G.S., Carrara, I., Nicolodelli, G., Milori, D.M.B.P. and De Pascale, O. 2016. 'Laser cleaning and laser-induced breakdown spectroscopy applied in removing and characterizing black crusts from limestones of Castello Svevo, Bari, Italy: a case study', *Microchemical Journal* 124: 296–305.

Shirakawa, M.A., Loh, K., John, V.M. and Gaylarde, C.C. 2012. 'Resistance of cyanobacterial fouling on architectural paint films to cleaning by water jet', *Current Microbiology* 64: 312–16.

Shirakawa, M.A., Gaylarde, C.C., Sahão, H.D. and Lima, J.R.B. 2013. 'Inhibition of *Cladosporium* growth on gypsum panels treated with nanosilver particles', *International Biodeterioration and Biodegradation* 85: 57–61.

Shirakawa, M., John, V., De Belie, N., Alves, J., Pinto, J. and Gaylarde, C. 2015. 'Susceptibility of biocalcite-modified fiber cement to biodeterioration', *International Biodeterioration and Biodegradation* 103: 215–20, doi:10.1016/j.ibiod.2015.04.003.

Silva, M., Rosado, T., Teixeira, D., Candeias, A. and Caldeira, A.T. 2015. 'Production of green biocides for cultural heritage: novel biotechnological solutions', *International Journal of Conservation Science* Special Issue 6: 519–30.

Sohlberg, E., Bomberg, M., Miettinen, H., Nyysonen, M., Salavirta, H., Vikman, M. and Itavaara, M. 2015. 'Revealing the unexplored fungal communities in deep groundwater of crystalline bedrock fracture zones in Olkiluoto, Finland', *Frontiers in Microbiology* 6: 573, doi:10.3389/fmicrob.2015.00573.

Szewczyk, B., Hoyos-Carvajal, L., Paluszek, M., Skrzecz, I. and Souza, M.L. 2006. 'Baculoviruses: re-emerging biopesticides', *Biotechnology Advances* 24: 143–60.

Tang, S., Xi, W., Cheng, Z., Yin, L., Li, R., Wu, G., Liu, W., Xu, J., Xiang, S., Zheng, Y., Ge, Q., Ning, K., Yan, Y. and Zhan, Y. 2016. 'A living eukaryotic autocementation kit from surface display of silica binding peptides on *Yarrowia lipolytica*', *ACS Synthetic Biology* 5: 1466–74, doi:10.1021/acssynbio.6b00085.

Tao, Y., Zhang, X., Au, D.W.T., Mao, X. and Yuan, K. 2010. 'The effects of sub-lethal UV-C irradiation on growth and cell integrity of Cyanobacteria and green algae', *Chemosphere* 78: 541–7, doi: 10.1016/j.chemosphere.2009.11.016.

Troiano, F., Gulotta, D., Balloi, A., Polo, A., Toniolo, L., Lombardi, E., Daffonchio, D., Sorlini, C. and Cappitelli, F. 2013. 'Successful combination of chemical and biological treatments for the cleaning of stone artworks', *International Biodeterioration and Biodegradation* 85: 294–304.

Urzì, C., De Leo, F., Bruno, L., Pangallo, D. and Kraková, L. 2014. 'New species description, biomineralization processes and biocleaning applications of Roman catacombs-living bacteria', in *The Conservation of Subterranean Cultural Heritage*, C. Saiz-Jimenez (ed.), 65–72. London: CRC Press.

van der Werf, I.D., Ditaranto, N., Picca, R.A., Sportelli, M.C. and Sabbatini, L. 2015. 'Development of a novel conservation treatment of stone monuments with bioactive nanocomposites', *Heritage Science* 3: 29, doi:10.1186/s40494-015-0060-3.

Villa, F., Vasanthakumar, A., Mitchell, R. and Cappitelli, F. 2015. 'RNA-based molecular survey of biodiversity of limestone tombstone microbiota in response to atmospheric sulfur pollution', *Letters in Applied Microbiology* 60: 92–102, doi:10.1111/lam.12345.

Watson, G.S., Green, D.W., Schwarzkopf, L., Li, X., Cribb, B.W., Myhra, S. and Watson, J.A. 2015. 'A gecko skin micro/nano structure: a low adhesion, superhydrophobic, anti-wetting, self-cleaning, biocompatible, antibacterial surface', *Acta Biomaterialia* 21:109–22, doi:10.1016/j.actbio.2015.03.007.

Young, M.E., Alakomi, H.L., Fortune, I., Gorbushina, A.A., Krumbein, W.E., Maxwell, I., McCullagh, C., Robertson, P., Saarela, M., Valero, J. and Vendrell, M. 2009. 'Development of a biocidal treatment regime to inhibit biological growth on cultural heritage: BIODAM', *Environmental Geology* 56(3–4): 631–41.

Zhu, T. and Dittrich, M. 2016. 'Carbonate precipitation through microbial activities in natural environment, and their potential in biotechnology: a review', *Frontiers in Bioengineering and Biotechnology* 4: 4, doi:10.3389/fbioe.2016.00004.

AUTHORS' ADDRESSES

- Christine C. Gaylarde, Department of Microbiology and Plant Biology, University of Oklahoma, 660 Parrington Oval, Norman, OK 73019, USA (cgaylarde@gmail.com)
- B. Otto Ortega Morales, Departmento de Microbiología Ambiental y Biotecnología (DEMAB), Universidad Autónama de Campeche, Campeche, México (beortega@uacam.mx)

Milton Keynes UK
Ingram Content Group UK Ltd.
UKHW051835251023
431338UK00004B/36

9 781909 492646